# Australia and the Origins of Agriculture

Rupert Gerritsen

BAR International Series 1874
2008

Published in 2016 by
BAR Publishing, Oxford

BAR International Series 1874

*Australia and the Origins of Agriculture*

ISBN 978 1 4073 0354 3



BAR Publishing is the trading name of British Archaeological Reports (Oxford) Ltd. British Archaeological Reports was first incorporated in 1974 to publish the BAR Series, International and British. In 1992 Hadrian Books Ltd became part of the BAR group. This volume was originally published by Archaeopress in conjunction with British Archaeological Reports (Oxford) Ltd / Hadrian Books Ltd, the Series principal publisher, in 2008. This present volume is published by BAR Publishing, 2016.

Printed in England

BAR
PUBLISHING

BAR titles are available from:

BAR Publishing
122 Banbury Rd, Oxford, OX2 7BP, UK
EMAIL info@barpublishing.com
PHONE +44 (0)1865 310431
FAX +44 (0)1865 316916
www.barpublishing.com

Dedicated to the memory of

*Eleanora (Nora) Gerritsen*

1921 - 2008

*"the history of cultivated plants is allied to the most important problems of the general history of organized beings"*

Alphonse de Candolle, 1884:viii

# Preface

In writing this work I have been conscious that I follow a great tradition, one that harks back to Herodotus, Polybius, Tacitus, Suetonius, Bede, Gregory of Tours, Gibbon and many others. It is certainly a project that I believe is consistent with that cultural tradition. At an individual level I am a product of my culture, and although this book is largely about another cultural tradition, that of Indigenous Australians, it really reflects some of the more significant concerns and interests that derive from my cultural heritage. In terms of those concern and interests, I hope it will be seen to be making a meaningful contribution to the attendant debates and questions that it endeavours to address.

One of the enduring and perhaps most important undercurrents in historiography is the issue of how one represents events from the past. While some have taken the position in recent times that historical writing is just one form of personal narrative it appears to me more than that, it is a distinctive type of writing that is anchored by real actions and events, by realities in the past. Accordingly, if one stops and reflects for but a moment, it is obvious that there is an almost overwhelming flood of sensory input from the world around us, second by second. Should we wish to convey that experience to another we normally, and probably unavoidably, resort to a 'shorthand' reality. We don't relate the detail of our perceptions of a room for example, we simply state we are in a room and any of the salient features of the room. To deal with the complexity of reality at even the most basic level the mind has many devices, such as the ability to categorise and to make behavioural generalisations, to enable us to process and communicate information. Another dimension or layer is then added to this, connecting to our inner being, our emotions and cognitions, when we try to interpret and give meaning to the features, regularities, and patterns of events that take place around us. This necessitates a "theory of mind", the ability to read, predict and interpret the motivations, reactions and intent of others, an innate and well-developed capacity in humans. Consequently we do not describe reality as such, be it past or present, we are simply modelling reality, formulating the model in words and then transmitting it using language.

In dealing with the broad sweep of history as I have attempted here, and in this context I include prehistory, another level of abstraction is introduced. Various paradigms characterising and interpreting changes in societies, or segments of societies, and the factors that influence those changes, have arisen over time. They are in fact still evolving, so that our understanding of the past not only reflects the concerns and perspectives of the present, but also the tools we currently have at our disposal to achieve that. Which of those tools provides the 'best' understanding of trends in history of course depends on which facet of history we may be seeking to understand. Within that framework some explanatory systems have greater resonance and are perceived as being more credible than others. This does not mean those explanatory systems are intrinsically or objectively superior as they still remain a subjective assessment, albeit one with the concurrence of a broad cross-section of those who have an interest in the subject matter. Another test of systems of explanation borrows from the ideal of the scientific paradigm, that they are able to accommodate and explain all data, or in this case the broadest range of information. Sometimes a Kuhnian type of paradigmatic revolution may occur, in which a whole range of unexplained or barely understood phenomena becomes explicable. Hopefully the material presented here may meet those tests, and our understandings may indeed move forward a little.

In motivational terms my interest in this topic is part of a broader project, a desire to push back the boundaries of knowledge. It is an ideal I have pursued in most of the research I have undertaken, driven by my intellectual curiosity, and the sense of achievement I experience when I am able to do that. Of course not all research need be of that nature, functional research for example has its place. From a personal perspective, however, it saddens me to see that many whose remit it is to pursue the goal of revealing new knowledge and understandings, lose sight of that and become bogged down in prosaic concerns and trivial matters. Perhaps I am being too idealistic in this regard.

This work challenges the stereotype that Indigenous Australians prior to the British colonisation of Australia were all nomadic hunter-gatherers. It also questions current mainstream explanations for the development of agriculture. Because of this it is likely to invite controversy. I accept that, it is a position I have willingly and knowingly placed myself in. I, in fact, hope there will be reasoned criticism and some degree of controversy as that is a healthy sign, a necessary part of the discourse that may lead to a deeper understanding of the issues I raise. All that I ask is that those who see fit to criticise ensure they have read and understood the content of this book.

Finally, I cannot conclude without thanking and acknowledging the many individuals from various walks of life who have helped me bring this work to its completion. Their assistance should not of course be construed in any way as an endorsement of the arguments I have put forward. The individuals concerned include: The library staff of the National Library of Australia, particularly the staff of the Petherick Reading Room and the Map Room, Emeritus Professor John Mulvaney, Professor Peter Veth, Professor Rolf Gerritsen, Professor Helen James, Professor John Parker, Dr Mary O'Dowd, Dr Nonja Peters, Dr Ian Keen, Dr Helen Lette, Dr Annie Clarke, Dr Michael Pickering, Dr Fiona Walsh, Judith Scurfield, Allen Mawer, Katharine West, Jamon Moon, Anita Jaroslawski, Margaret Saville, Jenny Ward, Rodrigo Aguilera, Lesley Hyndal, Bill and William Mallard, Ron Gerritsen, W. C. (Bill) Gerritsen [deceased], Eleanora (Nora) Gerritsen [deceased], the staff of the W.A. Department of Indigenous Affairs Library, Lyn Jones, Janet Buick, Shirley Fletcher, Bryce Moore, Patti Steele, Lado Shay, Jim Croft, Murray Fagg, Max Cramer, Lloyd Browne, Arienne Alphenaar of the Department of Environment, Water, Heritage and Arts Central Library, Peter Reynders and Rohan Reynders.

I trust that I have not overlooked anyone, and hope that you the reader will get the same enjoyment and intellectual exhilaration from this book that I experienced in creating it.

Rupert Gerritsen

31 May 2008

# Table of Contents

# Chapter 1.
# Considering the Origins of Agriculture

**Introduction**

The origins of agriculture are a matter of profound interest, not only in academic circles, but to many lay people. This is because the advent of agriculture as a particular form of food production, along with animal domestication, appears to have marked an unparalleled transformation in human affairs, the consequences of which are still being played out in the present day. Over time extensive investigations in a range of disciplines have been conducted in an endeavour to understand the phenomenon, investigations which have attempted to answer such basic questions as: what happened?; where did it happen and when?; how did it happen?; and perhaps most importantly, why did it happen at all? More recently it has been realised that developments in hunter-gather societies were more varied and complex than originally portrayed, and that different pathways and forms of "food production" were conceivable. Accordingly other patterns of development have come to be recognised, one in particular involving the formation of distinctive, complex, socioeconomic and sociopolitical structures. Societies exhibiting such characteristics are often labelled as "Complex Hunter-Gatherers". But this recognition has also made the question far more perplexing. Consequently, any broad explanatory framework must now account for more than just the development of agriculture. Part of the project has now become a struggle to find appropriate paradigms, identify common ground and promulgate sound explanations embracing all these diverse circumstances.

Research into the location and timing of the "agricultural revolution" over the last one hundred years or so has resulted in the determination, in at least proximate terms, of when and where such developments took place. In the process it has become apparent that it was not a discrete event but a circumstance that occurred in a number of different places, to some, virtually contemporaneously. Some degree of consensus would also appear to have been reached in characterising what happened - that certain groups managed to accomplish, unaided, the transition from hunting and gathering to farming (pristine development), apparently in the space of only a few centuries. The characterisation of that transition does depend, however, on the paradigmatic viewpoint. Usually such explanations are framed in anthropological or economic terms but the process can also be described, as Rindos has done, within a biological evolutionary framework. In Rindos' view a period of coevolution resulted in humans entering into a mutualistic relationship with plants.[1]

Questions as to how and why agriculture arose are more problematic. Numerous common factors, pathways and theories have been proposed in this regard, seeking to explain how agriculture arose, usually employing some form of process description or model. These, by and large, involve environmental and/or demographic factors, the botanical characteristics of particular plants and their interaction with humans, their integration into the human economy and the social dynamics that resulted when those plants became the economic base, the principal sources of sustenance, in the early agricultural societies. Often such discussions also seek to address the related issues of the domestication of plants and animals and the apparent synchronous development of domestication-based economies in different parts of the world. Debates in regard to domestication revolve around its significance, and the role of intentionality, while those encompassing synchronous development usually attempt to identify global or common factors that may have influenced local pathways.

It is within this context a number of researchers have drawn attention to the apparent absence of agriculture in Australia,[2] and the need to explain what I will call the "Australian Problem". In any initial consideration of the "Australian Problem", however, it is necessary to recognise that there are actually two supposed problems, one being the "failure" of agriculture to develop indigenously and the other, its "failure" to diffuse into Australia, despite contact with Indonesian (*Macassan*) agriculturalists or New Guinean horticulturalists. Although not always explicitly stated or recognised, significant differences probably exist in the factors and dynamics that led to the pristine development of agriculture, as opposed to agriculture that arose as a result of outside influences, as a result of cultural transfers. Consequently these two issues will be treated separately. In addition, a further question will be raised relating to the concept of Complex Hunter-Gatherers. While some, Lourandos,[3] Williams,[4] and Builth[5] in particular, have put forward extensive evidence and arguments in making a case

1 Rindos 1984.
2 e.g. White 1971:184-5; Harris 1977a:209; Rindos 1984:161-2; Williams 1987: 320; Price and Gebauer 1995:9; Hayden 1995a:297.
3 Lourandos 1983,1985:404-8,1988:153-4, 1993:72,75,78-80.
4 Williams 1987,1988.
5 Builth 2000a,b; 2002:302-6.

for the presence of Complex Hunter-Gatherers in Australia, their claims need to be more thoroughly tested, and this will be done in due course.

With these matters in mind, it is also my intention to employ the Australian situation to examine and test the validity of some of the frameworks, key arguments, and critical evidence, that have been put forward concerning the development of agriculture, animal husbandry and Complex Hunter-Gatherer economies. Such an undertaking is predicated upon the necessary assumptions that all humans have the same innate capacities,[6] and that some developments in Australia would probably have eventually paralleled those in other parts of the world, had Indigenous societies not been shattered by British colonisation. A corollary of these assumptions is the recognition that particular geographic, environmental, climatic, demographic and cultural factors, either singly or in concert, must have affected development in this continent. An exploration of these factors is, in fact, what constitutes the examination of the "Australian Problem" in both its forms.

While Indigenous Australians in their traditional circumstances were certainly not "fossil" versions of Paleolithic hunter-gatherers, there are strong structural similarities, or functional equivalences, principally in their subsistence economies, to other pre-agricultural societies known through the archaeological record.[7] As will become apparent in the arguments presented in forthcoming pages, developments in a number of "traditional societies"[8] in a number of regions in Australia were indeed following a trajectory leading to recognisable agricultural or supposed Complex Hunter-Gatherer economies. Consequently, if these arguments prove to be well founded, a valuable opportunity exists, given the availability of much recent ethnographic information and observations, to examine the relevant issues at a time when such developments were actually taking place, rather than as a post facto analysis relying exclusively upon archaeological evidence. From this, it is hoped, a sounder understanding of the origins of agriculture may well be obtained.

### Arguments and Evidence

In attempting to address some of the issues raised thus far, the approach I intend to take will be based on the assumption made earlier, that all humans have the same innate capacities. By extension I would claim developing hunter-gatherer and early "food production" societies have more in common with contemporary agricultural, industrial and post-industrial peoples than is often realised. This statement is not as banal as it might first appear and others have made similar observations.[9] Its significance lies in the inference that some of the difficulties that have been encountered to date may really be the result of a surreptitious assumption that hunter-gatherers really are somehow "different".[10] As a consequence limited, specious or inappropriate paradigms may have been applied on occasion to explain evolving behavioural patterns in hunter-gatherer societies, whilst well-developed and more appropriate ones have been overlooked.[11] Furthermore, as has been pointed out by Barnard and Woodburn, problems may arise in analysing hunter-gatherer societies because our culturally conditioned conceptual categories are not necessarily the same as those of the peoples being studied, and this may confound our efforts.[12] These views should not, however, be construed as a reworking of the issues of cultural relativism and power relationships in ethnography, typified by deconstructionism and post-modernism, which have coloured debates to a considerable extent in recent times, in a number of disciplines.[13] Those approaches are, of course, quite valid, and so are embraced in the philosophical underpinnings, "consensual reality" and "cognitive consensus", of this work.[14] Hence, maintaining awareness of cultural divergence must be considered as essential, but ultimately studies of this nature should not deny or obscure the universality of many human behaviours and qualities.[15]

It is for these reasons, and because we are dealing with issues relating to changes in peoples' modes of subsistence and production, that I will rely, at least in part, on the language of economics in my deliberations.[16] In essence the problem, in my view, involves the conditions, causes, dynamics and consequences of economic development in a "paleolithic" context. This position is similar to that advocated by Higgs and Jarman,[17] and Cook,[18] some time ago. Determining the role of environmental and demographic factors, social and political dynamics, or other circumstance, in the developmental process is, however, a critical consideration. Innumerable researchers have advocated local or global climatic changes as the trigger for the appearance of more complex paleoeconomies, be they based on agriculture, animal husbandry or some other form of intensified

6 Murray 1988:5-8.

7 Davidson 1989:75; Lourandos 1997:332-3,335.

8 By "traditional" I mean around the time of first contact with people from other lands and before the culture and economy was substantially altered by exogenous influences.

9 e.g. Barnard and Woodburn 1988:5; Plattner 1989:5; Butlin 1993:52-3,66.

10 See for example White 1993:207-9.

11 Garber 1987:356; Cashdan 1989:33; Hayden 1994:223; Isaac 1996:4; Kelly 1998.

12 Barnard and Woodburn 1988:10-11.

13 Murray 1988:9.

14 Gerritsen 2007a.

15 Barnard and Woodburn 1988:5.
This is what Plattner (1989:5) has termed "the psychic unity of humankind".

16 Notwithstanding the Formalist/Substantivist debate of the 1960s and 1970s (Polanyi 1968; Plattner 1989:13-15; Isaac 1993) which in my view set up a false dichotomy that simply "obfuscates" (Trigger 1985:193). While "embeddedness" may preclude a formal economic analysis, economic functions are still discernable and operate in pre-capitalist economies. Granovetter (1985) argues that "embeddedness" is found in modern economies and that pre-capitalist economies are not much different in this regard, and the problem is more one of perceptions.

17 Higgs and Jarman 1969:40.

18 Cook 1968:209.

subsistence. Similarly, population pressure has often been posited as the cause for such developments, Cohen's *The Food Crisis in Prehistory* being a classic exposition of this position. Alternatively many other scholars, such as Bender,[19] Lourandos,[20] and Hayden,[21] have argued that socioeconomic or sociopolitical dynamics provide the initial stimulus for economic development rather than a condition allowing or fostering further development. The challenge here is to determine which factor, or factors, model or schema is the most plausible, coherent and congruent with the data.[22] To achieve this, some means of discrimination is required to determine which of these complex and highly interpretative explanations has the greatest validity. The arguments and evidence for each theory will of course be discussed, but the ultimate intention will be to examine, as a critical test, the ability of each to "predict" or explain the geographical location and timing of early complex food producing sedentary economies where they have appeared in various parts of the globe in prehistory. Such a superficially simple evaluative technique is, I believe, in actuality a powerful tool for examining the coherence and cogency of these, or any other, relevant set of hypotheses. Naturally, further elaboration and a fuller exposition of this proposition will be entertained in the ensuing discourse, before any definitive conclusion is possibly reached.

Before concluding, some further clarification of the methodology and the nature of the evidence that will be exploited is necessary. I propose to employ an analytical framework which will involve investigations of a wide range of key exemplars, critical factors, notable developments and diagnostic situations, spread over many centuries and a number of continents. Informing this approach is the *Annales* framework, typified by Braudel and the concept of *longue duree*.[23] It is my intention, furthermore, to rely primarily upon historical ethnographic evidence in the arguments that will be advanced, producing an "historical ethnography".[24] When judiciously blended with archaeological evidence this becomes "reconstructive ethnography".[25] Each of these evidentiary domains, archaeological, historical and ethnographic, has its own limitations, of course, and this represents part of the challenge. Apart from preservational biases, for example, the element of chance in archaeological research may give quite erroneous or distorted impressions.[26]

Archaeological evidence may even be ignored or overlooked where it does not fit with preconceived notions.[27] It has also been pointed out on a number of occasions that archaeological finds are effectively meaningless unless they are interpreted in the context of some form of behavioural paradigm, usually based on ethnographic analogy.[28] By the same token, historical ethnographic evidence also has its limitations, such as those common to any work based on historical sources. Ethnography, moreover, effectively only represents in most instances a temporal "snapshot" of a particular group, and often misses private acts, uncommon or unspectacular but significant activities (such as women's gathering), and broader regional processes and diachronic trends.[29] In the employment of ethnography, unfounded assumptions may also be made that hunter-gatherers are "living fossils,"[30] that their societies are in stasis,[31] or that the ethnographic present can be projected far into the past, "naive historicism" as Ames terms it.[32] Failure to take account of these potential biases has led, on occasion, to the inappropriate employment of ethnographic data in attempts to reconstruct past hunter-gatherer communities.[33] In addition, Shipman has cogently argued that with the degree of variation in even the most basic traits in hunter-gatherer societies, there are significant dangers in extrapolating from any particular cultural circumstance.[34] Compounding this, the effects of outside influences on hunter-gatherer societies have, in some studies, either been ignored, not been recognised, or not given due weight, providing further scope for distortion.[35] Contact with state societies, commercial markets and mechanised transport have often been identified as major sources of such influences.[36] All these issues are relevant to any consideration of the "Australian Problem", particularly the latter.

In specific application to Australia it should be noted that relevant archaeological evidence is quite limited, especially in some of the regions that will come under consideration. Consequently, there will be a marked reliance on historical ethnography, keeping in mind the preceding qualifications. While Australia is often characterised as having been isolated from developments in other parts of the world this is not completely accurate.[37] In the first instance there has been a narrow corridor of communication with the rest of the world through New Guinea via Torres Strait and Cape York, open for much of the time since first human settlement

19 Bender 1978,1981.
20 Lourandos 1985.
21 Hayden 1990.
22 Murray 1988:11.
23 Braudel 1980; Knapp 1992:xv-xvi.
24 By "historical ethnography" I mean the synchronic reconstruction of the ethnography of a particular group at a certain point in time, in the context of Australian Indigenous studies, usually at time of contact. I see this as being distinct from the type of ethnohistory which constructs the history of a particular group over a period of time, usually in post-contact scenarios in Australia, this being couched in historical rather than ethnographic terms.
See Gerritsen (2007a) for a fuller exposition of this methodology.
25 See Gerritsen 2007a.
26 This has been demonstrated in a number of studies, for example in regard to the likely archaeological consequences of Hadza hunting (O'Connell, Hawkes and Jones 1988). See also Tainter (1978:128) and Meehan (1988).
27 Gerritsen 2001a.
28 Koyama and Thomas 1981:5; Flannery 1986:4; O'Connell, Hawkes and Jones 1988:114.
See also Allen (1996:138) for consideration of issues regarding ethnographic analogy with specific reference to the Australian context.
29 Wobst 1978:303-6; Yesner 1980:727.
30 Jones and Meehan 1989:131.
31 Yesner 1980:727; Blumler 1996:36; Mulvaney and Kamminga 1999:4.
32 Ames 1991:937. See also Hiscock 1999:101.
33 Kabo 1985:605.
34 Shipman 1983:31-2,43-4.
35 Mulvaney and Kamminga 1999:4; Schrire 1984:1-2; Plattner 1989:5-6; Widlock 2005:18-20.
36 O'Connell, Hawkes and Jones 1988:114.
37 e.g. Deetz 1968:283; Blumler 1996:36.

of Australia, about 46,000 or more years ago.[38] Other more powerful influences began to manifest themselves from the beginning of the 17th century with the arrival of Europeans (principally the Dutch followed by the British) and Indonesians (*Macassans*). These incursions had impacts well beyond their points of contact.[39] The *Macassans*, for example, appear to have stimulated the ceremonial exchange cycle in Arnhem Land in north Australia, hundreds of kilometres beyond their immediate sphere of influence.[40] Similarly we find valued European goods such as glass and axes up to 200 kilometres or more beyond the bounds of British colonisation, preceding the "frontier" by thirty years.[41] As in North America, smallpox epidemics also preceded the "frontier", devastating many populations, the extent of its demographic, social and economic impacts still the subject of considerable academic debate in Australia.[42] Whatever the specifics, it is nevertheless beyond question that there has been a progressive and profound alteration of Indigenous societies in Australia as external influences have increased. From an evidentiary point of view, where relevant, such influences and the changes that have arisen need to be recognised and taken into account.

In treating particular cases, areas, groups or regions it further needs to be recognised that the impact of exogenous influences was highly variable. Some may not have been affected by influences preceding the frontier, others dramatically so. When Aboriginal people first experienced contact with non-Aboriginal people, usually British explorers, "pioneers" and government officials, they were often afraid and uncertain, but at the same time curious about the strangers. Significant conflict usually followed once it became apparent the colonists intended to usurp their lands, resistance developed, and finally, in the face of defeat, attempts were made to adjust to the encroachments of the colonists and maintain some semblance of traditional lifestyles. In many cases this was succeeded by a process of disintegration, depopulation and near complete destruction of traditional life. The rapidity with which this happened was influenced largely by the degree of intrusion. Where intensive settlement by colonial intruders occurred the destruction of traditional life has been almost total and Aboriginal people have struggled, and still struggle, to maintain some semblance of their culture. Conversely, that process was much slower, much later or less intense in large parts of the arid and semi-arid zones and the tropical north of Australia. In some of these areas the land was perceived by the colonists to be of little economic value, resulting in minimal economic impact and low non-Aboriginal population densities. In some cases, such as in Arnhem Land in the north eastern part of the Northern Territory, deliberate decisions were made to restrict access by non-Aboriginal people and traditional culture has adapted and continues in a rich but modified form.[43] Consequently, in considering ethnographic evidence the interplay of all these factors and forces need to be weighed in determining the degree any specific item of evidence accurately reflects particular features of traditional, i.e. pre-contact, life. Where there is any suspicion that what was observed may have been a consequence or artefact of recent pre- or post-contact dynamics such evidence will be either qualified or discarded.

Finally one other matter needs to be clarified - what is meant by "Australia". Australia as a modern political unit includes not only mainland or continental Australia, Tasmania and offshore islands but almost all the islands in Torres Strait. The political boundary, in fact, comes within a few kilometres of the coast of mainland Papua New Guinea. A cline of intensification evident from the Pre-Contact Period (before 1606) across Torres Strait has previously been noted, including agricultural systems on some of the more northerly islands through to limited cultivation on the most southerly islands off the tip of Cape York.[44] By this definition of "Australia" agriculture was indisputably practiced in Australia in the pre-colonial era. But it would appear that various researchers, when they refer to Australia, presuppose a geographical definition that encompasses continental Australia, but which may also include Tasmania and offshore islands. For simplicity the term "Australia" will, henceforth, follow a geographic definition incorporating continental Australia, Tasmania and offshore islands, with the Torres Strait Islands being treated as a separate geographic entity.

[38] Johnson 2006:57-61.
[39] See for example Reynolds 1978.
[40] Thomson 1949:5-6; Clarke 2000:328-333.
[41] Morgan 1852:192; Hepburn 1853:62; An Old Hand [Hamilton] 1880:20; Curr 1886:1:78;2:375-6; Sharp 1952:69; Clark 1987:29-30.
[42] Ames 1991:937-8; See for example Butlin 1983,1993; Lourandos 1997:35-8.
[43] See Rowley (1972) and Clark (1998) for a more detailed exposition of this aspect of Australian history and the nature of the impacts.
[44] Harris 1979; Moore 1979; Barham and Harris 1985:247-8; Barham 2000:224.

# Chapter 2.
# People, Islands and Deserts

Approximately 55,000 years ago, with global climate cooling and sea levels falling, a land bridge existed between Australia and the lands to the north, now known as New Guinea. The latest hominin, *Homo sapiens*, had by then found their way out of Africa and established themselves in south west Asia. At some time after this they made their way into Australia, the first Australians, and also began to enter an icy Europe. But it was probably not until about 15,000 years ago that the first humans set foot in the Americas. Remarkably, once there, in a relatively short space of time, the first Native Americans populated the length and breadth of the two continents, developed agriculture and other forms of food production, generated chiefdoms, states and empires, and built some of the greatest cities on Earth, boasting architectural wonders the equal of any elsewhere. In both Eurasia and the Americas, preceding the advent of agriculture, other forms of specialised food procurement, and settled life, there was a period in which virtually all remaining vacant or sparsely populated ecozones were occupied, populations increased relatively rapidly, population densities increased, and people started to inhabit smaller and more sharply defined territories (increasing circumscription), or became more sedentary. Furthermore, "logistical" procurement became more pronounced, a wider range of food sources were exploited (the "broad-spectrum revolution"), more elaborate procurement and food processing technologies were employed, and subsistence organisation, as well as social and political structures, became more complex. In effect more intensive and increasingly specialised forms of subsistence evolved along with greater regional diversity in many facets of the culture. This phenomenon is known as "intensification".[1]

Such is the importance of this process in setting the stage for "the great leap forward" it has been argued we ought to be seeking to explain the origins of intensification rather than the origins of agriculture.[2] Intensification, as a global phenomenon, began in south west Asia around 15,000 BP [BP = Before Present], evolved into the Mesolithic or Epipaleolithic period, and culminated in agriculture between 11,500 - 9000 BP.[3] A similar period in Meso-, North and South America, leading into the Archaic, commenced around 10,000 BP, with early agriculture in the form of maize (*Zea mays*) cultivation apparent by 6250 BP.[4] This process was repeated in east Asia giving rise to rice and then millet farming between 11,200 - 8500 BP, perhaps earlier.[5] By and large the timing of intensification varied from place to place, region to region, but nevertheless it seems to have been "universal" in human populations.[6] Different economies developed, different plants came under human control, but in every case the same pattern appears to have emerged.

It has been argued that the process of intensification was also evident in Australia.[7] However, according to the proponents of this view, this transitional period in Australia did not begin to clearly manifest itself until 7000 - 5000 BP at the earliest.[8] It was still proceeding when British colonisation (from 1788) during the Contact Period (1606 – 1956) began to profoundly affect most traditional Indigenous societies. Parallels have been drawn between developments in Australia during this transitional period, known as the Australian Recent [the mid Holocene to the Contact Period], with the Mesolithic of south west Asia and Europe, the American Archaic, and the Japanese Jomon period.[9] One specific development, the seed-gathering economy of the *Paakantyi*,[10] has been labelled as "incipient agriculture".[11] Nevertheless the disparity in the timing of intensification in Australia relative to other parts of the world, especially

[1] Fagan 1995:155-9; Williamson 1998:143; Flood 1999:248; Weiss et al. 2004.

[2] Bender 1978:204.

[3] Price and Brown 1985:13; Unger-Hamilton 1989:101; Hillman and Davies 1990: 195,208-9; Anderson 1992:180; Allen and O'Connell 1995:vii; Fagan 1995:151-73; Anderson 1998:148-50,152; Willcox 1999:493; Bar-Yosef 2003; Bellwood 2005.
There is considerable debate about definitional and dating issues which will be discussed in subsequent chapters.

[4] de Tapia 1992:157; Fritz 1995:5; Smith 1997:932; Sherratt 1997: 275; Benz and Long 2000:460; Piperno and Flannery 2001; Matsuoka et al. 2002.
See n63 of Chapter 6 for further discussion of the dating.

[5] Zhimin 1989:646-7; Crawford 1992:9,13-15; Smith 1995a:127,130, 133-4,136; Underhill 1997:112-4,120,138; Higham and Lu 1998:870-1; Lu 1998:902,907; Pei 1998; Zhang and Wang 1998; Zhao 1998: 886,895; Wang et al. 1999; Shelach 2000:367-8,379; Toyama 2003:164-5; Yan 2003:152; Yasuda 2003a:134; C. Zhang 2003:189-90; Bellwood 2005:116.
See n44 of Chapter 6 for fuller discussion of dating and diagnostic issues.

[6] Price and Brown 1985:13; Allen and O'Connell 1995:vii; Sassaman 1992; Fagan 1995:155,157,159.

[7] Allen and O'Connell 1995:ix.

[8] Morwood 1981:42-3; 1984:343; Lourandos 1985:400-1; Smith 1986: 314; Lourandos and Ross 1994; Allen and O'Connell 1995:viii; Lourandos 1997:322-3.

[9] Lourandos 1997:23,333.

[10] Allen 1972,1974.

[11] Mulvaney 1975:242

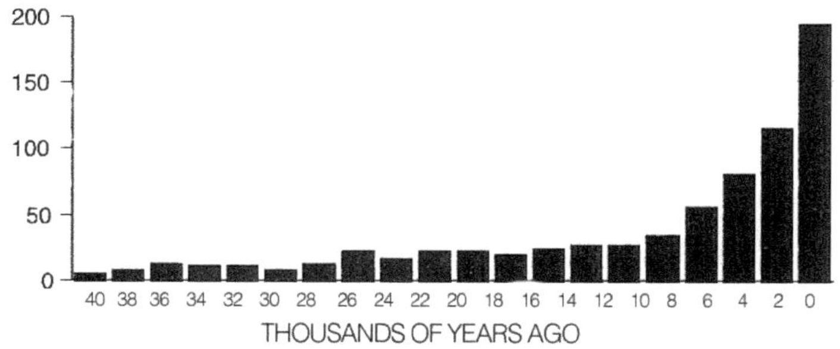


**Figure 1**: *Number of Rock Shelters Occupied Over Time in Australia (After Johnson 2006: Figure 8.1)*

the Americas,[12] requires some explanation. In some senses it is yet another facet of the "Australian Problem".

Intensification is a process that is not easy to precisely characterise or specifically define and will undoubtedly continue to provide a fertile field for academic speculation and debate. Its validity, in particular the necessity for "broad-spectrum procurement" to precede specialised modes of food production, has been challenged, particularly by Henry in regard to the antecedent Neolithic culture, the Natufian, of south west Asia.[13] Other controversies centre on appropriate archaeological methodologies for detecting patterns of intensification as well as, inevitably, its causes.[14] But before making any attempt to pass judgements on some of these issues a review of some of the Australian evidence may be instructive.

As noted earlier, some of the features of intensification included occupation of remaining ecozones, population increasing more rapidly and population densities also increasing. Often these trends were underwritten by the development and embrace of new subsistence strategies. In Australia the advent of the "Small Tool Tradition", the introduction of the dingo, exploitation of new plant foods, and the appearance of fish hooks, as well as the development of other, more elaborate, fishing techniques, have been singled out as some of the more conspicuous indicators of intensification through the adoption of new, increasingly specialised, subsistence regimes.[15] Determining prehistoric population and population densities, as a means of demonstrating more generally that intensification was under way, is a more problematic exercise. The Pre-Contact Period population of Australia, for example, has been a matter of debate for almost eighty years, with estimates varying by a factor of six. Attempting to ascertain population trends over millennia is even more difficult. One method involves analysis of the number and size of archaeological sites, and the frequency with which those sites were utilised. The results of a study of this type in south eastern Australia shows slow population growth initially (i.e. 50,000 BP to 10,000 BP), with a small spurt during the climate-optimum of the mid-Holocene, before increasing exponentially during the Australian Recent.[16] The most recent Australia-wide survey of the number of occupied rock-shelters (Figure 1) shows almost exponential growth from 8000 BP. Although the site data in these types of study could be interpreted as an artefact of the research methodology, or the result of local populations becoming more concentrated,[17] the consistency of the phenomenon in a variety of ecozones with different subsistence bases suggests otherwise.[18]

Paralleling this Australia-wide increase in populations and population densities was the permanent exploitation and occupation of almost all "marginal" ecozones, such as the most inhospitable desert areas, rainforests, wetlands and offshore islands. Previously these ecozones had either not been utilised or only utilised on a temporary basis when conditions were more benign. Using islands as an example, during the early Holocene many were created around the Australian coastline as a result of rising sea levels, which peaked at just above the present level around 8000 BP. Concomitantly local populations were either cut off, as in Tasmania, or retreated to mainland areas. On Kangaroo Island, a large island off the central south coast, the people of the "Kartan" culture became isolated and appear to have been unable to sustain themselves, dying out after several thousand years as the local climate deteriorated.[19] But on High Cliffy Island, part of the Montgomery Group, just 10 kilometres off the north west Kimberley coast, despite its small area (0.3 sq. km), exploitation was either continuous

[12] Allen and O'Connell 1995:viii.
[13] Henry 1989:17-18.
[14] Lourandos and Ross 1994:58; Williamson 1998:144-5.
[15] Luebbers 1978:ix-x,219,245-6; Lourandos 1985:401; Nicholson and Cane 1994; Holdaway and Stern:330,374-5; Lourandos 1997:22; Lourandos and David 1998:110; Williamson 1998:143,145; Flood 1999:248-9; Mulvaney and Kamminga 1999:44-7,230,316-7; Barker 2004.
See Hiscock (1986) and Freslov and Frankel (1999) for a contrary view, and Ulm (2002) contesting the validity of the evidence from south east Queensland.
[16] Lourandos 1997:224-5,Fig. 6.17.
[17] Bird and Frankel 1991:10; Johnson 2006:136-7.
[18] e.g. Luebbers 1978; Sullivan 1982; Hughes and Lampert 1982; Ross 1985; Lourandos and Ross 1994:58-9; Beaton 1995; Dortch 1997:19, 24,27,29,31; Allen 1998; Bowen 1998; McNiven 1998:78-9,87-8; Dortch 1999; Hall 1999; O'Connor 1999a,b; Morse 1999; Smith 1999; McNiven 2000:30; Lamb and Barker 2001; Ross 2001; David 2002:121-2,136,150.
David (2002:123,150) suggests the drop off in late Holocene sites are the result or a reluctance to date those sites. The likelihood of late Holocene sites, being nearer the surface, suffering destruction may be another reason for the apparent drop off.
[19] Lampert 1981; Bowdler 1995:945-6; Flood 1999:141-2; Mulvaney and Kamminga 1999:334-6

*Figure 2: Stone House Structure, High Cliffy Island˜North West Kimberley, Western Australia (After O'Connor 1999a: Figure 6.14)*

or commenced shortly after its separation from the mainland (6700 BP). The island had become permanently incorporated in the *Yawijibaya* subsistence pattern by 3000 BP and was being intensively exploited by 650 BP.[20] Permanent habitations, low stone huts or shelters, evolved as a manifestation of this increasing intensity of exploitation (Figure 2).

Yet other islands on the west and south west coasts of the continent, Barrow, Garden, Rottnest Islands and the Recherche Archipelago for example, were never visited following their abandonment, despite being within easy reach in some cases.[21] Elsewhere, along the south coast of Australia west of Port Phillip Bay, although lacking in coastal watercraft, at least some of the islands within swimming distance appear to have been visited occasionally.[22] Even so, Griffiths Island, only 100 m off-shore, was seemingly never visited.[23] This contrasts with North Keppel Island off the central Queensland coast which, although not visited for a couple of thousand years (4000 BP) following its separation from the mainland, nevertheless became regularly or permanently occupied within the last 1000 years.[24] Another apparently disparate example of island use was to be found around Tasmania. As shown in Figure 3, drawn principally from historical ethnographic data, virtually all islands around the coast were exploited on a permanent or seasonal basis, except those off the north east coast.

The islands that were exploited were visited by means of a distinctive type of Tasmanian "canoe" consisting of bundles of grass or bark lashed together, or simply by swimming. In the latter case it was usually women who tackled distances of up to 2.5 kilometres through vicious tidal rips.[25] It is thought the distances involved and the strength of the current was reason the north east Tasmanian islands were not visited.[26] While the Tasmanians were swimming and canoeing short distances other Aboriginal groups had managed to reach the Percy Islands 85 kilometres off the Queensland coast. This involved oceanic voyages of up to 27 kilometres.[27] To put this and the other evidence of island exploitation in perspective, consider firstly the number of

[20] O'Connor 1987:34; 1994:102; 1999b.
Although High Cliffy Island was only a small island it had a very large exploitation zone, with the Montgomery Reef covered an area of 300 sq. kms
[21] Nicholson and Cane 1994; Bowdler 1995:949.
[22] Bates 1985:252; Bowdler 1995:947,949
[23] Gaughwin and Fullagar 1995:43.
[24] Wyndham 1889:41; O'Connor 1992:50; Rowland 2002:68-9.
[25] Jones 1976:236.
[26] Jones 1976:256
[27] Rowland 1984.
Capt. Cook also reported "houses" on Lizard Island, approximately 30 kilometres from the mainland in far NE Queensland in 1770 (Cook 1893:321).

*Figure 3: Exploited Islands Around Tasmania*
*(After Jones 1976: Figure 7)*

islands being visited, regularly exploited, or permanently occupied over time, particularly in the last 3000 years,[28] as shown below in Figure 4.

In this instance the numerical increase in island usage can be treated as a primary manifestation of intensification. But how does one explain the disparate and highly variable pattern of island exploitation around Australia? Theoretical studies of islands globally identify distance, size and configuration (of islands as isolates, in groups or chains) as factors in the global pattern of island colonisation,[29] all applicable to Australia. To effect this sustained exploitation of islands, especially more distant ones, a number of innovations are required. Particular types of equipment employed in the exploitation of marine resources, such as canoes, fish hooks and bone tipped fish spears have been identified as the key elements in this process. These were in use in Australia, but not uniformly, other factors were clearly operating. For example, the average dating of the

[28] Barker 1996:35-7; Border 1999:136; Lamb and Barker 2001.

[29] Keegan and Diamond 1987.
Very few Australian studies have considered these factors. Bowdler (1995) is an exception.

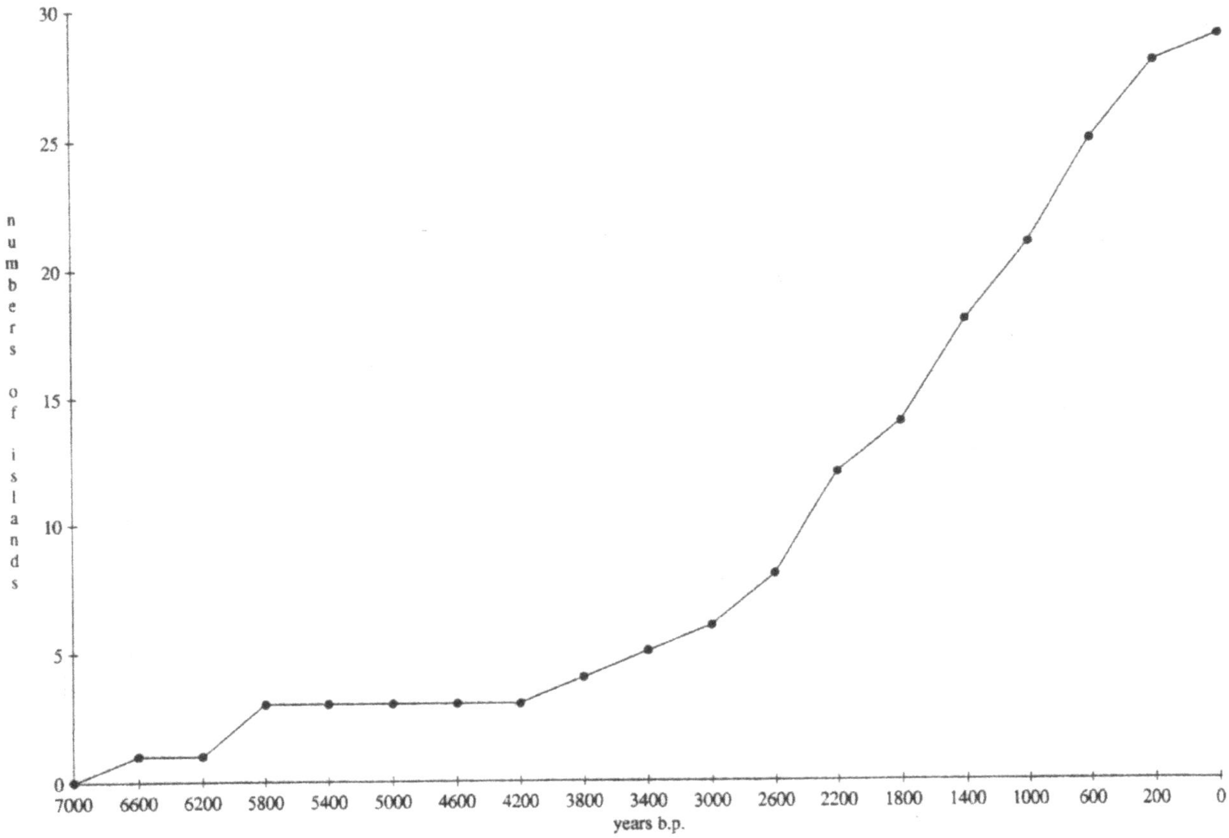


*Figure 4: Number of Islands in Use Over Time*
*(Bowdler 1995: Figure 6)*

commencement of exploitation in the Holocene of nine islands north of the Queensland border, where a crossing of more than 3 kilometres was required (beyond normal swimming distance), was 2853 BP. This contrasts with the archaeological evidence indicating only three islands where a similar crossing was required were visited south of Queensland, along the south east, south, west and lower north west coasts of Australia, and then not until 1467 BP on average.[30] Apart from Tasmania, the only islands permanently occupied were those off the north and north east coasts of Australia.[31] So a number of questions arise – what led to these differences; was the means developed specifically for the purpose of getting to and exploiting islands or was this just an extension of traditional subsistence activities; was it the product of the incremental adaptations or more efficient technology characteristic of intensification? - was it the result of increasing population pressure on the adjacent mainland areas?

One scenario, "resource pull" in the form of rich fishing grounds around islands, has been suggested as an incentive to develop fishing technology, such as fish hooks, and watercraft capable of reaching islands.[32] But this has been challenged on the basis that islands in some cases were exploited without the adoption of new technologies, or that fishing was often not a major subsistence activity on the adjacent mainland areas.[33] Some have tried to argue that climatic change, manifested in coastal regions as changes in sea level, had an important bearing on the development of coastal marine economies. According to this view sea levels rose considerably at the end of the Last Ice Age before stabilising about 8000 years ago. Only then, once new habitats and niches had evolved, did local populations adapt to take advantage of the opportunities offered, new techniques and technologies emerged, and the coastal marine economies became established, extending to coastal islands.[34] Apart from the contested proposition that marine resources were initially depleted and then built up again following stabilisation of sea levels,[35] this scenario is difficult to test as most of the evidence for preceding subsistence activities lies beneath the waves. Islands may

[30] Bowdler 1995:947,949; Gaughwin and Fullagar 1995:43. Eighty Mile Beach was the limit on the north west coast. Historical ethnographic evidence does, however, indicate that some of the southern and western Australian islands beyond 3 kilometres were being visited. For example see Bates n.d.a: Notebook 24 - Informant T. Carter. Recent research has established that Hook Island off the central Queensland coast appears to have been continuously occupied throughout the Holocene (Lamb and Barker 2001).

[31] Jones 1976:260; Campbell 1982:96; Rowland 1986:77,83; Bowdler 1995:954; Mulvaney and Kamminga 1999:320-2; O'Connor 1999a:1,116; Lamb and Barker 2001.

[32] Jones 1976; Sullivan 1982:15-16.

[33] O'Connor 1992; Gaughwin and Fullagar 1995:47.

[34] Binford 1968; Lampert and Hughes 1974; Yesner 1980; Beaton 1995: 801.

[35] Ulm and Hall 1996:55; Attenbrow 1999; Barker 1999:125; 2004:12-15.

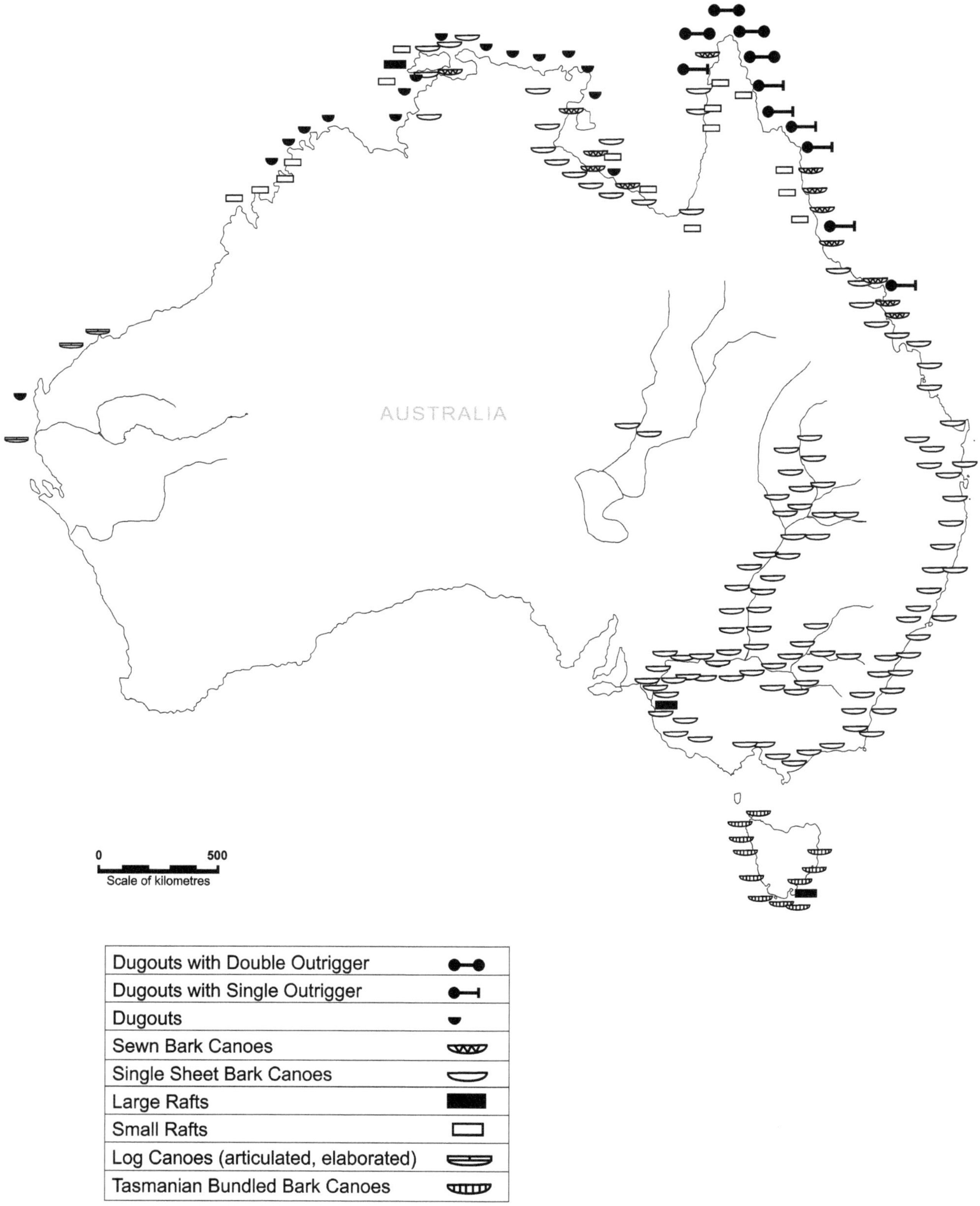


*Figure 5a: Distribution of Watercraft Around Australia*

**Figure 5a. notes**

1. While very extensive research has been carried out to determine this and the other distributions, the research cannot be considered exhaustive. Where the occurrence is well documented only limited or representative examples are used. Where there is questionable, contradictory or inconsistent evidence data has not been included on the distributional map unless there is some form of strong corroboration or overriding evidence.
2. These distributions are based as far as possible on observations and information collected from the time of historical European contact until the early part of the British occupation of each particular area, archaeological and ethnographic evidence, and oral traditions. The maps are indicative of the distributions as they stood around the time of British occupation. It needs to be recognised, however, that parts of the distributions appear to have changed (particularly in regard to watercraft in the Kimberley region) in the period between first European contact and British occupation.
3. In some instances there are words for canoes, fish hooks and fishing lines in local vocabularies, but these are not taken as being certain that these were in use traditionally unless there is some corroboration, as it may simply indicate knowledge of them or recent adoption (particularly fish hooks), not actual or traditional usage.
4. In some cases watercraft were not in common use, only being made intermittently or contingently for particular circumstances or occasions (such as when the rivers were running or there were floods). In these instances such watercraft were used simply as a means of transportation and they did not serve any other integrated economic or subsistence function. Consequently they have not been included in the distribution.
5. Some references are included as indicative of an absence of watercraft, fishtraps and so forth.
6. Nets used in hunting are not included in the net distribution.

Bates n.d.a: Notebook 24 – Informant T. Carter; Carstensz 1623: 39,41; Dampier:1697:464,468; Swaardecroon, Chasselijn and Craine 1705:170,172; Witsen 1705:175; de Haan 1756:92,94,97; Banks 1768- 71: 2:52-4,73-4,91,93,134; Cook 1768-71:304-6,308-9,312,339,396-7; Parkinson 1773:135,141,147; Flinders 1814:2:137,154,198; Oxley 1820:323,346; Pamphlett and Uniacke 1825:101,103-5,110,112,116, 118,121,124,126; King 1826:1:38,40,43-4,67,101;2:66-7,69,87-90, 137,200-12; Lockyer 1826-7:28/12/1826; Dawson 1830:79,247; Sturt 1833:2:201; Bennett 1834:1:115,242; Irwin 1835:23; Imlay 1838:232-3; Oakden 1838: 289; Strangways 1838:29; Mitchell 1839:1:283 et seq.; Perry 1839:40; Stephens 1839:75; Grey 1841:2: 65-264; Eyre 1845:2:263-4,310,314; Russell 1845:314; Stokes 1846: 1:79,80,112-3,173-5;2:15-16; Angas 1847:1:90; 2:230-1; Leichhardt 1847a:375; Austin 1851:7; Helpman 1851:268; Macgillivray 1852:45- 9; Henderson 1854:2:137,153; Austin 1855:xi; Phillips 1856; Brierly 1862a; Jukes 1862; Cooper 1862; Brierly 1862b; Capt. Jarman 1863:57; Stuart 1864:77; Martin 1865:265,286,288; Bennett 1867:552; Lord and Baines 1876:142; Smyth 1878:1:lviii-lix,407- 22;2:298; Wyatt 1879:169; Davies 1881:143; Gregory 1884a:41; 1884b:56; Jung 1884:119; Curr 1886:1:255,269,273,291,293,295, 335,356-7; 2:27,29,31,33,41,43,107,126,131,138,147,151,161,183, 200,225,289,309,313,334,339,391-5,415,418,421,423,425-7,429,431, 445-9,453,481,483,487-91; 3:7,22,33,35,41,43,48,51,53,57,63,75,77,81,83,89,93,99,101,105-9,125,128,133,135,137,141,143-4,149,151,197,211,215,221,227,229,233-7,241-2,247,261,267,277,288,291, 299,309,311,319,331,333,337,341,351,357,359,371,375,379- 81,383-5, 389,393,395-401,407-11,415,419,423-7,431-3,445,449-59,465,475-83, 487-95,501-15,519,529-35,548,551-63,583,587-9,639; Basset-Smith 1894:331; Lockyer 1888:590; Beveridge 1889:19,35,64-5,82,85-8,91; Lumholtz 1889:316; Wyndham 1889:40-1; Mann 1893:1:98-9; Stretton 1893:228; Boyd 1896:48,50-2,59; Willshire 1896:49; Saville-Kent 1897:8; Worsnop 1897:118-22; Fawcett 1898:153; Parker 1898:102; Roth 1899:152-9; Durlacher 1900:90; Hawker 1901:2:18; Brockman 1902:11-2,17-8; Hey and Roth 1903:11; Peggs 1903:361; Roth 1903: 65; Cary 1904:32,34-6; Spencer and Gillen 1904:680-2; Thomas 1905; Thomas 1906:83-6; Le Souef 1907:247; Thomas 1907; Eylmann 1908:379-80; Mathews 1908:66,68,72,75-81; Roth 1908; Roth 1910a; Spencer and Gillen 1912:2:483-4,348-51; Welsby 1913:38-40; Basedow 1914:303-5; Fitzpatrick 1914:50; Brown 1916; Love 1917: 32-3; Basedow 1918:189,193-4,219; Stuart 1923:22-3,60; Teichelmann and Schuermann 1923:162; Basedow 1925:161-2; McConnel 1930-1:104; Piddington 1932:344; Thomson 1934:226-7, 229; Davidson 1935; Kaberry 1935:412-3; Love 1936:8; Thomson 1936:71; Blount 1937:33,35; Anon. 1938:560; Davidson 1938:73; Thomson 1939a:121-3; 1939b:212,214; Hornell 1940:116-7; Berndt 1941:23-4,27-8; Bryant 1943:7; Black 1947:352-4,357-61; Thomson 1949:52,58-9; Thomson 1952; Warner 1958:458-60; Duncan-Kemp 1961: 7; Crawford 1965:7; Curr 1965:48,50-1,79; Duncan-Kemp 1968:124; Sharp 1968:351-2; Macknight 1972:304,309; Edwards 1973:9,10-14,18,26,30,39,65,68-70; Tindale 1974: 37,57,72,75,94,114,145,147,175-6,182,199,240-3,249,252,254,257,261-2,282; Akerman 1975; Dargin 1976:31; Macknight 1976:27,36; Craze 1977:22; Rowland 1980:3; Doran 1981:76,78-9; Lampert 1981:174; Ryan 1981:32-47; Gillespie 1984:4,6,31; McKellar 1984:13,47; Bates 1985:253,257-8, 284; Rowland 1986; Rowland 1987:40; Trigger 1987:80-1; Baker 1988:179-80; French 1989:31; Turbet 1989:51-5; Rathe 1990:48-9; Gott and Conran 1991:61; Nicholson and Cane 1991; Berndt, Berndt and Stanton 1993:14,83,85-6,118; Gerritsen 1994a:313n13; Bowdler 1995:955; Mitchell 1996:181,183; Clark 1998- 2000:4:49,50-1,53-54, 78,98; O'Connor 1999a:116; Barham 2000:228-9,239,241,243-5,248-9; Gerritsen 2001a:20-1,24-5; Barker 2004:32,38-42.

well have been previously exploited, only to be drowned, necessitating adjustment or re-establishment of the subsistence pattern as sea levels stabilised.[36] Furthermore, it fails to provide a cogent explanation for the pattern of island exploitation seen around Australia, and probably elsewhere. Similarly, arguments based on population pressure do not account for the pattern of island exploitation either and simply beg the question - what was it that led to the population pressure?

Perhaps this issue and the phenomenon of intensification can be more readily understood in terms of diffusion, endogenous development and greater economic specialisation. Consider, for example, the distribution of traditional watercraft, fish traps and weirs, net fishing and fish hooks in Australia where these elements of were recorded in the Contact Period, 1606 – 1956.[37]

[36] Barker 1991:106; Morse 1993; Nicholson and Cane 1994:115; O'Connor 1999a:123.

[37] It should be noted that these distributions were not static during the Contact Period and changes are discernible from different reports arising during separate contacts early in the Contact Period. In such cases the later reports, if still pertaining to the use of traditional materials in a traditional context, are used.

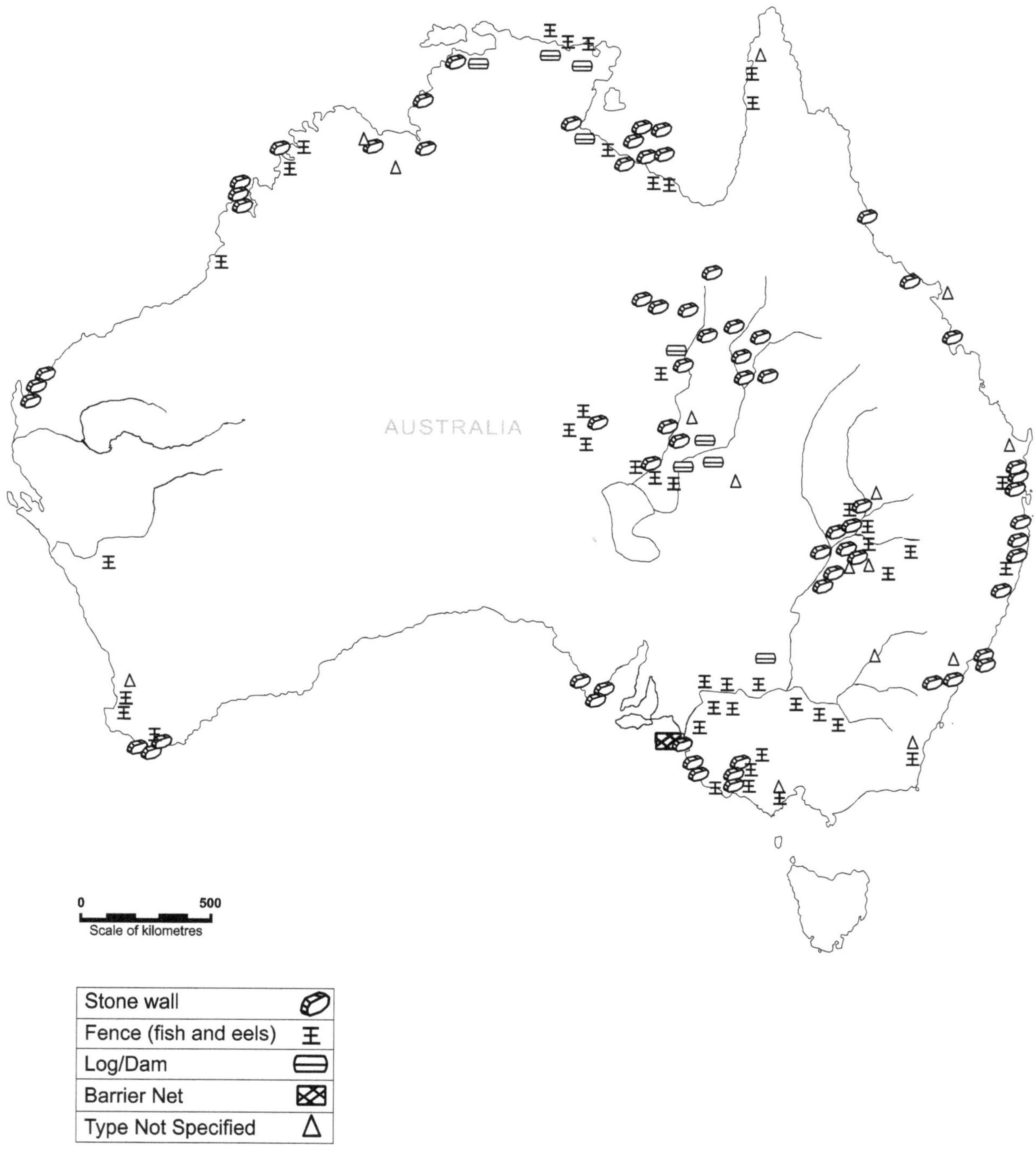


*Figure 5b: Distribution of Fish Traps and Weirs Around Australia*

**Figure 5b. notes**

Bates n.d.b:101; Dampier 1697:465; Dampier 1703:149; Vancouver 1798:1:38; King 1826:1:16,31; 2;96; Lockyer 1826-7:3; Nind 1832: 32-3; Sturt 1833:1:41,72; Anon. 1834:332; Armstrong 1836:793; d'Urville 1839:2:276,Pl.35,fig.3; Mitchell 1839:1:93; Robinson 1844a; Eyre 1845:2:253,301; Neill 1845:1:425; Angas 1847:1:112, 155,174; Mitchell 1848:103,113,121; Sturt 1849:1:105; Henderson 1854:2:136-7; Browne 1856:492; McKinlay 1861:69; Wills 1861:17; Wright 1861-2:85; Krefft 1862-5:368; Davis 1863:290; Stuart 1864: 252-3; Lang 1865:19-20; Lewis 1875b:13,26; Smyth 1878:1:202,234; Forrest 1880:29; Curr 1886:1:65,343;2:378;3:306; Douglas 1886:83; Beveridge 1889:81-2; Fraser 1892 :52; Stretton 1893:241; Kirby 1894:35-6; Stow 1894:12; Boyd 1896:57; Roth 1897:95; Worsnop 1897:100-2; Hall 1898:219; Mathews 1900-1:78; Roth 1901:23; Jackson 1902:14; Mathews 1903; Cary 1904:34; Petrie 1904:73-4; Anon. 1905:226-7; Parker 1905:109-10; Basedow 1907:23; Norton 1907:102; Banfield 1909:53-4; Kenyon 1912:110; White 1914:143; North 1916-17:126,128; Horne and Aiston 1924:64; Basedow 1925: 129; Ilbery 1927:25; Bunbury 1930:69,87; McConnel 1930-1:101; Daley 1931:25; Piddington and Piddington 1932:345; Hammond 1933: 46; Enright 1935; Love 1936:137-8; Thomson 1936:73; Anon. 1938: 560; Thomson 1938; Mountford 1939; Thomson 1939b:214,220; Mitchell 1949:149; Thomson 1949:19,27; Flecker 1951; Duncan-Kemp 1952:105,142,186,237; Thomson 1954:118; Durack 1959:94,96; Tindale 1962:302,304; Massola 1963:255; Crawford 1965:7; Curr 1965:79; Crawford 1968:56-7; Duncan-Kemp 1968:124,274-5; Massola 1968; Massola 1969:101-2; Gill 1970:30; McCarthy 1970:59; Allen 1972:51; Campbell 1973:9; Baudin 1974:178,487; Tindale 1974:18, 23,61-2,111,147,243,254,Col. Pl. 34; Stockton 1975; Akerman 1976; Dargin 1976:45; Dix and Meagher 1976; Dortch and Gardiner 1976: 265,274-7; Lourandos 1976:183-7; Northern Territory Museum Site Files 1976-83; Coutts, Witter and Parsons 1977a:23-4; Happ 1977: 36-7,39,43-4,47,52; Pretty 1977:313; Campbell 1978:122-3; Coutts, Frank and Hughes 1978; Lourandos 1980a:251-4; Campbell 1982; Coleman 1982:8; Watson 1983:43; McKellar 1984:47,67; Bates 1985: 251,253-5,257,325; Walters 1985a; Walters 1985b:51-3; Rowland 1986:79; Trigger 1987:79-80; Godwin 1988; McKelson 1989:37; Mulvaney 1989:19; Nicholson and Cane 1991; Mulvaney and Green 1992:347,376; Berndt, Berndt and Stanton 1993:96-7; Tacon 1993:224; Nicholson and Cane 1994:113; Attenbrow and Steele 1995; Lourandos 1997:47; Bowen 1998; Clark 1998-2000:2:145,161-3,178, 306; 3:48,55; 4:185,300; Border 1999:133; Dortch 1999:25,32; Flood 1999:200-1,254; Mulvaney and Kamminga 1999:34-5; O'Connor 1999a:13; Clarke 2002:154-6; Dortch 2002:13; Peron 2003:119-20.

Based almost exclusively on ethnographic reports, these distributions reflect to some extent the degree of specialisation in the exploitation of aquatic or marine resources. Each of these types of structure or forms of technology requires some level of labour investment. This investment is directly related to not only the size but the complexity of the structure or technology involved, the greater the size and complexity the greater the investment, either in terms of labour or trade goods. Presumably, in those regions where a greater number of more complex aquatic or marine exploitation technologies or structures were used, a greater and more sustained portion of daily effort was dedicated to procurement of aquatic or marine resources. In effect a greater commitment was being made to one type of subsistence economy.

Two of those regions where all four of the identified means of aquatic/marine/maritime exploitation appear, Arnhem Land and coastal Northern Territory, and Cape York and the east coast of north Queensland, are known to have been strongly influenced by "foreign" cultures. In the case of Arnhem Land and the Northern Territory, this is the accepted explanation for the appearance of dugouts with sails, these canoes initially being obtained in trade from the *Macassans*, seafarers from Sulawesi who had been visiting the area since around 1700, perhaps earlier.[38] Similarly, the appearance of dugouts with single or double outriggers around the tip and down the east coast of Cape York is attributed to Melanesian influences transmitted through Torres Strait. Other innovations such as fish hooks in these regions are also attributed to exogenous influences.[39] But at the other end of the scale and the continent were the Tasmanian "canoe rafts". As Tasmania had been cut off from the mainland by rising sea levels some time after 12,500 BP, their watercraft were presumably a purely local development. The fact that Tasmanian islands, such as Hunter Island, requiring a crossing of 2.5 kilometres, and Maatsuyker Island, 6.5 kilometres from the nearest landfall, were not visited until 2500 BP and 570 BP respectively suggests the distinctive Tasmanian canoe rafts had developed only in the last few thousand years.[40]

The Murray-Darling basin, the south east coast of Australia, and the north east of the Lake Eyre Basin are three other areas where subsistence was strongly oriented around aquatic or marine resources - fish traps, nets, fish hooks and watercraft forming part of the suite of exploitational hardware in those regions. They appear to have been separated to some extent from other regions where the full gamut of equipment was used by gaps of hundreds of kilometres. Although there is evidence of some use of watercraft around Moreton Bay (see Figure 5a) it has been argued that fish hooks and watercraft diffused down the east coast and that this gap in the distribution on the central east coast arose because either the ecology was not suitable in that part of the coast, or that the local people rejected their use for cultural reasons.[41] Both reasons are suspect however, Moreton Bay (Brisbane) for instance, where subsistence was based on intensive exploitation of marine resources using spears, fish traps and nets, was eminently suited to

[38] Thomson 1949:52,59; Macknight 1972; 1976:61-82,92,96-9,157-8,161; 1986:70; Mitchell 1996:186; Morwood and Hobbs 1997:198; Clarke 2000:325-332; Bulbeck and Rowley 2001:59-64; Gerritsen 2001b.

[39] Davidson 1935:69-70; Rowland 1987; Mulvaney and Kamminga 1999: 263,322; Barham 2000:241-249.

[40] Bowdler 1995:946,949; Flood 1999:207.

[41] Bowdler 1976; White and O'Connell 1982:151; Walters 1988:106-9; Nicholson and Cane 1994:113.

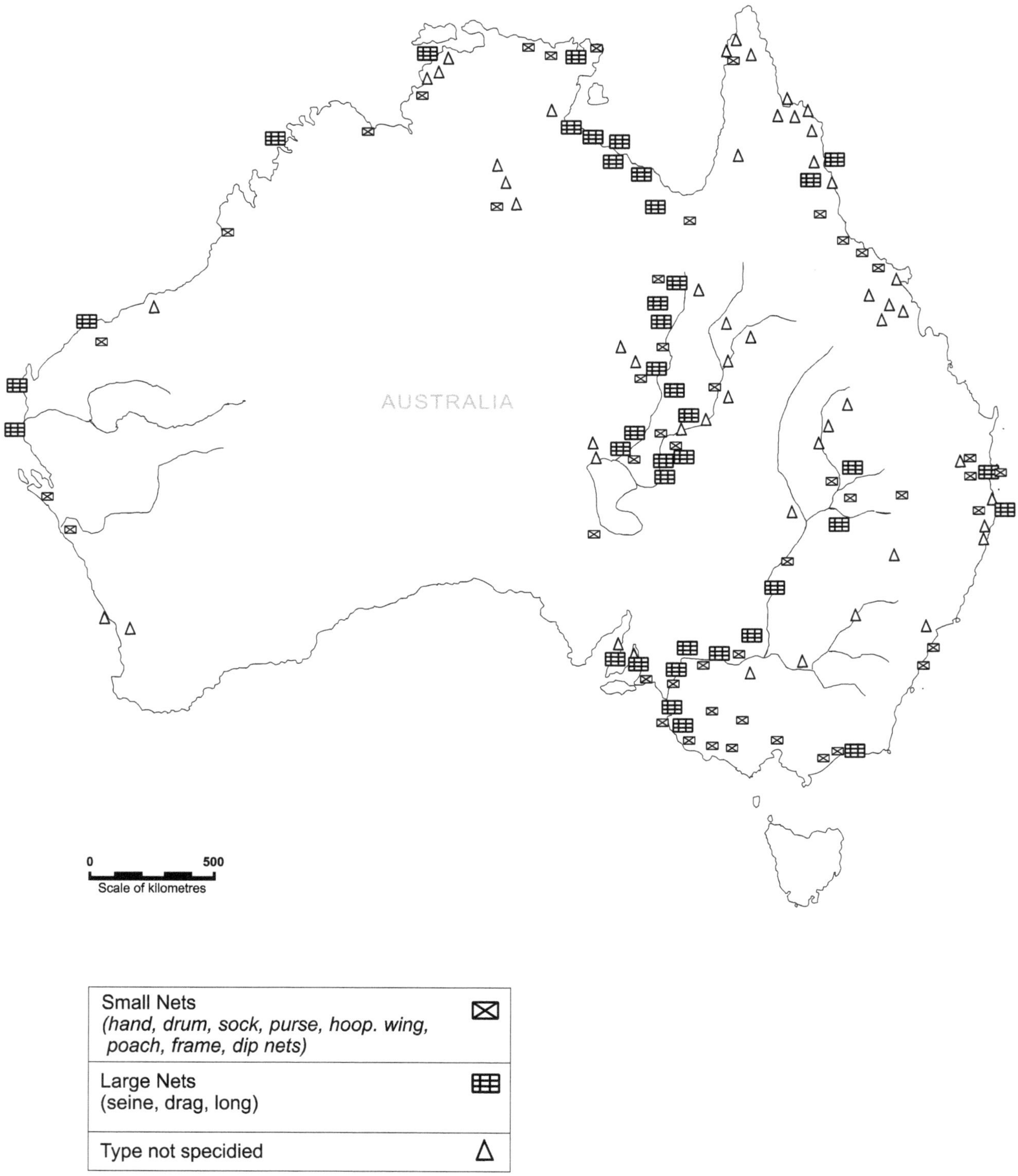


*Figure 5c: Distribution of Net Fishing Around Australia*

**Figure 5c. notes**

Bates n.d.c:70; Bates n.d.d:53; Bates n.d.e:3; Dampier 1697:465; Swaardecroon, Chasselijn and Craine 1705:170; Pamphlett and Uniacke 1825:100,107,113-4,119,122; Sturt 1833:1:89-90,93,106; Imlay 1838:232; Mitchell 1839:1:100,287,305; 2:128-9; Perry 1839 :40; Eyre 1845:2:251-3,261,311-12; Angas 1847:1:101; Mitchell 1848:113,119,367; Sturt 1849:1:106-7,261,270; 2:62,68; Henderson 1851:2:122,136; Phillips 1856:270; Gregory 1858:4; Howitt 1861:11; King 1861:4; McKinlay 1861:21,46; Brahe 1861-2:89; Landsborough 1862:101; Krefft 1862-5:361,366,368; Davis 1863:113,203,219-20; Stuart 1864:62-3; Oldfield 1865:269,274,295; Gason 1874:34; Lewis 1875b:13-14,26,42; Hic et Ubique 1877:13; Hodgkinson 1877:217, 521; Smyth 1878:1:142,234,389-90; 2:305; Andrews 1879:85; Bennett 1879:312; Meyer 1879:192-3; Schurmann 1879:219; Wyatt 1879: 177; Tolmer 1882:1:222; Sanger 1883:1224; Gregory 1884a:167; Gregory 1884b:58,71-3; Jung 1884:114; Palmer 1884a:284; Walcott 1884: 98; Curr 1886:1:251,270,288,291,299,330,343; 2:18,24,144,183,191, 201,331,341,377-8,401,418,432,465,471 ,497; 3:19,22-3,97,252,306, 335,342,354,365,544,548,570,580; Beveridge 1889:19,73,76,78-80, 83; Lumholtz 1889:331; Watkins 1890-1:44; Kirby 1894:34; Stow 1894:12; Wells 1894:518-9; Gason 1895: 169,172; Gregory 1896:44; H.P. 1897:81; Parker 1897:124; Roth 1897:94-5,127; Worsnop 1897:90-1; Coghlan 1898b; Fawcett 1898:153; Durlacher 1900:37-8, 95-6,111,160; Hawker 1901:2:18; McLellan 1901:258; Meston 1901: 63; Withnell 1901:21-2; Roth 1903:47,67; Clement 1903:3,7,16, fig.20; Petrie 1904:66-8,73-4; Parker 1905:107-10; Thomas 1906: 92-3,95; Basedow 1907:24; Gill 1907-8:228; Banfield 1908:165; Eylmann 1908:375,382; Banfield 1909:49; Spencer and Gillen 1912: 2:466; Fitzpatrick 1914:42; Hall 1917:24; Banfield 1918:184-5; Teichelmann and Shuermann 1923:154; Horne and Aiston 1924:33,62-3,176; Basedow 1925:129-30,fig.3; Bennett 1927a:410; Bunbury 1930:87; McConnel 1930-1:102; Davidson 1933; Duncan-Kemp 1934:112,121-2, 258; Kaberry 1935; Thomson 1939b:214,220; Harvey 1943:110-12; Johnston 1943:298; Thomson 1949:19,27; Lamond 1950; Williams 1950:7; Flecker 1951:2; Tindale 1951; Duncan-Kemp 1952:126,236-7; Morey 1952:40; Durack 1959:94,96; McArthur 1960: 109,113; McCarthy and McArthur 1960:152,158; Duncan-Kemp 1961: 41,118,231; Crawford 1965:6,8; Curr 1965:110,131,171; Cleland 1966:127; Duncan-Kemp 1968:3,124; Carmichael 1973:23; Tindale 1974:18,61-2; Bowdler 1976:249-50; Happ 1977:53-4; Salvado 1977: 159; Moore 1979:150-1; Kerwin and Breen 1981:288-93; Reuther 1981:III,119/1700; Eyre 1984:156-7; Vanderwal and Horton 1984: 110-1; Bates 1985:250-1,256-7; Jenkin 1985:14; Walters 1985b:50- 3; Rowland 1986:78; Satterthwait 1986:39; Colley and Jones 1987:70-1; Trigger 1987:78-9; Marchant 1988:150; French 1989: 35; Turbet 1989:49; Rathe 1990:47,56,73; Nicholson and Cane 1991; Berndt, Berndt and Stanton 1993:28,95-100,562-6,Pls.11,16; O'Connor 1993:227; Bulmer 1994:49; Nicholson and Cane 1994:113; Attenbrow and Steele 1995; Flood 1999:200; Mulvaney and Kamminga 1999:355; O'Connor 1999a:126; Clarke 2002:151-154,157,159; 2003: 150; Tolcher 2003:14.

watercraft and fish hooks.[42] Whereas there is no direct evidence to support the cultural explanation for the gap, it is apparent from the few dated examples of fish hooks that fish hooks were diffusing from the north while being independently invented in the south.[43] Fish hooks appear in north Queensland on Keppel Island (1070 BP)[44] and Whitsunday Island (800 BP)[45] at about the same time as they appear at locations 2000 kilometres away on the south coast of New South Wales, such as Durras North (700 BP)[46], northward to Bass Point (570 BP)[47] and Port Stephens (490 BP).[48] The gap had simply not been "closed".

Another gap of hundreds of kilometres in the distribution of fish hooks and watercraft existed between the New South Wales/eastern Victorian region and the Murray-Darling Basin, suggesting that this too was a locus of independent invention. A similar conclusion could be reached in regard to the fishing infrastructure and technology found in the north east Lake Eyre Basin. The presence of dugout and log canoes as well as fish hooks (the only instance of fish hooks in the western part of Australia) and nets (for catching dugong and turtles) in a region centred on the northern Gascoyne coast of Western Australia might be considered a further example of independent invention.[49] However, certain lines of evidence suggest that these features might have arisen as a result of the impact of Dutch sailors marooned in the region in 1712.[50]

In summary it would seem foreign influences played a significant role, through technology transfers, in maritime intensification in northern Australia, and perhaps the Gascoyne-Shark Bay region, while in other parts of Australia parallel autochthonous developments were taking place. This processes involved what has been termed "niche shifting", in coastal regions from "coastal economies" to "marine economies" for example.[51] While it is assumed that this has happened across much of northern Australia, it is also evident in Tasmania.[52] For much of the last few thousand years, the Tasmanians were pursuing predominantly terrestrial and coastal economies but then in certain areas they began to incorporate more islands and islands further from the coast in their subsistence round. Hunter Island, for example, became part of the pattern of exploitation about 2500 BP with Maatsuyker Island's seals becoming an integral food source from 570 BP on.[53] Effectively the Tasmanians were "niche shifting",

42 Hall 1982; Walters 1985a,b.
It would seem bark canoes were in use to some extent in Moreton Bay and the rivers of the adjacent coastal region. See Pamphlett and Uniacke (1825:103-5,110)
43 Mulvaney and Kamminga 1999:263; Gerritsen 2001a:24-5.
44 Rowland 1980:13; 1981:67; 1983.
45 Barker 1991; Lourandos 1997:210; Barker 2004:42,145.
46 Lampert 1966:94; Lourandos 1997:210; Mulvaney and Kamminga 1999: 292.
47 Bowdler 1976:254; Hughes and Djohadze 1980:17.
48 Dyall 1982:55; Sullivan 1987:104.
49 Phillips 1856:270; Oldfield 1865:269,274,295; Rathe 1990:47,56, 73; Gerritsen 1994:176-180; 2001a:25.
50 Gerritsen 1994a:180-1,252-60.
51 As defined by Gaughwin and Fullagar 1995:39. Their characterisation is limited in my view and I would suggest 2 further distinctions - a maritime economy that exclusively exploits the coastal marine zone and adjacent islands, and a pelagic economy that is exclusively based on the exploitation of island marine zones and the open oceans.
52 Jones 1984; Vanderwal and Horton 1984.
53 Jones 1984; Vanderwal and Horton 1984; Nicholson and Cane 1994; Bowdler 1995:946,949.

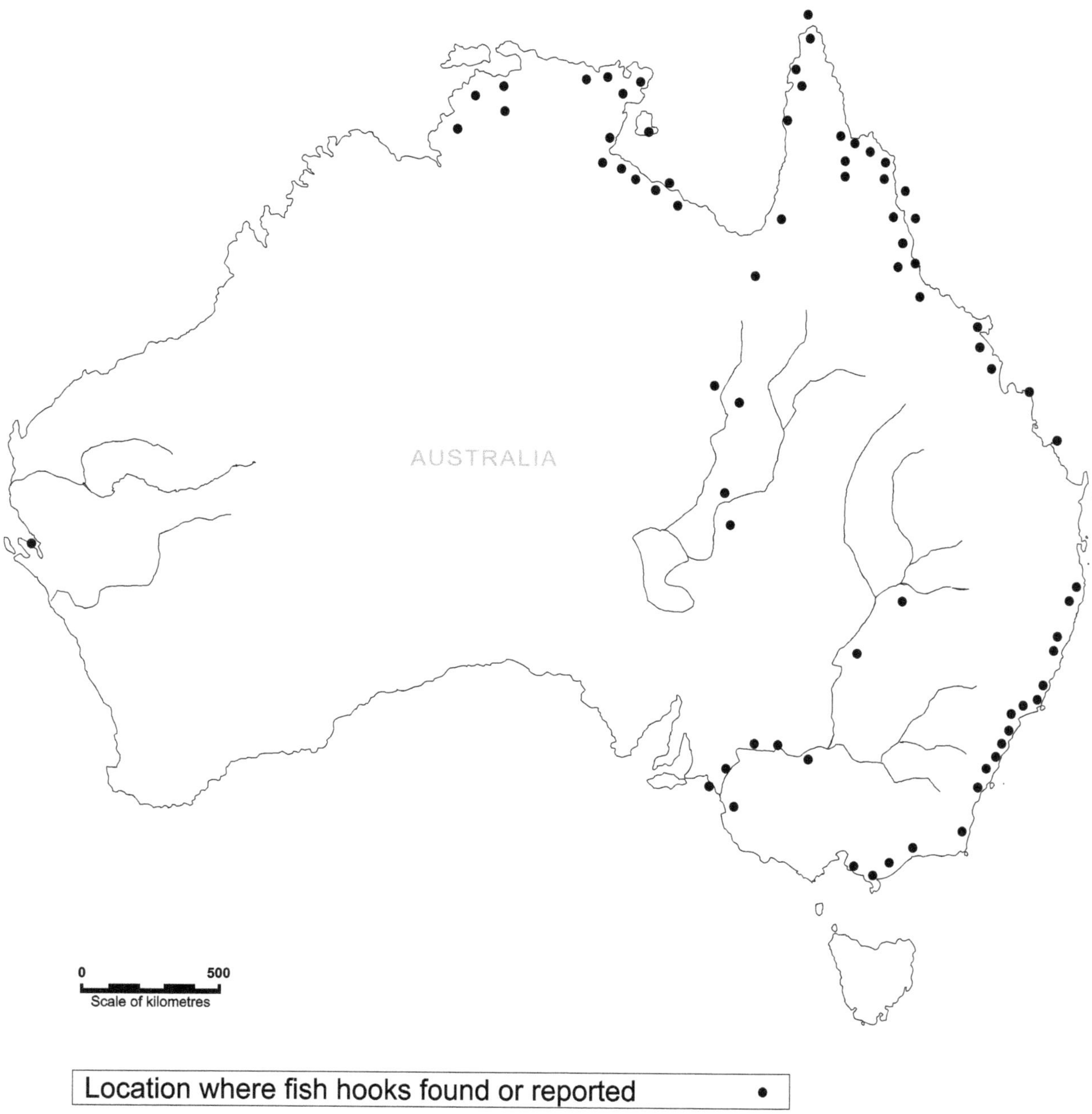


*Figure 5d: Distribution of Fish Hooks Around Australia*

**Figure 5d. notes**

Fusiform bi-points, otherwise known as *muduks* or fish gorges, have not been included in the fish hook distribution.

Dampier 1697:465; de Haan 1756:92; Banks 1768-71:2:131; Oxley 1820:176; King 1826:1:37,43,354; 2;14,67; Williams 1839:61; Grey 1841a:1:378; Schurmann 1844:73; Oldfield 1865:274; Ridley 1878:253; Meyer 1879:192; Schurmann 1879:219; Wyatt 1879:171; Jung 1884:114; Palmer 1884a:284; Curr 1886:1:299; 2:201,334,341, 378,418,426,432,443,471,494; 3:284,348,407,548; Beveridge 1889: 73,86,88,91; Bicknell 1895:143; Durlacher 1900:162; Roth 1901:20- 1; Roth 1903:47; Petrie 1904:57,73; Roth 1904:33-4; Spencer and Gillen 1904:677,figs 239,240-2; Basedow 1907:25; Fountain 1907: 289; Banfield 1909:59; Fitzpatrick 1914:45; Teichelman and Schurmann 1923; Williams 1923:61; Basedow 1925:132; Kaberry 1935: 414; Love 1936:136; Thomson 1936:73-4,Plates IX,XI; Thomson 1939b:210; Lamond 1950:169; Massola 1956; McArthur 1960:113; McCarthy and McArthur 1960:184-5; Duncan-Kemp 1961:211-2; Duncan- Kemp 1968:124,275,319; Hardy 1969:12; Farwell 1971:67; Macknight 1972:305; Bowdler 1976; Dargin 1976:25; Happ 1977:58,67; Rowland 1980:4,13; Rowland 1981; Dyall 1982:55,59; Rowland 1983; Bates 1985:250-7; Trigger 1987:79; Marchant 1988:150; Walters 1988; Lawson 1989:57; McKelson 1989:37; Barker 1991; Nicholson and Cane 1991; Penney 1991:86; O'Connor 1993:227; Nicholson and Cane 1994:113; Lourandos 1997:210-11; Attenbrow, Fullagar and Szpak 1998; Border 1999:133; Flood 1999:200; Mulvaney and Kamminga 1999:263,292; O'Connor 1999a:13,126; Gerritsen 2001a; Clarke 2002:150-1; Barker 2004:42,145.

developing marine economies in some places. But this was not possible without the watercraft capable of journeying to islands such as Maatsuyker Island. Which again raises the question, was there deliberate development of the seafaring technology with the aim of exploiting the untapped seal colonies of islands such as Maatsuker Island, or was it an evolutionary process involving a series of small incremental developments allowing groups such as the *Needwonne* to discover the seal colonies? It would be premature to answer to such a question at this point, before giving further consideration to intensification.

A significant manifestation of intensification, at the opposite end of the spectrum to the marine environment, is the move beyond the final frontier in the desert, the details of how and when also being a matter of much debate. Although humans had been present here for possibly up to 55,000 years, occupation of the desert areas, which cover two-thirds of Australia, probably did not, on current evidence, begin until about 26-34,000 years ago.[54] When climatic conditions were at their most extreme during the Last Glacial Maximum (19,000 – 17,000 BP) some areas, such as parts of the Western Desert, which had only been sparsely inhabited, were probably completely abandoned for a time.[55] It has also been argued that the harshest areas, the "Barrier Deserts" such as the Great Sandy and the Great Victoria Deserts, were not populated at all, or only sparsely, until the last 5000 years.[56] How this came about is, as ever, subject to some debate, though there are areas of broad agreement as well. Seeds are a significant source of subsistence in the arid zone, having reputedly "underwritten" the occupation of such areas.[57] Various types were collected and ground, it would appear, from the earliest times, as attested to by the discovery of 30,000 year old grindstones at Cuddie Springs in inland northern New South Wales.[58] But these and other early grinding implements were generalised tools, used to grind various seeds, tubers and root foods. Only in the last 5000 years or so did specialised seed grinding implements evolve from the earlier multi-purpose grindstones.[59] Other adaptations, such as maintenance of conservation areas to fall back on in hard seasons, deep wells, improved hafted woodworking tools and more open social networks, providing greater flexibility in a highly variable environment,[60] also appear to have played a role in the colonisation and ongoing occupation of the arid zone. Accompanying these developments was the evolution of broad-spectrum procurement, although the timing of this is still an open question, and limited population growth, albeit at a very low rate.[61]

Explanations for the development of specialised seed grinding and the exploitation of remaining niches in the arid zone encompass climate change, population growth and sociopolitical dynamics. M. A. Smith, for example, originally proposed that this facet of intensification was triggered by a decline in precipitation 3-4000 years ago, following a more benign period during the mid-Holocene. In his view population rose in the arid zone during the wetter period, only to suffer considerable resource stress when conditions deteriorated, with specialised seed grinding an adaptive response to this situation.[62] Population growth in the context of more complex social relations has been invoked by Jane Balme to explain arid zone intensification, climatic change not playing any significant role in this.[63] As an alternative, or addition, to this Veth has also placed developments in the arid zone in the context of more complex social relations, in the form of extensions to social networks. But he points to a range of technical innovations beside seed grinding, such as deep well construction, as another important contributing factor.[64] Finally, extension of social networks or alliances, is given overarching primacy by Lourandos and Ross, in their view this factor triggered economic demand and subsequent innovations.[65] Unfortunately all these theories have been subject to rigorous scrutiny and, it has been argued, all have their shortcomings. None, according to Edwards and O'Connell, "account for the apparent mid-Holocene expansion in diet breadth anywhere in Australia."[66] So what is the explanation? Perhaps a new theory is required, but before this can even be contemplated other aspects of traditional life in Australia must be considered for the picture to be a little more complete.

[54] Veth 1989,1993; Allen and O'Connell 1995:viii; Edwards and O'Connell 1995:774; Hughes and Hiscock 2005:10.
Smith (1986,1989a) argues that all desert areas of Australia were occupied from the earliest times, albeit only thinly populated, while Hiscock and Wallis (2005) have recently proposed an alternative "desert transformation" model, whereby groups altered their subsistence base as the climate worsened so that episodic exploitation of the barrier desert, and human presence, continued.
[55] O'Connor et al. 1998:20; Thorley 1998; Hiscock and Wallis 2005: 43-47.
[56] Veth 1987,1989; Veth et al. 1990:48-9; Veth 1993:103-114; 2005.
[57] Smith 1986,1989b; Edwards and O'Connell 1995; Allen 1998; Smith 1998:136.
[58] Fullagar and Field 1997.
[59] Smith 1986, 1989b; Edwards and O'Connell 1995:774.
[60] Gilmore 1932:168; Strehlow 1965:143; Cane 1984; Veth 1987,1989; Veth et al. 1990; Veth 1993:105-6,113; Fisher 1997:20.
[61] Edwards and O'Connell 1995:774-80; Fullagar and Field 1997.
[62] Smith 1989b:314.
[63] Balme 1991.
[64] Veth 1987,1989; Veth et al. 1990.
[65] Lourandos and Ross 1994:57.
[66] Edwards and O'Connell 1995:779; Allen 1998.

# Chapter 3.
# People and Plants in Australia

**Planting and Cultivation in Australia**

Deliberate planting on a systematic basis is a hallmark of early agricultural endeavours, be it in south west Asia, Mesoamerica, China, or wherever. By any normal criteria it is an absolutely essential component because, without planting, agriculture is not possible. Consequently, comprehending the dynamics that were involved in such a step, and the transition that evolved from it, is fundamental to our understanding of the origins of agriculture. In regard to Australia the 19th century theorist de Candolle, in his work *Origin of Cultivated Plants*, emphatically declared that:

"*The lowest savages know the plants of their country ... but the Australians and Patagonians ... do not entertain the idea of cultivating them.*"[1]

This view, usually expressed in less pejorative terms, became, and still remains for many, an accepted fact. But, contrary to de Candolle's pronouncement, planting by Aboriginal peoples in various parts of Australia was by no means uncommon, as will be documented below. And, as will become apparent in later chapters, agriculture was indeed practiced in parts of Australia by Aboriginal people in a traditional context.

The first instance in which it was claimed Indigenous Australians engaged in planting arose in March 1658 when a shore party from a Dutch ship "Emeloordt", searching for 68 sailors stranded two years earlier on the central west coast of Western Australia following the sinking of the "Vergulde Draeck", reported to the Captain:

"*On land we found much low scrub and in some places also seeding land which they burn off in some places, arable or seeding land, but we saw no fruit but some herbs that had a fragrant scent.*"[2]

This was in all likelihood an erroneous report. What the crew members from the "Emeloordt" had probably encountered was the practice of firing scrub and grasslands, a common feature of Aboriginal subsistence strategies, often referred to as "firestick farming". In their naivety they had interpreted this as land clearance in preparation for planting.

1 de Candolle 1884:2

2 Major 1859:83.

The earliest recorded example of planting by Indigenous Australians in Australia was first observed in the 1850s, by noted explorer and surveyor A. C. Gregory. According to Gregory the *Nhanda* of the Victoria District, 300 kilometres further north on the central west coast of Western Australia:

"*never dug a yam without planting the crown in the same hole so that no diminution of food supply should result.*"[3]

leading R. H. Maiden to comment in 1889 in *Useful Native Plants of Australia*, that this yam:

"[*Dioscorea hastifolia*] ... *is the only plant on which they* [Aboriginal people] *bestow any kind of cultivation, crude as it is.*"[4]

Was this just a peculiar, isolated occurrence? It would appear not, for at least two other varieties of yam, *Dioscorea bulbifera* [*D. sativa*] and *D. bulbifera* var. *elongata* [*D. elongata*], were planted in Australia in traditional circumstances, on Cape York and its adjacent islands in Queensland. This has been documented on a number of occasions in a variety of places - on east Cape York among the *Umpila*;[5] at Lockhart River and islands offshore;[6] and among the *Walmbaria* and *Mutumi* of Flinders Island and the adjacent mainland[7] - and west Cape York among the *Wik-Mungkan*[8] and possibly the *Tjungundji*.[9] Similar activity has also been noted on Groote Eylandt, off the Arnhem Land coast, with "women digging up yams and replacing part of the stem in the hole,"[10] as well as at Kalumburu

3 Gregory 1886:23
Gregory, who had carried out exploration and survey work in the area in the late 1840s and early 1850s originally made the claim in 1882 (though this was not published until 1887), specifically stating:
"The natives on the West Coast of Australia are in the habit amongst other things of digging up yams as a portion of their means of subsistence; the yams are called 'ajuca' in the north and 'wirang' in the south. In digging up these yams they invariably re-insert the head of the yams so as to be sure of a future crop, but beyond this they do absolutely nothing which may be regarded as tentative in the direction of cultivating plants for their use." (Gregory 1887:131)

4 Maiden 1889:22

5 Hynes and Chase 1982:40.

6 Harris 1977b:437.

7 Hale and Tindale 1933-36:113; Tindale 1974:96; Tindale 1977:349.

8 McConnel 1957:2; Campbell 1965:208.

9 Campbell 1965:208.

10 Tindale 1974:96

in the very north of Western Australia.[11] These examples quite possibly arose through a diffusional dynamic given their proximity to and/or contact with peoples such as the *Macassans* and Torres Strait Islanders who already practiced cultivation. The fact that all sites on Cape York lie within an area designated as being subject to [Torres Strait] "Island Cultural Traits" supports this inference.[12]

The same explanation may also apply to a further two, possibly three, species that were reportedly planted within the region subject to Melanesian influence. Hynes and Chase report the *Pama Malnkana* planted coconuts (*Cocos nucifera*), each tree being owned by the planter,[13] while the *Uradhi*, according to Crowley, cultivated the sweet potato (*Ipomoea batatas*) prior to contact with Europeans.[14] The *Wik-Mungkan* also reputedly planted another, unidentified, tuber named as "*Koothai*".[15]

But what of practices beyond the areas subject to the aforementioned outside influences? Again a variety of sources indicate that other Australian groups may have been undertaking planting in their traditional context. Following Gregory's account of the planting of yams on the west coast, a further claim was made in 1906 that Aboriginal people had been undertaking planting. N. W. Thomas, in a largely derivative work on Australia, made the extraordinary statement that:

"*The cultivation of purslane (*Portulaca*) seems to be a well-established fact. It is grown like melons on slightly raised mounds; before the seed vessels are ripe, the plant is cut, turned upside down and dried in the sun; then the seed vessels are plucked and rubbed down and the seed collected. Many pounds weight can be collected in a day, even when there is no cultivation.*"[16]

Rather than being a well-established fact, Thomas's assertion is the only instance in which it has been claimed *Portulaca oleracea* was cultivated in Australia.[17] Commonly known as *munyeroo*, it was certainly heavily utilised, a staple, in parts of Australia,[18] as well as being cultivated in many other parts of the world.[19]

Following Thomas's suspect claim, however, instances of planting by particular Aboriginal populations began to appear more frequently. It was reported that Aboriginal groups from south east Queensland germinated the very large seed cones of the "bunya pine" (*Araucaria bidwilli*), according to the informant they were "all sprouting". However the intention here appears to have been not to grow these, but to eat them, by preserving them in swamps, waterholes or creek beds for two or three months.[20] The bunya pines and their nutritious cones, found principally in the territory of the *Waka Waka* and *Gubbi Gubbi* peoples in the Blackall Ranges and Bunya Mountains of south eastern Queensland, were the focus of gatherings of at least 700 people from the local and adjacent regions, some groups travelling over distances of up to 320 kilometres or more.[21] This normally occurred every third year, when all the bunyas usually bore fruit synchronously.[22] Small groves of bunya pines were individually owned and inherited and it has been suggested that the bunya seeds were indeed planted.[23] While there is some indication that groups frequenting the area may have played some role in the "pines" local distribution, perhaps through "sprouting" leading to "incidental planting",[24] others have offered the opinion that deliberate planting was unlikely as these conifers took a lifetime or longer to reach maturity.[25]

Mary Gilmore, better known in Australia as the noted poet Dame Mary Gilmore, was next to claim, in the 1930s, that Aboriginal people, in this instance the *Wiradjuri* of central New South Wales, had been engaging in planting and husbandry. Recalling a number of incidents she allegedly witnessed in her childhood in the early 1870s, either at Wagga Wagga on the Murrumbidgee River or, more likely, at Cowabbie Station, 65 kilometres northwest of there, Gilmore recounted:

"*It was here that I first saw more or less extensive planting of seeds by the lubras* [women]. *This was done at the direction of the chief, who, when the earth was cool enough to walk upon* [following firing], *went first to one shrub and then another, testing and examining the burnt capsules, and where these had been destroyed directing the attention of*

11 Crawford 1982:31,44-5. In this instance the top of *Dioscorea bulbifera* var. *elongata* [*Ganmanggu*] and var. *rotunda* [*Gunu*] was replanted.
12 Crowley 1983:306,Map 4.
13 Hynes and Chase 1982:40.
14 Crowley 1983:312.
15 Campbell 1965:208.
This term could not be found in the Dictionary and Source Book of the Wik-Mungkan Language (Kilham et al. 1986) though it may be a misspelt abbreviation of *may kuthal*, the bulbous root of "swamp grass" (Kilham et al. 1986:79). Alternatively Harris (1977b:433) recorded *kuthay* as a Dioscorea sp. At Lockhart River in east Cape York, so *Kuthai* may simply be applied to any form of bulbous root (See Kilham et al. 1986:71,70).
16 Thomas 1906:113.
17 Thomas provided a very select bibliography (one page, 1906:251) with a statement indicating that a comprehensive bibliography would be published later (1906:xi-xii). Nothing in the select bibliography could be found supporting Thomas's claim and no trace could be found of a comprehensive bibliography as a subsequent publication. The level of details suggests that the information may come from a primary source, perhaps a correspondent, but there is no confirmation of this. Consequently the statement could not be verified and must be treated circumspectly.
People in Central Australia did, nevertheless, sometimes place "specially constructed stone rings" to contain these plants (Latz 1995:250)

18 Allen 1974; Smith 1986:29.
19 It was also exploited in North America, among the Iroquois and Hurons, as well as by peoples in the southwest U.S. and California, as well as being cultivated in Europe and parts of Asia chiefly as a vegetable (See USDA 1936:24;Byrne and McAndrews 1975; and http://www.fao.org/docrep/t0646e/t0646e0t.htm:3.).
20 Petrie 1904:23; Daley 1931:20.
21 Meston 1895:82; Petrie 1904:11-13,15-16,23; French 1989:18-9.
Bunyas were also apparently found in the territory of the *Dalla* and *Jarowair* people as well (French 1989:18).
There are claims of up to 20,000 people attending bunya feast (French 1989:19) but such claims are not generally accepted.
22 Mathew 1910:93-4.
23 Petrie 1904:16; Hyam 1940:117,
24 Walker 1977:13
25 Bennett 1930:13-14.

*the women to them. The women gathered fresh ones from untouched shrubs, and planted them where the destroyed ones stood. The use of ammonia was apparently known, for when the seeds were in the scratched but still warm earth, the little boys were requisitioned to damp them. Grass-seed was gathered, a heavy kind, and I helped in this. When a lot, still in 'the ear' had been collected it was rubbed by hand, and then by a turn of the wrist the bark container into which it was shaken tipped left and right, and the grain lay clear at one end, with the husks in a little pile at the other .......... The separated seed was sorted, the unsound or small rejected and the best planted in the burnt area but very lightly covered. It was not scattered, it was put in small pinches so that if some failed there would still be enough for a tussock. How I remember the grass planting so well is that being a child I thought all seed was the kind wanted, and gathered ripe and unripe, and chiefly the sort that rolls on a sickle-shaped terminal. This, I was told, was not what was required, that the wind would plant this; that it was what did not run, and then catch in the earth, that had to be planted by hand.*

*I had often seen the blacks set individual seeds as well as replace, where the plant grew, those of what they ate. But the former was usually done where they themselves first made a small fire of twigs in order to prepare the earth by heating it. Whenever they gave me the fruit of the ground-berry, as we called it .................... I was invariably asked for the seed, which was immediately planted beneath the growth from which it came. I remember once I had an exceptionally large berry, and without thinking spat the seed out when the pulp was eaten from it. Because it was a large seed the lubra I was with scolded me for waste, hunting the grass till she found it, so that it could be put back from where it originally came from and a strong plant grow from it .............. Then there was their usage in the matter of the quandong-tree [Santalum acuminatum]. They looked to see which of the stones or nuts were male or female, and planted accordingly. When a grove was in flower they brought branches from another grove to fertilize the blossom.*"[26]

Gilmore's account includes some highly significant claims. Not only does she insist that she witnessed the planting of at least four different species of plant, only one of which can be clearly identified, she proffers evidence of deliberate fertilisation, pollination and seed selection. This last mentioned behaviour, if true, is most remarkable in light of the debates that have raged regarding intentionality in the domestication of plants by early agricultural groups in other parts of the world. Before continuing, however, the issue of the accuracy and veracity of Gilmore's recollections must be considered.

While there is no direct corroboration of the planting and other activities mentioned by Gilmore,[27] this is not unusual as most accounts of planting in Australia rely upon a single report. Despite the level of detail in her account, her lack of specificity in terms of exact time and place nevertheless make even indirect corroboration difficult. The only specifics of this nature relate to the cross-pollination of the quandong where, in contrast to the other parts of her evidence, she later named two places where she had seen this done, "Yarrengerry" and "near Bethungra".[28] Certainly she was in the region during the period in question, her family, particularly her father, were highly sympathetic to local Aboriginal people,[29] and there is evidence that she played with Aboriginal children in their camp on Houlaghans Creek near Wagga Wagga.[30] She maintains she had spent six weeks with the *Wiradjuri* on the banks of the Murrumbidgee River in the charge of "the chief women of the tribe"[31] when recovering from poisoning by arsenic "intended for them."[32] Although nothing is known of the reputed poisoning, it is known she had two serious accidents as a child, the second, when she was only six years old, requiring an extended period of convalescence.[33] This could possibly be the period she alludes to when recounting the episodes of planting. However the textual context points to Cowabbie Station as it follows a discussion of Aboriginal firing practices. Cowabbie is situated on the grasslands of the open plains in contrast to the Murrumbidgee floodplain, consisting of open woodland, with the traditional Aboriginal economy having a strong riverine orientation.[34] Views differ as to Gilmore's veracity, opinions ranging from there being, "a generalized truth in her writings,"[35] to, "one of those fantasies that she invented, largely because they added more colour and intensity to her narratives."[36] Gilmore's writings often provide a highly romanticised portrayal of Aboriginal people,[37] at times in terms that are not consistent with established anthropological understandings, such as frequent references to "chiefs".[38] Events she describes elsewhere, such as massacres and atrocities committed against Aboriginal people in the region, appear to relate to

[26] Gilmore 1934:221-2

[27] The only corroboration offered by Gilmore is a claim (1935:v) that she had received a letter which "endorsed" her observations regarding the replanting of fire damaged plants.
[28] Gilmore 1934:222
[29] Read 1983; Wilde 1988
[30] Cusack, Moore and Ovenden 1965:28.
[31] Gilmore 1934:183
[32] Gilmore 1935:173
[33] Wilde 1988:46.
[34] Flood 1988:274.
[35] Read 1983:26.
[36] Wilde 1988:289.
[37] e.g. "The Waradgery Tribe" in 1932:172-3
[38] Although she frequently uses this term, she shows an awareness that they were not "chiefs" in the more conventional sense by acknowledging they had "no autocratic power of life and death" (1935:227). She may not have necessarily been misleading in this regard either, as George Augustus Robinson, Chief Protector of the Port Phillip Protectorate at the time, appears to have encountered a man named "Wol.lut.but" in southern New South Wales in the early-mid 1840s, who he described as "chief of all the natives of this country, regulates their fights and disputes" (Clark 2000:184). And Howitt (1904:303) alludes to the existence of hereditary headmen, called *Bidja-bidja*, in the southern *Wiradjuri*, the people from the area Gilmore refers to.

incidents that took place before she was born.[39] It has been suggested that, having heard of these events when young from people directly conversant with them, she transposed herself into the story.[40] There is also evidence that her father, perhaps others, may have been the real source of at least some of her information on Aboriginal matters.[41] Although I am inclined to this view, Gilmore's evidence must, unfortunately, be treated with great caution, if not rejected outright.

Following the suggestion that an isolated grove of cabbage palms (*Livistona australis*?) at Orbost in Victoria may have been planted,[42] there were no further reports of planting by Aboriginal people for over thirty years. Then a spate of reports appeared in the 1970s and 80s, many by professional anthropologists and archaeologists, beginning with the sowing of *kurumi* near Lake Darlot in the Western Desert of central Western Australia. According to Dix and Lofgren's informants traditionally *kurumi* "seeds were carefully scattered in the cracks of the clay pan to ensure that after heavy rains a bountiful crop would be produced."[43] The planting of *kurumi*, tentatively identified as *Tecticornia arborea*,[44] appears to have only been undertaken at only one particular claypan associated with a sacred stone arrangement.[45] Two years later, in 1976, Kimber reported having observed two men in Central Australia undertake planting of no less than four different species, "native fig" [*Ficus* sp.], *yella* ["bush potato", *Ipomoea costata*], *moolyardi* [*Acacia dictyophleba* - for spears] and finally Sturt's Desert Pea [*Clianthus formosus*], an adornment signifying a successful hunter.[46] These examples may, however, have been an adaptation to modern circumstances as the two individuals involved lived in fixed settlements outside their country, and all the plants involved had, apart from their functional and symbolic qualities, strong spiritual associations with their traditional lands.

While there may be doubt that the episodic planting observed by Kimber was a traditional practice, this is not the case with the broadcast seeding of "bush tomatoes" (*yalijarra* - *Solanum diversiflorum*), first noted by ethnobotantist Fiona Walsh in the same period. In this instance the *Mardu* of the central north of the Western Desert, in the north western region of Australia, traditionally broadcast the seeds over burnt ground near their camp prior to the rains.[47]

A short time later yet another example of planting in Central Australia, in this instance *Nicotiana* sp., was observed by another ethnobotanist Peter Latz, who, like Kimber, was uncertain whether this was a traditional practice or a "recent innovation".[48] Certainly the dried leaves of *Nicotiana* spp. were traditionally widely consumed in Central Australia, there being at least four *Nicotiana* species known to have been ingested by Australians for their nicotine content.[49] It is worth noting here that *Nicotiana tabacum*, the species used in modern tobacco, was originally widely grown and domesticated by Native Americans,[50] and it has been argued that psychoactive plants, "intoxicants", may be among the first to be grown extensively and domesticated.[51] Although the planting of *Nicotiana* in Central Australia is the only documented instance of this, the practice could possibly have been a traditional one and more widespread than Latz's evidence suggests. For example, when exploring the upper Lyons River in the central west of Western Australia, F. T. Gregory, brother of A. C. Gregory, recorded in 1858:

*"... we halted for the night in latitude 23°59'39". Tobacco here grew to sufficient size for manufacture, occupying many hundreds of acres of the best land.*"[52]

Shortly after Latz's observations were originally published another example of sowing, involving two species and apparently undertaken by a wide range of groups, came to light in inland Australia. In this case *ngardu* (*Marsilea drummondii*) and grass seeds, most likely *Panicum decompositum* or one of its relatives, was propagated by broadcast seeding, reportedly a traditional activity. This used to happen, according to Kimber's informant (Walter Smith), at a "good camp ... or might be a good soak or something, good sandhill ..."[53] These were seeds that were heavily utilised in the interior of Australia, playing a major part in the subsistence economies of a broad region that will be discussed later.

Finally, in terms of such accounts, D. E. Yen reported the planting of two *Ipomoea* species, *I. costata* and *I. polpha*, in an unspecified part of the desert region of Australia.[54] The cultivation of the first of these, *I. costata*, the "bush potato", was noted earlier.[55]

---

39 Read 1983:26.
40 Read 1983:27.
41 A note in Gilmore's handwriting and signed "M G" appears at the bottom of p.170 of Copy No.7 of Under the Wilgas (Gilmore 1932) indicating the Aboriginal legend "The Willi-wa Man No.2" originally came from her father, that there had been some confusion, and that it may not have been an Aboriginal legend at all. At various points she nominates her father (1932:168; 1934: 163-4) and refers others in general terms (1934:160). She also cites her uncles as a source for inordinately large gatherings (5,000+)at the Brewarrina fish traps (1934:192).
42 Hyam 1940:17.
It would appear that this was actually first suggested in 1907 (Le Souef 1907:80).
It is more likely this grove was a remnant from a period when the climate in this part of Australia was warmer and wetter as they also appear in eastern New South Wales.
43 Dix and Lofgren 1974:74
44 In all likelihood it was *Tecticornia arborea* as Tindale (1974:145) reports this species being collected and stored for months in kangaroo skin bags in that area, though he called it *buli buli*. A related species, *Tecticornia verrucosa*, was a staple in central Australia (Smith 1986:29).
45 Dix and Lofgren 1974/1090
46 Kimber 1976:146-7,149.

47 Walsh 1990:34. See also Tonkinson 1991:53.
48 Watson 1983:40; Latz 1995:63.
49 Gerritsen 1994a:195-6; Latz 1995:62-64,230-36.
50 Keeley 1995:265.
51 Hayden 1995a:282,294.
52 Gregory 1884a:46 (31 May 1858).
This was the first documented British colonial exploration of this region.
53 Kimber 1984:16-17.
54 Yen 1989:60.
55 Kimber 1976

Summarising, it can be seen that there is direct observational or ethnographic evidence detailed here of at least seventeen distinct plant species that were probably planted by Aboriginal Australians.[56] In fact there are more, as will become apparent. Although some of these accounts, particularly Gilmore's, may be suspect, the evidence of planting would appear to be incontrovertible. Certainly a proportion of these examples may have arisen as a consequence of influences from peoples to the north of Australia, Dutch pre-colonial interaction or modern circumstances. But there is good evidence that planting of at least six, perhaps seven, of these species was the outcome of purely indigenous developments.[57] Furthermore, there may have been other situations in which planting was practiced but not recorded. The brevity of small-scale planting means that it could easily have been missed and this was quite likely in relation to the 19th century, where much of our information for large parts of southern Australia depends upon the observational powers of amateur ethnographers from that era. The stereotype, typified by de Candolle's pronouncement at the beginning of this section, that Australia's Aboriginal people didn't engage in planting, cultivation or agriculture may also have compromised the observational acuity of these 19th century ethnographers. It is noteworthy in this context that all the valid examples recorded in the 20th century occurred in areas where traditional or quasi-traditional societies were still persisting (i.e. in northern Australia, central Australia and the Western Desert). Another interesting observation relates to the distribution of occurrences. All seven plantings identified as purely indigenous developments occur in the arid and semi-arid zones of central Australia, although this may simply be an observational artefact. While not included in the enumeration, the Gilmore examples are, however, consistent with this category to some degree, coming from a semi-arid zone. A similar pattern of seed planting occurring predominantly in arid and semi-arid regions has also been observed in California in relation to traditional Native American practices.[58]

**Other Forms of Interaction With Plants**

Studies of Aboriginal interactions with plants and the environment in Australia were hampered for a considerable period by an inability to appropriately conceptualise the relationship, until Hynes and Chase developed the notion of "domiculture".[59] This characterised the interaction as a form of whole or total environment management leading to a "domesticated environment" and has great validity.[60] The people involved controlled, to an extent, the total environment through manipulation of segments, niches or elements of it, producing in Hynes and Chase's view, "the conditions for plant domestication."[61]

One of the principal tools in this approach was the extensive use of fire, often called "firestick farming", involving controlled burning of selected areas. Firing, which required a detailed knowledge of the environment, allowed easier capture of small game, encouraged new growth, created more diverse plant and animal communities, and also facilitated movement through the country.[62] Natural productivity was increased to some extent by this practice, perhaps by as much as 500-900%.[63] This has led some to deem it to be a form of cultivation.[64] Aboriginal peoples, however, saw their role in this not as cultivators, however, but as curators of the natural environment.[65] With the cessation of their direct involvement in their country, and the firing of it, a traditional custodian observed that it "[be]come wild now. No-one look after him,"[66] a point supported by numerous instances in Australia of grasslands reverting to dense bush and scrub following dispossession.[67]

Numerous other practices can be alluded to in keeping with the domicultural approach. For example, temporary dams were constructed in channels flowing out of jungle wetlands around the Roper River in the Northern Territory of Australia

[56]
1. *Dioscorea hastifolia,*
2. *Dioscorea bulbifera,*
3. *Cocos nucifera,*
4. *Ipomoea batatas,*
5. *Ipomoea costata,*
6. *Ipomoea polpha,*
7. *Tecticornia arborea,*
8. *Acacia dictyophleba,*
9. *Solanum sp.*
10. *Nicotiana sp.,*
11. *Marsilea quadrifolia/drummondii,*
12. *Santalum acuminatum,*
13. *Clianthus formosus,*
14. "native fig" (*Ficus* sp.),
15. Unidentified shrub,
16. Unidentified grass (probably P. decompositum or other Panicum sp.)
17. Unidentified "ground-berry".

"Koothai" (possibly a Dioscorea sp. or "swamp grass") is not included as it cannot be identified with certainty as a separate and distinct species from those already listed. The Thomas evidence on *P. oleracea*, as well as the claims the bunya and the cabbage palm were planted, are not included either as they are either not certain or based on eyewitness accounts.

[57] *Ipomoea costata, Ipomoea polpha, Tecticornia arborea, Solanum* sp., *Nicotiana* sp., *Marsilea drummondii* and *Panicum decompositum* or other similar *Panicum* spp.

[58] Keeley 1995:265.

[59] Hynes and Chase 1982

[60] Hynes and Chase 1982:38.
This concept highlights one of the difficulties of using "domestication" as a diagnostic criteria for agriculture. Human intervention along the lines described, particularly firing, probably results not only in changes to the form and distribution of plant communities but also probably in the morphogenetic structure of the species. Consequently some species may be "domesticated" prior to planting and cultivation, such as described by Ladizinsky (1987) with pulses (*Lens* sp.) in south west Asia, and then undergo further morphogenetic changes when systematically planted and harvested.

[61] Hynes and Chase 1982:49.

[62] Yen 1989:58; Jones 1969:226-7.
The extent of these practices has been called in to question by Horton and others (Horton 1982; 2000:70-101; Collins 2006:329- 48). Horton has argued that the impact of firing was "unique local effect" (1982:256), and I am inclined to share this view, that it only had local or at most regional impacts, and was not a generalised phenomenon as is often assumed.

[63] Mellars and Reinhardt 1978:261; O'Connell, Latz and Barnett 1983: 99.

[64] O'Connell, Latz and Barnett 1983:99.

[65] Jones 1990:32.

[66] Hynes and Chase 1982:41.

[67] Jones 1969:228.

at the end of the rainy, "wet", season to maintain water levels and thus encourage preferred bird and plant life.[68] Activities directed toward enhancing productivity by manipulation of plant communities, as well as animals, insects, grubs and worms can also be considered to fall within the domicultural paradigm. Burning of cycad palms (*Cycas* sp.) to increase nut production,[69] similar treatment of the trailers of the "bush potato" by the *Kukatja* of the northern part of the Western Desert to promote tuber production,[70] and again with rushes ("flag" - *Typha augustifolia*?) in southern Western Australia, to "improve it,"[71] are further instances. Other interesting examples include the stimulant "*pituri*" (*pitjuri - Duboisia hopwoodii*) from southwest Queensland and the "yam-daisy" (*Microseris lanceolata*) formerly found extensively in western Victoria.

*Pitjuri*, a plant containing nicotine and nicotine related compounds, when prepared, was extensively traded over large areas of central Australia, reaching into southern Australia.[72] Such was the demand for *pitjuri* that Aboriginal consumers reportedly would "give everything they possess for it,"[73] a significant observation in light of earlier comments that intoxicant plants may have been the earliest cultivars. While *pitjuri* was not planted, as far as is known, the older branches were certainly burned to encourage growth of more potent new shoots. The second exemplar, the prolific yam-daisy, was a very important component of subsistence in south eastern Australia, especially in Victoria. Known as *murnong*, explorer Major Thomas Livingstone Mitchell noted in south west Victoria 1836 that, "in many places the ground was quite yellow with the flowers,"[74] and Chief Protector Robinson reported on one area of central northern Victoria where there had been "millions of murnong or yam, all over the plain"[75] in the early colonial period (1840). Digging to obtain *murnong* loosened the soil, a condition encouraging further propagation, leading one researcher to describe it as a form of "natural cultivation".[76] Alas, soil compaction and grazing resulting from the introduction of sheep and cattle by the colonial invaders has made the yam-daisy a rarity these days.[77]

Activities intended to enhance the productivity of individual plant species were not limited to manipulation of the plants themselves but could also include their preferred environments. The *Iliaura* and *Wanji* of the eastern Northern Territory, for instance, blocked up of small creeks and channels so that they overflowed and flooded the surrounding plains, promoting the growth of grasses such as *P. decompositum*.[78] Similarly a dam 330 ft. [100m] long and 6 ft. [1.8m] high, first reported in 1969, had been constructed at some time on the shallow Bulloo floodplain in the arid northwest of New South Wales, ostensibly to promote the growth of grass and seed plants such as *Panicum* sp., *Portulaca oleracea* and *Echinochloa* sp.[79] A similar intervention carried out on Cape York entailed simply clearing the ground around favoured species springing up near camp sites and erecting small barriers to protect them from playing children.[80]

Beyond this generalised approach, more specific interventions were also undertaken, including some of the examples of planting discussed earlier, as well as "selective propagation", an interaction which results in the propagation of the plant, without actual planting. Illustrating this is the careful harvesting of yams by breaking off the tuber, leaving the top and trailer behind, resulting in regrowth. This was a deliberate action, with the harvesters quite conscious of the consequences of the method,[81] and was first noted in the late 1850s in the *Watjandi*, a northern *Nhanda* group.[82] The *Watjandi*, unlike their southern neighbours, did not engage in utilisation of *D. hastifolia* to such a significant extent, their country being much drier and containing only isolated patches of alluvial soils. Traditionally the *Anbarra Gidjingali* of Arnhem Land, Northern Territory also harvested yams [*manbandi - D. transversa*] in this manner, as did the *Pama Malnkana* of east Cape York [probably *D. bulbifera*].[83]

One other form of behaviour resulting in the propagation of plants is what is known as "incidental dispersal", "incidental planting" or "incidental domestication"[84], the bunya pine cones discussed earlier being a possible example. A number of other occurrences have been noticed in Australia, typified by the proliferation of "wild tomatoes" and *yukidjirri* berries over a wide area as a result of the *Walbiri* of the Western Desert transporting "dampers" made with these.[85] A consistent association, one the *Anbarra Gidjingali* in Arnhem Land were quite conscious of, between old camp sites and endemic fruit trees, particularly *Eugenia suborbicularis*, is also thought to be the result of incidental dispersal. More recently watermelons, a modern addition to the *Anbarra Gidjingali* diet, have also started appearing around camp sites.[86] Lastly, the frequent appearance of a

[68] Campbell 1965:206.
Similarly the Owen Valley Paiute in eastern California built dams and irrigation ditches up to 6 kilometres long in wetlands to promote the growth of wild hyacinth, nut grass and spike rush (Harlan 1995:13; Pringle 1998:1447).
[69] Lourandos 1997:48
[70] Cane 1984:88.
[71] Grey 1841a:2:294.
[72] Watson 1983; Berndt, Berndt and Stanton 1993:116.
[73] Bancroft 1879:7
[74] Mitchell 1839:2:272.
[75] Clark 1998-2000:1:135.
[76] Gott 1982:65; 1987:45.
[77] Gott 1983:12-13;1987:40,45.

[78] Tindale 1974:102; 1977:347.
[79] Rowlands and Rowlands 1969; Jones 1979:143.
Such dams may well have had other additional functions, as will be discussed in Chapter 5.
[80] Hynes and Chase 1982:41
[81] Jones 1975:23.
[82] Oldfield 1865:277
[83] Jones 1975:23; Hynes and Chase 1982:40.
[84] Rindos 1984:156.
[85] Kimber 1976:145.
In recent times the *Alyawara* introduced *S. chippendalei* south of its normal range through the discard of the seeds of this fruit when bought back to camp in cars (Peterson 1979:173).
"Dampers" are flat, unleavened bread-like food in Australia.
[86] Jones 1975:24.

plant, "*dillines*" [*dilanj – Nitraria billardieri*], on and around the occupation mounds found in western Victoria and on the Murray River is strongly suspected to be the result of incidental planting. Bearing large berries containing a stone, "the natives," according to Beveridge, "are extremely fond of these berries, and to this fondness may be attributed the fact of their prevalence on the cooking mounds."[87]

**Human-Plant Interactions**

A variety of Australian examples of human-plant interactions have been presented in the preceding sections. While not necessarily typical of Aboriginal cultures in Australia they do represent a range of such interactions. Similar interactions have, of course, been observed in other supposedly incipient and early agricultural economies. The point has been made elsewhere that planting should not be separated "from the antecedent activities of intervention,"[88] as indicated by plant remains in the archaeological sequence in eastern North America showing a gradualistic, incremental development toward agriculture.[89] Having arrived at a similar conclusion, Jarman *et al.* had previously attempted to characterise the process by developing a typology of intensification reflecting the growing concentration on particular plants, with subsistence becoming increasingly oriented toward those plants and the control of them.[90] Their classification included:

1) Casual gathering: "resources exploited simply because they are in the vicinity"
2) Systematic Gathering: "particular plants or groups of plants become the focus of regular and systematic economic activities" with "some husbanding of resources"
3) Limited Cultivation: "productivity is encouraged and regulated ... by transplanting, tillage, weeding, irrigation and so on."
4) Developed Cultivation: "propagation of crops under human control"

While this classification has value in terms of identifying some sub-types of subsistence economies it is probably too general to account for the Australian evidence. It also offers little insight into how people might shift from one category to another through intensification, particularly how the shift from Systematic Gathering to Limited Cultivation might be achieved. Casual Gathering is also probably based on a misconception, as the Australian evidence indicates there was usually an underlying resource conservation strategy, to avoid over-exploitation, in the selection of areas and specific plant communities to be exploited, particularly in women's gathering activities. What might appear to be casual gathering is often an informally programmed systematic exploitation of diverse resources.[91] This is the essence of the domicultural approach.

Considering their Systematic Gathering category, Higgs and Jarman's appear to assume that where there is intensive exploitation of particular plant resources that these plants would eventually be planted. But, apart from some examples which will be discussed later, there is little evidence that planted species prior to planting were necessarily gathered any more systematically, or exploited more extensively, than most other local resource.[92] The dynamics that led to plants being planted, and which ones assume economic importance, I would suggest, are more complex than Higgs and Jarman envisage. Moreover, certain Australian groups, such as the *Paakantyi* of the Darling River valley, harvested extensive stands of seed plants on a systematic basis, perhaps husbanding them to some degree but, according to current understandings, they never planted these.[93] Hence it is difficult to see what predictive power this categorisation has.

Lastly, where there was "limited cultivation", (i.e. planting) no obvious cases of tilling, weeding or irrigation were noted, a point that will be elaborated on a little later in relation to such activities in the evolution of agriculture elsewhere. Furthermore, the validity of including transplantation is questionable, as the Australian evidence shows that it was a feature in only two, perhaps three cases, where limited planting was traditionally carried out, with the *Umpila* and the Lockhart River people of Cape York, for example, transplanting yams on offshore islands.[94]

An alternative classification of human-plant interactions, from an evolutionary biology perspective, has been promulgated by David Rindos, providing for three categories of "domestication"[95]:

(1) Incidental Domestication:
  a) wild plant distribution "that appears to be the result of human feeding behaviour"
  b) "cultivated plants that have portions of their gene pool incidentally dispersed or protected"

(2) Specialised Domestication: "intensification of certain tendencies present in incidental domestication."

[87] Beveridge 1889:35. See also Gott and Conran (1991:34).
[88] Smith 1995b:211.
[89] Yarnell 1985:698-9.
[90] Jarman, Bailey and Jarman 1982:53-4.

[91] Chase 1989; Yen 1989.
The main form of what I would consider to be Incidental Gathering was undertaken by men out hunting and "snacking" on berries and other "consumables" they passed by in the course of their activities. See for example Oldfield (1865:271) and Hiatt (1968: 210) and fuller discussion on p.88 of Chapter 6.
[92] The nearest approach to Systematic Gathering in this regard was with the planting and harvesting of yams on Flinders Island where it was reported, "the yam is abundant, hundreds of these deep holes are evident in a comparatively small area." (Hale and Tindale 1933-6:113)
[93] Allen 1972;1974
[94] Hynes and Chase 1982:40; Harris 1977b:437.
The yams planted by the *Walmbaria* of Flinders Island were probably also originally transplanted from the mainland. Evidence will also be presented in the next chapter that the Nhanda transplanted their yams as well.
[95] Rindos 1984:154-166.

(3) Agricultural Domestication: "establishment and refinement of the systems of agricultural production"

Rindos also introduces the concept of "replacement planting", the "maintenance of the plant species in a pre-existing niche,"[96] which he contrasts with "agricultural planting", defined as "colonisation by the plant of ... areas ... created by human disturbance of the existing ecology."[97] Many criticisms have arisen in regard to the Rindos model, his work being characterised by some as "biological determinism" or "biological reductionism".[98] By reducing the role of humans to "obligate dispersal agents,"[99] some consider Rindos is attempting to eliminate the social or cultural element in the process of domestication and the development of agriculture,[100] as well as possibly denying human intentionality in this.[101] His employment of the term "coevolution" has also been challenged,[102] while the appropriateness of the term "incidental domestication" for a phenomenon others consider to be simply incidental planting or dispersal is another problematic area.[103]

Basically Rindos, in endeavouring to provide an explanation for the domestication of plants, tried to identify processes leading to this domestication. Consequently he proposes that humans initially affect plant distribution through incidental dispersal, and the creation of the disturbed habitats favoured by "weedy" plants, around camp sites. This pattern of interaction is exemplified by occurrences such as native fruits trees around camp sites in Arnhem Land and of particular plants around camp sites, such as *yukidjirra* berries, where seeds have been dropped following harvesting, preparation or consumption. Following that, humans further aid the dispersal of plants by preferentially assisting or protecting them (e.g. the *Iliaura* blocking up creeks) as well as by planting them in other niches (initial agricultural planting). Further intensification of these processes finally leads to full-scale agriculture where people significantly alter the local ecology and facilitate the plants' colonisation of these areas. Humans, in the course of this, move from being opportunistic to specialised feeders.[104] But there appears to be limited explanatory value in Rindos's schema. No account is taken of domicultural activities or selective propagation, there are few examples of incidental planting among the cases of planting in Australia, and none of the planted species appear to have been incidentally dispersed prior to their planting.

A later attempt to classify the relationships between humans and plants in terms of an "evolutionary continuum of people-plant interaction"[105] created a typology based on forms of "production" engaged in by hunter-gatherers and early farmers. Harris's basic schema provides for:

(1) Wild Plant Food Procurement: burning vegetation, gathering/collecting, (foraging) protective tending.
(2) Wild Plant Food Production: replacement planting, ransplanting, weeding, (with minimal tillage) harvesting, storage, drainage/irrigation.
(3a) Cultivation: land clearance, systematic soil tillage (with systematic tillage).
(3b) Agriculture: cultivation of domesticated crops.

Such an ordering, in my view, is flawed in certain respects. A major defect in this conception lies in the insistence that specific optimisation behaviours such as weeding, drainage and/or irrigation and tillage accompany particular types of plant interactions. The evidence pertaining to the activities of Indigenous Australians, as well as elsewhere, does not appear to bear this out. Furthermore there appears to be a lack of clarity in the distinctions made between categories. "Harvesting", for example, is associated with the planting behaviours of Wild Plant Food Production but not Wild Plant Food Procurement where people are just "gathering/collecting". This begs the question of whether the *Paakantyi* of the Darling River were engaged in Wild Plant Food Production or Procurement? Including storage in the second category is also questionable as this is usually considered to be socioeconomic adaptation rather than a human-plant interaction. In a sense the problem lies in Harris's approach which, like Higgs and Jarman's, is really a classification of types of "productive economies" and not just of human-plant interaction.

All the models that have just been outlined seek to describe and classify in some rational and consistent way a range of human-plant interactions in conjunction with optimisation behaviours that eventually become an integral part of any mature agricultural system. Part of the difficulty in the preceding examples lies in generalising in regard to optimisation behaviours - clearing, tilling, weeding, fertilising, watering, protecting and so forth - when there are no systematic studies to support those generalisations. As will be pointed out in the next chapter, the evidence indicates that at least some of these activities may well be quite late developments, refinements adopted when agriculture had already become the dominant means of food production. Compounding this difficulty are attempts to then correlate such activities with particular types of plant interactions, without a clear body of evidence demonstrating such correlations. Undoubtedly, in general terms, a continuum of increasingly close or intense human-plant interactions is perceptible, and at some point various optimisation behaviours are adopted. But these are not necessarily embraced synchronously or in conjunction

96 Rindos 1984:161.
97 Rindos 1984:161.
98 Rosenberg 1990:405; Flannery 1986:16.
99 Rindos 1984:159.
100 Chase 1989:47; Rosenberg 1990:400; Pearsall 1995:192.
101 Blumler and Byrne 1991:27; Riley 1991:44.
102 Blumler and Byrne 1991:26n8.
It has been argued that as it is only plants that are evolving (i.e. changing genetically), and not the humans involved in the process, the application of the term "coevolution" is inappropriate (Blumler and Byrne 1991:26n8). The relationship can probably be more accurately described as a case of "predator-prey mutualism" (Pearsall 1995:159).
103 Price and Gebauer 1995:17.
104 Rindos 1984:158-9.

105 Harris 1989:17

with particular types of plant interactions and may simply be parallel processes driven by different dynamics. Consequently, it may be more productive to consider each process independently and initially focus exclusively on human-plant interactions.

This consideration begins with the observation that human-plant interactions can be treated as a continuum in which human strategic behaviours range from generalist to specialist, reflecting an increasing involvement in, and manipulation of, the life cycle and reproduction of subject plants. In other words, a scale of intervention is discernible in human-plants interactions in which people can be seen to have increasingly specific and direct involvement with plants. Employing this principle a typology of human-plant interactions can be developed, as illustrated below:

(1) General Selective Encouragement: Activities that enhance the productivity of particular ecosystems and/or plant communities e.g. firestick farming.

(2) Specific Selective Encouragement: Activities that promote the productivity of individual plants or discrete communities of a particular species e.g. burning *pitjuri* bushes or rushes.

(3) Incidental Propagation: Activities which do not necessarily involve any specific intervention but which result in the propagation of economically useful plants in frequented locations e.g. camp following fruit trees in Arnhem Land.

(4) Selective Propagation: An intervention that results in the propagation of a species of plant without actual planting e.g. selective propagation of yams by the *Gidjingali* and *Watjandi*.

(5) Replacement Planting: Deliberate planting to ensure a particular individual plant, or group of that species, being utilised is replaced e.g. replanting of yam tops by *Wik-Mungkan*, replanting of *I. costata* and *I. polpha*.

(6) Translocational Planting: Carriage and planting of seeds or propagules to a new location (usually outside the normal habitat of that species) e.g. *Umpilla* transferring yams to offshore islands.[106]

(7) Productive Planting: Planting that increases the population, and often the density and/or the size of the niche, of the exploited plant species in a particular locale, up to a limit determined by the extent of the locale and human needs e.g. *Nhanda* yam planting practices.

In all the cases alluded to so far, and underlying the typology that has been outlined, it would appear the principal concern of the groups undertaking respective activities is the utility of the exploited plants. This applies whether it is a desire to have access to suitable spear trees, an assured supply of "native tobacco" or plentiful and reliable sources of food. As interventions become more specific there tends to be an increase in total output and productivity, especially with food plants. Genetic changes in plants may, and do, occur at any point on the continuum, the nature and extent of those changes being dependent on such things as the type of interaction, level of exploitation and the genetic and ecological characteristics of the plant in question. Domestication, in the sense of genetic changes in plants making them either partially or wholly dependent on humans for their reproduction, would appear to be more likely to arise as the degree of intervention increases. With so many variables involved, however, one cannot be rigidly prescriptive in this regard.

Finally, it should be pointed out, the scale of human involvement in the life cycle and reproduction of plants does not cease with Productive Planting. Breeding is an additional level of interaction where the genetics of the subject plant are manipulated to produce new varieties or species with improved utility. In the last century or so this has become a more refined process as the underlying mechanism has become better understood. Further to that, another more interventionary strategy has recently arisen, the Genetic Modification of plants. Again this is carried out, by inserting a desirable genetic characteristic into the particular plant species, with a view to increasing its utility, in this instance by operating at a biomolecular level. Consequently these last two items should be added to the typology of human-plant interventions:

(8) Breeding: Activities that intentionally manipulate the reproduction of exploitable plants, resulting in new varieties or species e.g. many current crops

(9) Genetic Modification: Activities that manipulate the biochemistry and genetic structure of particular species resulting in the introduction of desired characteristics e.g. pest resistant cotton

In conclusion it is important to stress the continuities involved. From this perspective I reiterate the point made earlier, that even the most traditional generalist hunter-gathers have much in common with "space-age" or "information revolution" societies. Stripped of sociocultural variation the approach is the same throughout, at least in regard to plants, it is just the inherent possibilities of the situation in which each finds themselves that is different.

[106] Translocational planting has been noted at the Pre-Pottery Neolithic A (PPNA) site of Mureybit and Abu Hureyra in SW Asia where einkorn was transferred from elsewhere, probably south east Turkey (Flannery 1986:15; Heun et al. 1997:1313), and the Kumeyaay of southern California (Smith 1995a:17). Translocational planting, of yams, in a non-agricultural context has also been noted among hunter-gatherers in Africa (Harris 1977a:213).

# Chapter 4.
# A Case of Indigenous Agriculture in Australia

**Preamble**

In 1606, when Europeans, as far as is known, first made landfall in Australia and had contact with Indigenous Australians for the first time,[1] the process of intensification had been in progress in Australia for at least five millennia. At about the same time as the Europeans became aware of the existence of the Great South Land the *Macassans* also began visiting parts of its north coast. Meanwhile, significant Melanesian influences continued to impact on Aboriginal cultures in north and north eastern Australia through the Torres Strait corridor, although possibly disrupted by the drowning of the Sahulian land bridge in the mid-Holocene.[2] All these dynamics, be they endogenous or exogenous, appear to have led to distinctive developments in particular parts of Australia. The first of these to come under consideration relates to the *Nhanda* and northern *Amangu* groups of the central west coast of Western Australia.[3]

As stated earlier, the received wisdom is that there was no indigenous agriculture in Australia, a preconception I will strongly dispute. But before deliberating on the possibility of this indigenous agriculture in Australia a digression is necessary in order to clarify what is meant by "agriculture". The characterisation and definition of "agriculture", describing what is, in effect, a complex array of behaviours and relationships, is highly variable and at times quite confused.[4] Agriculture, according to Reed is "the totality of the human practices involving those living energy traps which man plants, breeds, nurtures, grows, guards, preserves, harvests, and prepares for his own use."[5] Rindos treats agriculture as a form of plant colonisation of a well-defined area created by human disturbance,[6] while Yarnell describes agriculture as "intentional husbandry practiced by humans."[7] MacNeish sees it as the "planting of multipropagators (i.e. seeds) or cultivars in relatively large plots or fields,"[8] but according to Meadows it is simply the "practice of growing crops ... Anyone who sows or harvests domesticated plants is involved in agriculture."[9] While it would appear from this that there is little definitional consensus, in behavioural terms the principal elements appear to be, after Redman and Harris,[10] propagation (sowing, planting, usually of "domesticates"), husbandry (soil preparation, maintenance of moisture and fertility, exclusion of competitors, protection from predators), harvesting and finally storage. The scale or areal extent of these activities would also appear to be a significant factor, though not always explicitly stated, as some of the preceding examples show. This, in fact, is often one of the criteria employed to distinguish between agriculture and horticulture,[11] as typified by MacNeish's concept of horticulture as "cultivation that emphasises the planting of individual domesticates or cultivars in relatively limited plots."[12] But, how ever one views it, clearly planting is absolutely essential to any consideration of agriculture.

The common requirement that the planted cultivar be "domesticated" is highly problematic, however.[13] Certainly many, but not all, cultivated food plants appear to undergo morphological [phenotypic] changes and/or speciation, some quite quickly, once they are systematically planted, and in so doing may become human-dependent.[14] Whether a cultivated plant does become domesticated may depend,

---

1 When the "Duyfken", under the command of Willem Janszoon mapped about 300 kilometres of the western coastline of Cape York.

2 See Clarke 2003:183-86 for a listing of some the forms of influence.

3 There has been some difficulty over the years determining who occupied what territories in the region, and this is still a matter of debate (Blevins 2001:ix,1-7,146,154; Gerritsen 2004:85- 8). Here the area from 50 kilometre south of the Murchison River to central and western Shark Bay is assigned to the *Watjandi Nhanda* group (Murchison River) and an unnamed group at Shark Bay. In the Victoria District, Hutt River to Irwin River, I assign the *Nhandakorla Nhanda* to the northern part and the *Amangu Nhanda* (northern *Amangu*) to the southern part, although the southern group is often considered on linguistic grounds to be a separate group just called the *Amangu*. Unless specifically indicated otherwise *Nhanda* will be used in a generic sense in application to the Victoria District groups.

4 Scarry 1993:7; Harris 1996:3; Leach 1997:145-6.
An example of one end of the range of definitions is Pryor (1983:94), who equates agriculture in its "widest sense" with "food production" but then includes animal domestication in this as well as plant domestication.

5 Reed 1977:19.

6 Rindos 1984:161.

7 Yarnell 1985:698.

8 MacNeish 1992:11

9 Meadows 1996:392.

10 Redman 1978:91; Harris 1989:22.

11 Bronson 1977:25-6; Leach 1997:146.

12 MacNeish 1992:10

13 See Helbaek 1970:195; Higgs and Jarman 1972:7; Jarman 1972:15, 20,23; Wilke et al. 1972:204; Coursey 1976a:400n5; 1976b:72; Reed 1977:19; Redman 1978:92; Bailey 1981:2; Nishida 1983:306,310,318; Sowunmi 1985:127; Yen 1985:323; Ladizinsky 1987; Harris 1989:19; Shipek 1989:159; White 1989:157; Zohary 1989:369; Hillman and Davies 1990:156,166; Harlan 1992a:25; Ingold 1996:21-22; Rindos 1996:211; Spriggs 1996:525; Sherratt 1997:275; Anderson 1998:145; Willcox 1999:487-492; Benz and Long 2000:460; Cauvin 2002:51-61; Asouti and Fairbairn 2002:182.

14 It should be noted that even some species that are classically human dependent, such as wheat, can still in fact survive in the wild without human assistance (See Jarman 1972:23; Canadian Food Inspection

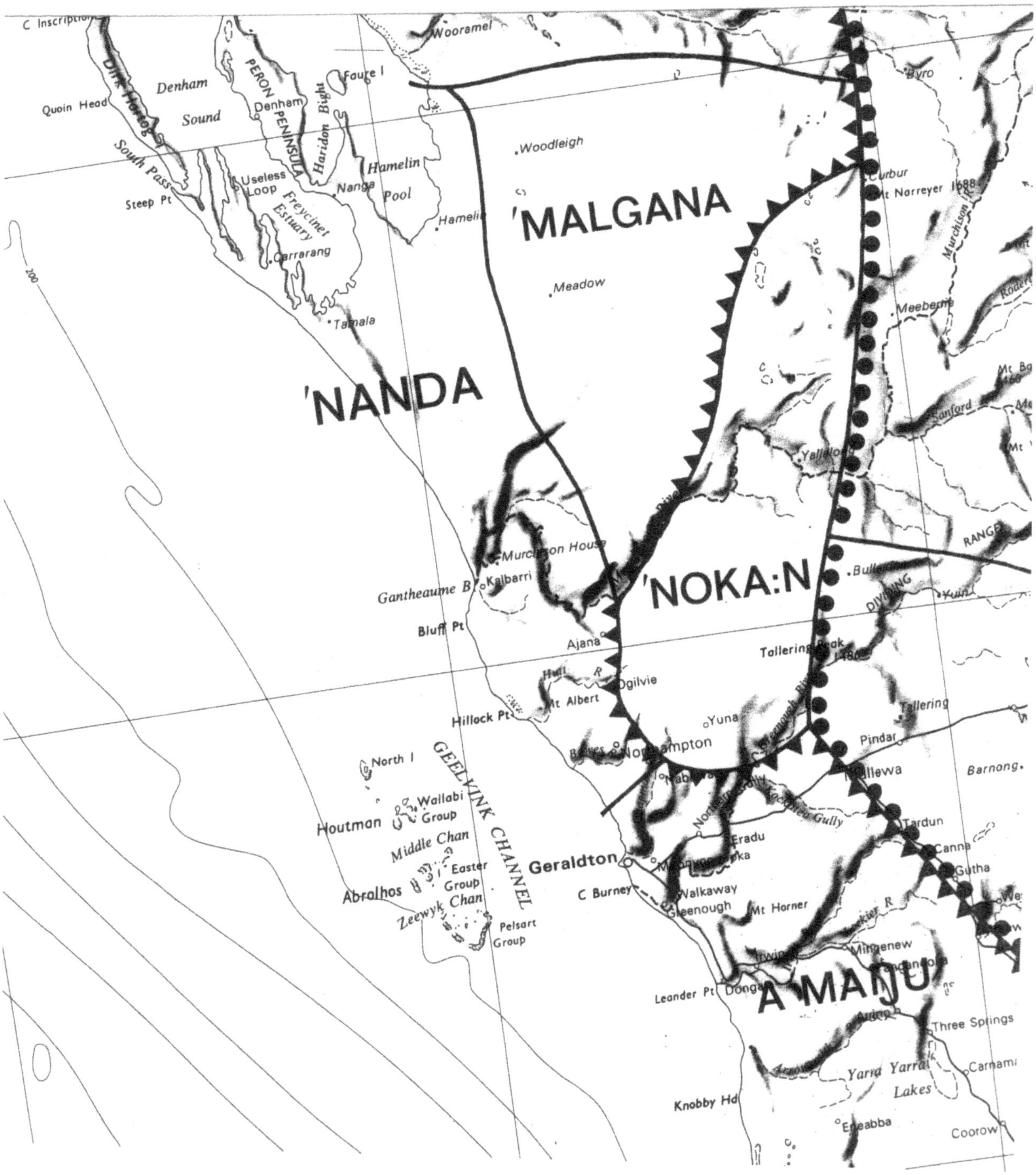


*Figure 6: Nhanda and the Central West Coast*
*(Tindale 1974)*

however, in part on the harvesting method as well as its genetic responsiveness.[15] And it should be noted that intensive harvesting, without planting, may also produce some of those changes.[16] Furthermore, domestication itself does not seem to be essential to systems otherwise characterised as agricultural, with some engaging in what has been termed "pre-domestic agriculture" or "pre-domestic cultivation" by Hillman and Davies, Willcox,

Agency Plant Biotechnology Office:http://www.inspection.gc.ca/english/plaveg/bio/dir/dir9901e.shtml#B1:5).
Moreover, many species of "domesticated" yam are cultivated forms of wild species (that continue to grow in adjacent areas), and simply revert to the wild form if cultivation ceases.

[15] Hillman 1996:194; Sato 2003:146.

[16] Ladizinsky 1987; Chikwendu and Okezie 1989; Hillman and Davies 1990:159; Harlan 1992a:25; Higham 1995:150; Evans 1996:63.
This may explain the occurrence of "domesticated" squashes (*Cucurbita pepo*) in Mesoamerica (Harrington 1997; Roush 1997; Smith 1997), thousands of years before maize and agricultural lifeways became

Cauvin and others.[17] Such seems to have been the case as in south west Asia, where current evidence indicates wild, undomesticated or only partially domesticated cereals were being grown well into the Neolithic, the period when it is usually accepted agriculture was becoming, or had become, established.[18] Rice (*Oryza sativa*) is another case in point. It had been cultivated for at least 3000 years as "ancient cultivated rice" before becoming "domesticated" by the end of the Chinese Middle Neolithic,[19] yet few would deny that a well developed agricultural economy was in effect in that period. Broomcorn millet (*Panicum miliaceum*), another Neolithic crop from China, does not appear to have ever been fully domesticated.[20] So, while domestication may be a common consequence, and strong indicator, of systematic cultivation, it is not a necessary condition for systematic, larger scale sowing and harvesting of plants. Certainly it may signify the subject species genetic response to those practices, as a biological marker signifying in its strictest sense possible entry into a fully commensal, human-dependent, relationship.[21] But it is nonsensical to define a human activity on the basis of how a plant may respond genetically when that response may take time and can be quite variable, depending as much as anything on the genetics of the particular species. Consequently it is, in my view, more rational to define agriculture just in terms of the human behaviours involved, that it can be demonstrated in particular cases that there is an on-going commitment to providing a significant proportion (say in excess of 30%) of the food supply through systematic planting.[22] Domestication can then be simply treated as an indicator, one that can be diagnostically tested, in ascertaining that systematic planting and agriculture may have been taking place, particularly in archaeological contexts where planting is not otherwise directly detectable.[23,24]

Plant husbandry, another element commonly incorporated in definitions of agriculture, involves an array of behaviours, which the evidence appears to indicate were not developed or applied synchronously. In effect husbandry is a series of refinements introduced over time, probably as adaptations aimed at improving the productivity of the crops being produced.[25] Decrue planting (sowing on muddy or flooded flats that dry out) or broadcast seeding is often noted in early agricultural endeavours involving seed crops,[26] with no clearing, or at most simply burning, and no deliberate tilling of the soil.[27] Measures to protect the crop, such as weeding, are also seemingly later developments, as may be the case with watering and irrigation.[28] In the European Neolithic, measures to retain or enhance fertility were not initially practised in some areas, and probably did not appear until the Bronze Age.[29] Even the physical storage requirement is not considered essential in agricultural systems, it being a frequent practice with vegetatively reproduced staples, such as taro, sweet potatoes and yams, for the tubers to be left in situ and harvested as needed.[30]

In summary it would seem that the characterisation of agriculture should include as a minimum: propagation and harvesting of plants on a systematic basis, of sufficient extent to provide a significant proportion of the food supply for the people undertaking those activities. It should also involve a strategy to conserve the resultant output through storage or other means. Non-essential complementary features should include activities intended to maintain or enhance productivity (i.e. husbandry), and the expectation that the cultivar is likely to have undergone, or be undergoing, some morphological change.

### The Case for Indigenous Agriculture in Australia

The notion that traditional Aboriginal Australians, prior to British colonisation, never grew crops, tilled the soil,

established there (Benz and Long 2000:460,464; Fritz 1995:5). Support for the contention that squashes were not grown in the period in question comes from the observation that it was not translocationally planted in the Tehuacan Valley for a further 2000 years (Lamberg-Kalovsky and Sabloff 1995:253), as would be expected if it were being cultivated. See also n61 of Chapter 6.

17 Hillman and Davies 1990:166,199-201; Colledge 1998:121; Willcox 1999:494; Colledge 2001:150; Cauvin 2002:51,53-59,214,217.

18 Hillman and Davies 1990:159,166,208; Kislev 1992:87-93; Bar-Yosef and Meadows 1995:66; Bar-Yosef 1998a:167,168-9; Anderson 1998: 147; Willcox 1999:478-80,489-93; Colledge 2001:146-7,150-1; Cauvin 2002:53-59,214,217; Holmes 2004:29-30.

19 Pei 1998; Zhang and Wang 1998; Sato 2003:147; Yuan 2003:161; W. Zhang 2003; Bellwood 2005:116,164.
Sato (2003:147) points out that "the classification between cultivated and wild rice is not always clear". See also Yasuda 2003a:130 and W. Zhang 2003.
According to Bellwood (205:150) harvested rice would have taken on domesticated characteristics after 5 generations.

20 Harlan 1995:154; Lu 1999:61; Crawford 2006:80-1; Lee et al. 2007.
There are other examples of cultivation for long periods before genetic changes occur, such as *Setaria* and corn in Mesoamerica (Flannery 1986:15; Lamberg-Karlovsky and Sabloff 1995:250), and sorghum in northern Africa, taking perhaps 6000 years to become domesticated (Wetterstrom 1998:34,37- 8,42).

21 Higgs and Jarman 1972:7; Coursey 1976a:387,391; Bronson 1977:27; Unger-Hamilton 1989:101; Kislev 1992:87-93; Bar-Yosef and Meadows 1995:66; Bar-Yosef 1998:167,168-9; Willcox 1999:489.

22 The qualification is also based on the observation that domestication may occur before the scale of planting reaches a point where the plant in question is providing a significant proportion of the food supply. See p.75 in respect to maize.

23 Unger-Hamilton 1989:89.

24 It should be noted, however, that domestication usually results in an improvement in productivity as the subject species develops "desirable" characteristics in response to the human selection regime (See Jarman 1972:19).

25 Helbaek 1970:194; Bailey 1981:8; Stark 1981:352; Pryor 1983:114- 5; Smith 1995:21; Cauvin 2002:20,57.

26 Harlan 1989:95; Shipek 1989:162; Hillman and Davies 1990:210; Anderson 1992:180,186; Smith 1995a:113; Pringle 1998:1447.
Decrue planting is also the norm in the Nile Valley, the farmers planting following the annual floods (Childe 1934:51; Wetterstrom 1998:34)

27 Boserup 1965:24; Stark 1981:352; Anderson 1992:180-1; Harlan 1995:13; Smith 1995:21; Anderson 1998:145,151; Ibanez 1998:140-142.
Of course "slash and burn" agriculturalists do little more than clear an area and plant a crop for one to several years. See for example Lu's (1999:136) description of corn and millet agriculture in parts of China until recent times.

28 MacNeish 1972:82; Anderson 1992:181,186,206; Smith 1995a:21; Balter 1998:1443; Anderson 1998:151.

29 Boserup 1965:24; Jewell 1981:226; Reynolds 1992:385; Bogucki and Grygiel 1993:408; Bakels 1997.

30 Coursey 1972:218;Ingold 1983:570n; Degras 1993:57.

planted or sowed is an old one. It was indeed one of the rationalisations employed to justify the expropriation of Australia, often repeated in the ethnographic literature of the 19th century.[31] It is also a common preconception or unquestioned assumption among modern scholars. However, one confounding case in particular, which arose as a consequence of an accident of history a little over 370 years ago, may warrant closer attention.

On the 16 November 1629 two Dutch mariners, 18 year-old cabin-boy Jan Pelgrom de Bye and 24 year-old soldier Wouter Loos, were deliberately marooned on the shores of the west coast of Australia, at a place now believed to be the mouth of the Hutt River,[32] 450 kilometres north of Perth, Western Australia. These hapless seafarers were abandoned on the orders of Captain Francisco Pelsaert for their part in a bloody mutiny. Five months earlier, in the early hours of 4 June 1629, the Dutch ship "Batavia" had been wrecked on Morning Reef in the Wallabi Group of the Abrolhos Islands, 60 kilometres from the Western Australian coast. All but about 40 of the 316 passengers and crew appear to have eventually managed to make their way to nearby islands, coral specks in most instances, by boat, swimming or clinging to flotsam. However, water was in critically short supply, so Captain Pelsaert and about 40 of the crew sailed in the ship's yawl to the continental mainland to search, without success. Rather than return to the beleagued survivors, Pelsaert next decided to head for the Dutch fort at Batavia (modern-day Djakata) in Java for help. And so, in an epic voyage, they sailed the open boat 2,400 kilometres back to Indonesia, hoping to effect a rescue. In their absence an horrific mutiny, the Batavia Mutiny, took place in which 125 men, women and children were murdered.

The mutineers had been secretly conspiring even before the "Batavia" was wrecked. With Pelsaert gone, the mutineers, led by Undermerchant Jeronimus Corneliszoon connived to have body of Dutch and French soldiers abandoned on one of the islands, West Wallabi Island, so as to remove any source of opposition. They thought the soldiers would die of thirst or hunger there, but in fact they prospered, finding water and living on bird's eggs, seals and tammar wallabies. The soldiers were soon joined by some stragglers who had escaped from the mutineers. They informed the soldiers of the mutineers' murderous rampage. Because the mutineers planned to capture any rescue ship and become pirates they decided it was necessary to eliminate the soldiers and stragglers, known collectively as the Defenders. Consequently the mutineers launched two attacks against the Defenders, both quite ineffective. In the course of the conflict the Defenders even managed to capture Corneliszoon and so the mutineers elected a new leader, Wouter Loos. Loos then launched another, much more effective attack, on the Defenders. They were about to overwhelm them when, on 17 September, Pelsaert miraculously reappeared in the rescue ship, the "Sardam". The leader of the Defenders, Webbie Hayes, raced to warn Pelsaert of the mutineers plot to seize his ship, and the mutiny was immediately put down and the mutineers rounded up. Following a series of inquisitions in which judicial torture was freely used, 8 mutineers, including de Bye, were sentenced to be hanged. Due to lack of evidence Loos escaped that sentence. Pelsaert and the ship's council decided in the end to maroon Loos and de Bye (because of his youth) on the Australian mainland.[33] Pelsaert subsequently issued the castaways, the first Europeans known to inhabit Australia, with orders, which read (in part):

"*.... to put ashore there or here, to make themselves known to the folk of this land by tokens of friendship. Whereto are being given by the Commandeur* [Pelsaert] *some* Nurembergen [cheap wooden toys and trifles], *as well as knives, Beads, bells and small mirrors, of which shall give to the Blacks only a few until they have grown familiar with them.*

*Having become known to them, if they then take you into their Villages to their chief men, have courage to go with them willingly. Man's luck is found in strange places;*"[34]

In order to assist the reluctant "pioneers" Pelsaert provided them not only with the aforementioned articles but also, on the day they were put ashore, "... a Champan [flat-bottomed boat] *provided with everything* [my emphasis]..."[35]

Exactly two centuries later (1829) the European colonisation of Western Australia began in earnest when the British established a permanent settlement, the Swan River Colony, on the site of the present-day state capital, Perth. For a decade the colony was largely restricted to an area within a radius of 80 kilometres around Perth, during which time the Indigenous populations were pushed aside and then "pacified" when they began to resist. Tentative explorations beyond this area took place during this period, but in 1839 an ambitious expedition was planned to examine the coastal lands between Perth and Shark Bay, 880 kilometres to the north. Led by Lt. George Grey (later Sir George Grey, Governor of South Australia, New Zealand and then the Cape Colony) the expedition soon foundered when, at Shark Bay, they were caught in a terrible cyclone [hurricane] - the

[31] e.g. Lang in Eyre 1845:2:298; Curr 1886:1:36,78; Worsnop 1897:2.

[32] Gerritsen 1994a:271-287; Playford 1996:237-42; Gerritsen, Slee and Cramer 2005; Gerritsen 2007b,2008.

[33] At the time the continent was *Terra Incognita* to Europeans, unknown apart from the shadowy outlines of parts of the north, south and west coasts.

Numerous accounts of the Batavia Mutiny have been written, for example:

- Drake-Brockman, H. 1963/1982 *Voyage to Disaster*. London: Angus & Robertson.
- Godard, P. 1993 *The First and Last Voyage of the Batavia*, Perth: Abrolhos Publishing.
- Gerritsen, R. 1994 *And Their Ghosts May Be Heard*. South Fremantle: Fremantle Arts Centre Press, pp.16-30,271-287.
- Dash, M. 2002 *Batavia's Graveyard* London: Weidenfeld and Nicholson.

[34] Pelsaert 1629:230

[35] Pelsaert 1629:237

"eye" passing directly over them. Having lost most of their stores and equipment Grey decided to return immediately to Perth in the two remaining boats. Continuing south for some 300 kilometres Grey attempted to put in at a place known as Gantheaume Bay, the embouchure of the Murchison River and Wittecarra Creek. Unfortunately a large and unexpected surf close to shore swamped the two boats, leaving Grey and his party 500 kilometres from "civilisation" with little food and no water. The party resolved at that point to walk overland back to Perth, experiencing great hardship, but accomplished this remarkable feat with the loss of only one life. In the course of his travails Grey kept a journal, later published as part of *A Journal of Two Expeditions In North-West and Western Australia, 1837-39*, in which he recorded at length the events that took place and his observations along the way. Several days after their mishap at Gantheaume Bay, as the ragged band of explorers approached the Hutt River, Grey began to make some interesting observations. About 18 kilometres from the Hutt River valley, in a valley leading into it, Grey recorded in his journal:

"*April 4 1839: we soon fell in with the native path we quitted yesterday; but now became quite wide, well beaten and differing altogether by its permanent character, from any I had seen in the southern portion of this continent ... And as we wound along the native path my wonder augmented; the path increased in breadth and its beaten appearance, whilst along the side of it we found frequent wells, some of which were ten and twelve feet* [3-4 m] *deep, and were altogether executed in a superior manner. We now crossed the dry bed of a stream, and from that emerged upon a tract of light fertile soil quite overrun with* **warran** [original emphasis] *plants* [the yam plant - *Dioscorea hastifolia*], *the root of which is a favourite article of food with the natives. This was the first time we had seen this plant on our journey, and now for three and a half consecutive miles* [5.6 kms] *traversed a piece of land, literally perforated with holes the natives made to dig this root; indeed we could with difficulty walk across it on that account whilst the tract extended east and west as far as we could see. It is now evident that we had entered the most thickly-populated district of Australia that I had yet observed, and ... more had been done to secure a provision from the ground by hard manual labour than I could believe it in the power of uncivilised man to accomplish. After crossing a low limestone range we came upon another equally fertile* **warran** [original emphasis] *ground .....*

The next day, as they reached Hutt River, Grey continued:

*April 5 1839: ..... The estuary* [actually a salt marsh, Hutt Lagoon] *became narrower here, and shortly after seeing these natives, we came upon a river running into it from the eastward* [Hutt River]; *its mouth was about forty yards* [36 m] *wide, the stream strong but the water brackish, and it flowed through a very deep ravine, having steep limestone hills on each side ... Being unable to ford the river here, we followed it in a SE direction for two miles* [3.2 kms], *and in this distance passed two native villages, or, as the men termed them, towns - the huts of which they were composed differed from those in the southern districts, in being built, and very nicely plastered over the outside with clay, and clods of turf, so that although now uninhabited they were evidently intended for fixed places of residence.*"[36]

Over the next 130 kilometres, as Grey and his party traversed the Victoria District, they encountered a series of small fertile river valleys (the Bowes, Chapman, Greenough, Irwin and Arrowsmith Rivers) each conforming to the pattern observed at Hutt River in terms of the intensive utilisation of the *warran* plant (it was actually known locally as *ijjecka* or *adjukoh*). Two further villages were sighted, at the Bowes and Greenough Rivers. Grey estimated the latter "contained at least a hundred and fifty natives"[37]. All the elements of Grey's observations were subsequently borne out over the following 12 years, as exploration continued and British settlements became established in the Victoria District [Hutt River to Irwin River], by a variety of explorers, government officials and pioneers including Lt. Helpman (Explorer), A. C. and F. T. Gregory (Surveyors, Explorers), James Drummond (Government Botanist, Pioneer), Mrs Brown (Pioneer), Phillip Chauncy (Surveyor), John Septimus Roe (Surveyor-General), William Burges (Government Resident), Edward Cornally (Shepherd) and Dr Robert J. Foley (Pioneer). Lt. Helpman, for example, provided a detailed description of the type of huts referred to by Grey:

"*We here met with the first native hut; it was well plastered outside and the timber which formed it was about 6in.* [15cm] *thickness, about 6ft.* [1.8m] *high inside and capable of holding ten persons easily.*"[38]

Captain Stokes and his party from the famous "Beagle" (Darwin was not on this voyage) even slept in one during a land reconnaissance in the Geraldton area in 1841:

"*We noticed their winter habitations substantially constructed and neatly plastered over with red clay ... Some neighbouring wigwams of superior structure gave us snug quarters for the night.*"[39]

[36] Grey 1841a:2:12,19
I believe I have identified the site of this first village, as well as some of the others in the region, through field investigations (Gerritsen 2002). The site of this first village appears to cover an estimated area of almost 4 hectares (Personal Observation 29/4/1998,21/5/1999) having an estimated population of 290(Gerritsen 2002:12).
[37] Grey 1841a:2:37
[38] Helpman 1846:9
[39] Stokes 1841:2-3.
How Stokes came to the conclusion these dwellings were only inhabited in winter is a moot point as he was only in the area for about one and a half days in early summer (1846:2:385-93). He probably just assumed this was when they were usually occupied because of their substantial construction. However winters here are quite mild (minimums never fall below zero, av. maximum 18 deg.C) compared to further south where far less substantial or protective shelters were constructed. Protection from rain

Perhaps of greater significance were other comments made by some of the vanguard of the colonial invasion. William Burges, the first representative of the Colonial government in the district, reported in 1851 that:

"*They* [the *Nhanda* people] *seem very little addicted to hunting and very few of them are even expert at tracking a Kangaroo. This may result from the great variety of edible roots, particularly the A-jack-o or warang which grows here in great abundance.*"[40]

a view echoed by Helpman:

"*They are a fine race of men but seem to depend* ***entirely*** [my emphasis] *upon warran and gum, of which they have great abundance. Very few, even of the women, had kangaroo skins*"[41]

Finally A. C. Gregory, as mentioned earlier, a noted Australian explorer, provided the telling observation mentioned in the last chapter regarding the *Nhanda* people of the Victoria District:

"*Agricultural science seemed to have made some progress, as they never dug a yam without planting the crown in the same hole so that no diminution of food supply should result.*"[42]

with the conclusion supported by Maiden in 1889 in *Useful Native Plants of Australia*:

"[*Dioscorea hastifolia*] ... *is the only plant on which they* [Aboriginal people] *bestow any kind of cultivation, crude as it is.*"[43]

The significance of these passages lie in the references to the *warran*, the yam plants growing "as far as we could see", or, "tracts of land several square miles in extent"[44] [calculated to cover an area of 15-16 sq. kms[45]], evidence of their propagation, the paths, the wells, the huts and the villages - all bespeak an agricultural society exhibiting a high degree of sedentism.[46] Gregory's information is unequivocal and clearly indicates propagation was in evidence, while Grey elsewhere observed "these circumstances all combined to give the country an appearance of cultivation, and of being densely inhabited, such as I had never before seen".[47] Other comments, particularly Helpman's and Burges', regarding the degree of dependence on the yams, if read in conjunction with Grey and others on the extent of the yam grounds and the presence of villages, strongly suggests that this propagation was carried out on a systematic basis, to an extent where it provided a major proportion of local subsistence.

Dr Foley, in 1851, referring to the Chapman River valley, noted "You can see for miles, and it being dug for *warrang* or *agacs*, as they call it here, you can see the nature of the soil."[48] At nearby Wizard Peak, Bynoe, the surgeon from the "Beagle", had noticed during their reconnaissance a decade earlier that "the native yam seemed to grow in great abundance."[49] Again Grey, in describing the Arrowsmith River valley reported, "the whole of this valley is an extensive warran ground".[50] Grey's remark, that the ground at Hutt River was "literally perforated with holes the natives made to dig this root", was often repeated by other observers, pointing to intensive harvesting.[51]

Unfortunately, apart from one minor study carried out in 1914,[52] no published archaeological investigations have ever been carried out in the Victoria District, necessitating a reliance on historical ethnographic sources. These limited sources do not contain any information on any husbandry that may have been undertaken by the *Nhanda* of the Victoria District. However it could be inferred that at Hutt River at least the plants may have been watered. Grey, it will be recalled, stated that, "we found frequent wells, some of which were ten and twelve feet [3-4 m] deep, and were altogether executed in a superior manner". He described them in another account as being "of a depth and size quite unknown in the southern portions of the continent."[53] The climate in this semi-arid region is of Mediterranean type, with hot, dry summers and mild, wet winters, and an annual

also appears an unlikely explanation. This area receives about 463 mm p.a., about half that of Perth, usually in winter from heavy falls lasting only an hour or two. Curiously Stokes' reference to "their winter habitations substantially constructed" is omitted from the later account.

40 Burges 1851a:6

41 Helpman 1849:3.

42 Gregory 1886:23

It will be recalled that Gregory, who had been in the district at various times between 1849 and 1854, originally made the claim in 1882 (though not published until 1887), in which he specifically stated:

"*The natives on the West Coast of Australia are in the habit amongst other things of digging up yams as a portion of their means of subsistence; the yams are called 'ajuca' in the north and 'wirang' in the south. In digging up these yams they invariably re-insert the head of the yams so as to be sure of a future crop, but beyond this they do absolutely nothing which may be regarded as tentative in the direction of cultivating plants for their use.*" (Gregory 1887:131).

43 Maiden 1889:22.

44 Grey 1841a:1:292.

45 See Gerritsen 1994a:298n19.

46 See Gerritsen (2002) for an analysis of the sites, form of the structures, size of the settlements, degree of sedentism and the settlement pattern. Information has recently come to light which appears to confirm the conclusion in that work that the dwellings were hemispherical in their morphology (Vivienne 1901:337).

Furthermore Rev. John Wollaston (Burton and Henn 1954:144) recorded in his diary (1 March 1853) that he had learnt from Government Botanist James Drummond Snr. and his son John Drummond (policeman stationed in the Victoria District) that in the Victoria District, "Their families are more numerous and less wandering in their movements."

47 Grey 1841b:44.

48 Foley 1851.

49 Bynoe in Stokes 1846:2:389.

50 Grey 1841a:2:54.

51 e.g. Helpman 1846:9-10; Roe 1847:26.

Rev. Wollaston (Burton and Henn 1954:144) also recorded in his diary (1 March 1853) that he had learnt from the Drummonds that in the Victoria District "the 'Warrang', a kind of Yam, greatly abounds and grows to a large size".

52 Campbell 1914.

53 Grey 1841b:44.

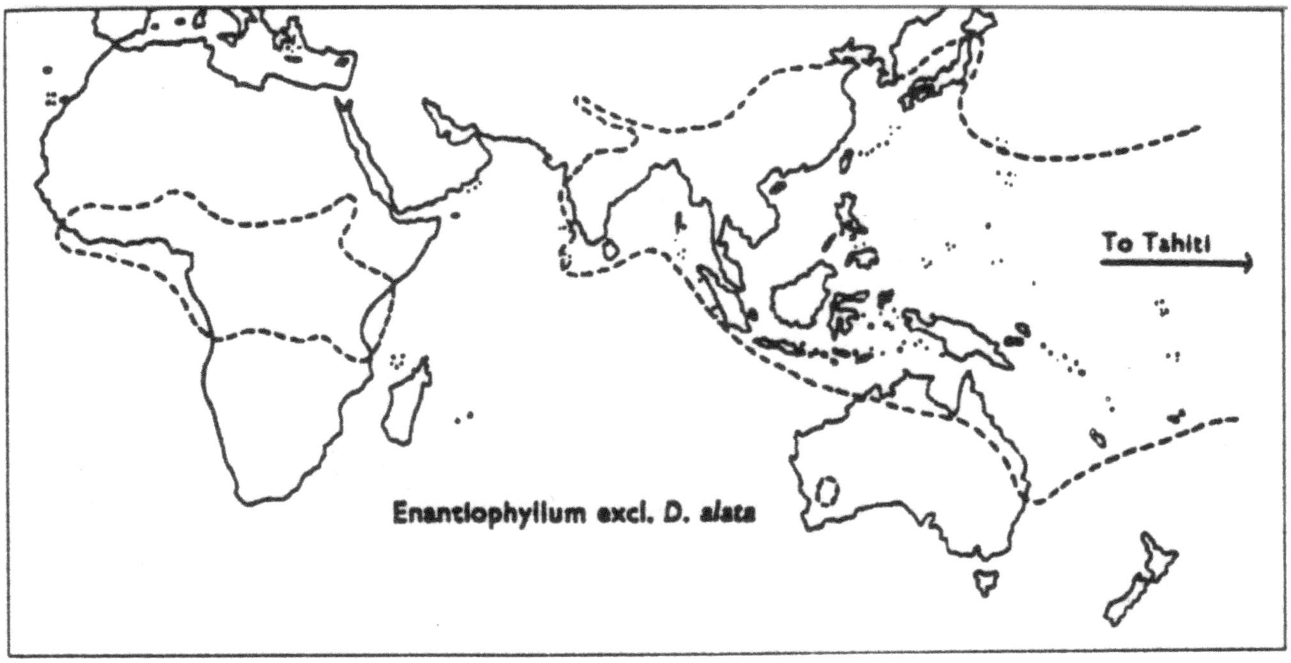


*Figure 7: Global Distribution of Enantiophyllum*
*(Courtesy of the Linnean Society of London)*

rainfall of 463 mm [18.5 in]. Wells of the type described by Grey normally only appear singly in the arid zone in Australia, an adaptation to the "more marginal areas of the desert lowlands."[54] Following an exhaustive search of the ethnographic literature, historical and contemporary, regarding the occurrence of a number of these type of wells in close proximity, only two comparable cases could be found. One was an account from northern Australia by explorer Ludwig Leichhardt of "some large wells, ten or twelve feet deep [3-4 m], and eight or ten feet [2.4-3 m] in diameter, which the natives had dug near the Zamia groves, but they were without the slightest indication of moisture."[55] The other was from the Irwin River valley, again in the Victoria District, where Lt. Helpman reported "several wells of considerable size dug about 8 feet [2.4 m] deep."[56] Given that there are dozens of permanent springs in the Hutt region,[57] Grey actually commenting on passing springs "every few hundred yards" and a "chain of reedy fresh water swamps" on 3 and 4 April, 1839,[58] as well as the presence of various watercourses (Hutt River and a number of creeks), it is difficult to account for the wells. The most likely explanation was that they were being used to water the yam fields and this contention is supported by their adjacency to the yam fields as well as by the similar occurrence of grouped wells in the Irwin River valley.

The yam plant referred to earlier, the *warran* or *ijjecka*, *Dioscorea hastifolia*, is a species of yam that is in fact one member of the tropical genus *Enantiophyllum*, described as " ... the most tropical yams".[59] Yet in Western Australia its distribution was almost exclusively encompassed by the South-West Botanical Province,[60] ranging from the wetter Mediterranean climate just south of Perth, to the prolific Victoria District, with isolated occurrences in the fringe of the coastal desert region of southern Shark Bay, reportedly "abounding" on Salutation Island [1.5 sq. kms] in that locality when the British reconnoitred there in 1851.[61] Around Perth the tuber grew to the thickness of a man's thumb,[62] "to a very large size,"[63] in the Victoria District and the thickness of a man's thigh at Shark Bay.[64] The global distribution of the *Enantiophyllum* genus, excluding the cultivar *Dioscorea alata*, is illustrated in Figure 7 below:

It can readily be seen that there are two main areas of distribution, in Africa and Asia/Oceania, separated by the deserts of the Middle East, with an isolated outlier on the central west coast of Western Australia. Australia, however, was completely isolated from the Paleocene [65 Million

54 Veth 1989:83;1993:106.
55 Leichhardt 1847b:403-4.
Martin (1865:281) reported another possible example, in this instance "native wells" which may have been 3-4 m deep, spaced about a mile apart inland from Roebuck and La Grange Bay in the south Kimberley region of Western Australia.
56 Helpman 1846:13-14.
Grey (1841a:2:39-41) also reported two separate wells of "considerable depth" and "great depth" in the Irwin River valley on 9 April 1839, even though there was a spring in the vicinity. Mrs Brown (1851:3) also reported a well (dimensions were not mentioned but it was deep enough for their dogs to fall in) near Mt. Hill which lies between the Greenough and Irwin Rivers.
57 Australian Topographic Survey Map 1741
58 Grey 1841a:2:10,13

59 Alexander and Coursey 1969:410.
60 Beard 1984; Keighery 1990:64.
61 Helpman 1851:273; Roth 1903:48.
Isolated specimens have recently been reported from the "northern range end" on North West Cape (Keighery and Gibson 1993:55,73).
62 Moore 1884:74
63 Burges 1851a:6
64 Oldfield 1865:277

Years Ago] until the mid-Miocene [20 MYA],[65] the period in which the *Dioscoreaceae* were evolving, probably in south east Asia or Africa,[66] virtually ruling out the possibility that *Dioscorea hastifolia* has evolved independently in Western Australia. If this is the case then it can only have reached its distribution in Western Australia by natural migration from the north of Australia, or with human assistance. Some have speculated that *D. hastifolia*, or its ancestral species, could have migrated by coming directly across the Great Sandy Desert,[67] or by "extending progressively down the coast" to reach the west coast and then "evolving into a distinct species over ... perhaps several million years".[68] Natural migration such as this is extremely improbable, however. By the Paleocene Australia had broken away from Antarctica and was heading north, before beginning to collide with the Asia/Pacific region in the mid-Miocene. In so doing it entered the desert belt now stretching across the middle of the continent. Consequently for *D. hastifolia* to reach the west coast of Western Australia it would have had to cross this desert belt. Given the fact that the deserts of the Middle East acted as an effective barrier dividing the African and Asian distribution of *Enantiophyllum* yams from the Pliocene [15 MYA] until after 2000 BP, when sea-borne traders and immigrants began transporting them around it,[69] this seems highly unlikely. Moreover, the present distribution of the genus east of Indonesia, including northern Australia, reputedly has only arisen in the last two to four millennia.[70]. As the "barrier dunefields" of the Great Sandy Desert have formed a biogeographic barrier for at least 35,000 years,[71] and a coastal route requires direct carriage or diffusion through "one of the largest arid coastal zones in the world,"[72] incorporating at least 200 kilometres of coastal desert on the upper west coast, migration from northern Australia seems highly implausible. Furthermore, if *Enantiophyllum* yams had only reach northern Australia sometime in the preceding two millennia, then they could not have evolved naturally "over ... several million years" into a distinct species in Western Australia as postulated by Playford. The absence of *D. hastifolia*, or any other species of yam, in any intervening areas, or in other parts of southern, western and eastern Australia, would appear to confirm this conclusion, that it did not migrate to the west coast of Western Australia. Consequently it must have arrived by some other means.

It will be recalled that when Pelsaert marooned Loos and de Bye in the aftermath of the Batavia Mutiny they were given "... a Champan *provided with everything,*" and presumably this included food. Pelsaert, it should also be noted, had returned two months earlier on his rescue mission in the "Sardam", having come directly from Java, where the ship had been requisitioned and provisioned before sailing. With numerous species of yam from the *Enantiophyllum* genus found in Java, and it being a common and extensive practice to use yams in victualling ships in the 17th century because of their keeping properties,[73] it is quite probable that the two mutineers were put ashore with a supply of yams. The obvious inference from this is that they introduced the yam to the *Nhanda*, along with the knowledge of its means of propagation, and perhaps some other cultural innovations such as the huts mentioned earlier. From that point on, however, all further developments were essentially of Indigenous provenance.

This proposition is supported by detailed arguments that have been put forward contending that a significant foreign influence was discernible in the Victoria District and the surrounding region at the time of British "settlement" there in 1849. This arose not only as a result of the impact of the abandoned mutineers, but probably the presence of at least part of a group of 68 Dutch castaways, marooned as a result of the sinking of the *Vergulde Draeck* further south in 1656.[74] One manifestation of this was the physical appearance of the *Nhanda*, A. C. Gregory, for example, reported that when exploring in the Hutt River region in 1848 he came across a tribe whose "colour was neither black nor copper, but that peculiar colour that prevails with a mixture of European blood."[75] These people Gregory wrote elsewhere had "light flaxen hair, the eyes approaching the colour of the same,"[76] and similar comments were made by others on numerous occasions in succeeding years.[77]

Mythological evidence has also been interpreted as evidence of the Dutch presence and incorporation into the traditional societies of the *Nhanda* and other peoples in the region. One example is the *Kooranup* concept, a set of beliefs based on the idea that there existed a land over the sea, in this case the Indian Ocean, where the spirits of the deceased went.[78] As one colonist, E. R. Parker, recounted:

"*there is the idea prevalent amongst the sea-coast tribes in the vicinity that their ancestors originally came from the west, and that their dead return thither, and that when the first Whites* [British colonists] *came in their ships*

65 Briggs 1987:67,78
66 Coursey 1967:34;1972:217; 1976b:70.
67 Burkill 1958-61:Fig 5,330,334.
68 Playford 1996:227
69 Burkill 1958-61:334; Coursey 1967:16-17,33; 1972:226-7; 1976b:70-1.
70 Alexander and Coursey 1969:417; Coursey 1976b:71.
This explains the presence of species such as *D. bulbifera* and *D. Transversa* in the humid tropical and sub-tropical coastal regions of northern and north eastern Australia.
71 Veth 1995.
72 Veth 1995:733.
The harsh barrier dunefield ecozone was only colonised and occupied by Indigenous Australians in the last 5000 years, while the current climate regime has been in effect for the last 7000 years. See Wyrwoll et al. 1986:208-10; Veth 1993:8-9,103-14.

73 Burkill 1966:827; Coursey 1967:16-18; 1972:227; 1976b:72; Hancock 1992:254; Hahn 1995:114.
74 It is conjectured that surviving members of a group of 68 sailors stranded on the coast following the wreck of the "Vergulde Draeck" in 1656 also made their way into the region (Gerritsen 1994a:232-46).
75 Gregory 1886:23.
76 Gregory quoted in Bates n.d.f:31.
77 See Gerritsen 1994a:71-73 for other examples
78 Gerritsen 1994a:66,71,145-50,160-1,169,171.

| English | Nhanda* | Dutch | Notes |
|---|---|---|---|
| Bad | Gooraa'ee (N) | Goor, goorheid | (Dutch – Nasty, nastiness) |
| Cold | Koon'dhetha (A) | Koudachting, koudheid | (Dutch – coldish, coldness) |
| Coughing | Oondoonda mok (A) | Aandoen ademtocht | (Dutch – gasping) |
| Hawk | Kirkenjo (W) | Kiekendief | (Dutch – the hawk-like kite) |
| Many/plenty | Boola (W) | Boel | (Dutch – a lot, a whole lot) |
| Wind | Windhoo (A) | Wind | |

*Table 1.*

| English | Nhanda* | Dutch |
|---|---|---|
| Aged | Oop'baija (A) | Oop bejard |
| Bowl, wooden | Bat.tje (W) (large oblong dish) | Badje (small bath) |
| Clod of earth | Turpa (W) | Turfje, turf (peat, turf) |
| Digging stick | Wippa (A,N,W) | Wipje (short plank) |
| Old Man | Wingja bardo | Wijze baard ('wise beard') |

*Table 2*

* Designated Nhanda dialects: A = Amangu; N = Nandakorla; W = Watjandi

*they believed them to be their ancestors risen from the dead.*"[79]

For a time the initial wave of British colonists were in fact known as the *djanga*, the dead, returned spirits, by Aboriginal people in western coastal areas, and accorded the respect normally extended to long lost relatives.[80] A variant found in the Victoria District, known as the *Moondung* Myth, gave the moon as the final resting place for souls, who were later reincarnated as the spirits of babies. According to well-known amateur ethnographer Daisy Bates:

"*The first half caste child whom Cornally* [shepherd Edward Cornally] *remembers seeing in Champion Bay* [Geraldton, early 1860s] *was also stated by the mother and mother* [sic – brother] *to come from the moon and when asked why the child was a different colour they stated that this was not uncommon amongst them.*"[81]

Nhanda social organisation was highly unusual as well, being based on reciprocal localities and explicit consanguinity rather than the sections or moieties which characterised social organisation in most other parts of Australia. Marriage partners had to come from groups associated with different localities or places, and they were not allowed to be any closer than *wadjira*, or cousins. This may have come about, as I originally suggested, as a result of the cultural influence of the castaways once they had been absorbed by the *Nhanda* and other local groups.[82] However, it could also be a result of the increasing sedentism and circumscription, as outlined earlier and in other research.[83] This may have led to the formation a new set of social conjugal relations based on a new form of social organisation.

Another discernible influence was a linguistic one. An example is the *Moondung* Myth, in which the spirits resided in the moon, the suffix *-ung* means 'belonging to'. In Dutch the 'moon' is 'maan', which here is associated with the explanation that their paler colour was a result of the paler light of the moon.[84] Many such linguistic coincidences and close correspondences are apparent in historical records of *Nhanda* and its dialects. For example see *Table 1.*

Other close correspondences are also apparent if one takes account of sound transpositions as well. One example is the Dutch 'j', which sounds as a 'y' to English speakers. It was presumed that the *Nhanda* heard it that way as well, so words of Dutch origin may have entered the language modified by that or very a similar sound transposition. An examination of *Nhanda* word lists recorded by British colonists in the 19th century indeed revealed other words of Dutch origin if allowance was made for the Dutch 'j' to 'y' transposition. For example see *Table 2.*

A full analysis of the *Nhanda* dialects, based on a 937 word/600 item vocabulary built up from early historical records, was been carried out, looking for direct correspondences as well as those arising not only from the 'j' to 'y' transposition but a number of other hypothetical regular sound transpositions. This analysis concluded that

79 Parker 1886:339.
80 See Gerritsen 1994a:163-168.
81 Bates n.d.g:65.
82 See Gerritsen 1994a:144-150 for a fuller discussion.
Note that the key in Fig. 3 on p.147 has been inadvertently reversed in publication in most copies of the 1994 edition and all of 2002 edition.

83 Gerritsen 2002.
84 In the Gascoyne version (450 km north) of the Moondung Myth the *moondung* are explicitly described as 'being like white people' (Bates n.d.h:87)

16% of the *Nhanda* language was of Dutch derivation.[85] It has been argued that the Dutch influence also explains some of the numerous unusual linguistic features in *Nhanda*, such as the appearance of the word-initial 'kn-' phoneme and 'initial phoneme dropping'.[86] While this linguistic evidence is considered controversial,[87] it is nevertheless difficult to dismiss the proposition that there was at least some degree of Dutch influence on *Nhanda*.

An issue that has arisen from this analysis is the degree of Dutch language influence in *Nhanda*, its extent is a little surprising given the small numbers involved. However, the fact that the Dutch castaways, with their radically different appearance and ways, were the first foreigners the *Nhanda* had ever had contact with, may have given them a disproportionate influence. If they did fully assimilate with the *Nhanda* then the degree of their language influence may have been further enhanced. However, it may also be another indication of the adoption of an agricultural way of life. In this scenario the abandoned mutineers only needed to directly influence a few families in the Hutt River area, with that local population exploding and expanding as a result of their cultivation of yams, a process known as "demic expansion." Consequently, this founder population would have carried their distinctive linguistic influence with them, as has happened with the development of agriculture in other parts of the world, resulting in a much broader impact.

The critical point, however, is that this introduced yam was probably already a "domesticate", although yams often do not become "true cultigens".[88] Presumably, once introduced, it underwent further speciation, either through founder-induced speciation, known as the "founder effect",[89] becoming a naturalised form of the original species or, as happens when domestication occurs, as a result of the selection regime applied by its *Nhanda* cultivators.[90] The exact genetic relationship between *D. hastifolia* and other species of *Enantiophyllum* yams has not as yet been determined. Once introduced it was carried to other localities, such as mentioned previously, presumably through trading, gifting and payments of yams, a common occurrence if early colonial accounts are a valid guide.[91]

In conclusion it can be seen that, by the minimum criteria established earlier, agriculture was being practised in the Victoria District, with evidence of propagation, and harvesting on a systematic basis and of sufficient extent to provide a major component of the *Nhanda*'s subsistence.[92] Furthermore, there is some evidence of possible husbandry (watering) as well as utilisation of what appears to be a domesticate in cultivation, with in situ storage being employed.[93] By extension the features of the *Nhanda* economy are consistent with most of the definitions of agriculture discussed earlier. Other indirect lines of evidence, such as a higher population density, large permanent or near-permanent dwellings and settlements, seemingly a high degree of sedentism, an atypical sexual division of labour, concepts of fixed land tenure, even specific artworks, are consistent with and support this conclusion.[94] We will return later to the Victoria District to re-examine developments there in terms of theories on the origins of agriculture, socioeconomic development and sociopolitical evolution, but will now consider other matters.

---

85 Gerritsen 1994a:113-30; Gerritsen 1994b.
Five historical sources (1840-1908) were employed, being temporally closest to the local language at the time of British colonisation and therefore least likely to suffer some form of language "disturbance". The decline of the *Nhanda* was such that now only a small number of individuals remain who can claim descent from the original population. Paralleling this has been a catastrophic loss of cultural and linguistic information (See Blevins 2001a; Gerritsen 2004).

86 Initial phoneme dropping is a feature in which equivalent words in *Nhanda* to those found in other Aboriginal languages, particularly the Kardu and Nyoongar subgroups, are almost identical except the initial phoneme has been 'lost'.

87 See Anonymous 1995; Gerritsen 1997; Blevins 1998; Gerritsen 2001b; Blevins 2001a,b; Gerritsen 2004; Blevins 2006.

88 Coursey 1976b:387,391.

89 Hancock 1992:117-121.

90 What constitutes "domestication" in yams is an open question but would seem to include selection for greater size, shallower rooting, reduced spines, less bitterness and lower toxicity (Chikwendu and Okezie 1989:349; Evans 1996:74). "Domestication" of cultivated yams can happen very rapidly, within three to eight years (Coursey 1972:225; Chikwendu and Okezie 1989; Dumont and Vernier 2000).

91 Oldfield 1865:236; Drummond 1842:719; Roe 1847:32.

92 Hobhouse, Wheeler and Ginsberg (1915:20-1) conceded that "technically" practices such as the *Nhanda* people's were the beginning of agriculture, but then discounted the notion because they claimed there was insufficient evidence of "clearing, digging and planting". However they were probably unaware of the extent of the areas apparently being cultivated by the *Nhanda*.

93 In situ storage is assumed because of the extent of yam harvesting frequently noted by nineteenth century explorers and colonists, and the absence of any evidence of other forms of storage in the historical ethnographic literature. However, Oldfield (1865:230) mentions stockpiling of yams in preparation for the "Caa-ro Ceremony" by the *Watjandi*, a non-agricultural northern *Nhanda* group, so it is possible above ground storage was practised but not recorded.

94 Gerritsen 1994a:86-7,139-41,185; 2002.

# Chapter 5.
# Sowing the Seeds

The traditional subsistence economy of the *Paakantyi* people of the Darling River region in western New South Wales has been nominated at various times as the prime Australian example of "incipient agriculture".[1] The researcher who originally drew attention to this region, Harry Allen, claimed that the people here met Kent Flannery's pre-adaptation criteria for a proto-agricultural society, of broader-spectrum procurement (i.e. utilisation of a wider variety of food sources), the employment of ground stone technology and the development of storage facilities.[2] But it was the large-scale harvesting of native millets and other seed-bearing plants by the *Paakantyi* that was seen as one of the most notable feature of their subsistence economy. It was these practices that Diamond has conjectured were "most likely to have evolved eventually into crop production."[3]

Incipient or proto-agriculture, as the terms suggest, relates to developments preceding a commitment to an agricultural lifeway. But what exactly does this entail? Originally Braidwood saw it as a "period of experimentation with plants and animals,"[4] but for others, such as Higgs, plant domestication was the critical threshold. Consequently he took the position that incipient agriculture occurred, "when certain plants were grown but not changed in form," pointing to the "Natufians", the precursor culture to the Neolithic of south west Asia, as an exemplar.[5] But, apart from the difficulties identified previously arising from an insistence on domestication as a critical indicator of agriculture, it now appears that wild, undomesticated or only partially domesticated cereals were being grown well into the Neolithic, the period when it is believed agriculture was becoming, or had become, established as an integral part of the lifeways of people in the Levant.[6] An alternative definition proposed by MacNeish, in which "planting of individual domesticates or cultivars in relatively limited plots,"[7] was seen as fulfilling the basic criteria for incipient agriculture, suffers from two defects, the inclusion of domestication as a benchmark and the notion that only small numbers of plants are involved. Furthermore, definitions such as these have also been criticised by Harris as being "vague" and "deterministic".[8] To overcome this problem others, rather than attempting to define incipient agriculture, have tried to identify necessary conditions or common factors evident in proto-agricultural societies. Employing this approach Hole described five preconditions including:

1. high quality, abundant and storable food sources,
2. highly seasonal availability of such foods,
3. the inability to escape this high seasonality,
4. the yield of these sources could be increased by some modification,
5. ability to stockpile commodities such as dung, bitumen or obsidian.[9]

It was further argued by McCorriston and Hole that the ability to gather and process plant foods in bulk, store them, ensure their preservation and distribute them as needed was also required.[10] Others, such as Hayden, have focussed on broader socioeconomic features. He pointed to a number of "consistent variables" in proto-agricultural societies such as:

1. high levels of sedentism,
2. storage,
3. high population densities,
4. processing and harvesting technologies,
5. suitable potential domesticates.[11]

Another approach, emphasising the ecological characteristics of regions in which agriculture has developed has been proposed by Richard MacNeish. In this, his five "necessary conditions" included:

1. environments with layered ecozones,
2. multiple food resources not exploitable from a base,
3. harsh seasonality,
4. intense interaction with ecologically similar regions,
5. potentially domesticable plants in one or more ecosystems.[12]

1 Mulvaney 1975:242; Flood 1999:265.
2 Allen 1972:95.
3 Diamond 1997:311.
4 Redman 1978:71.
5 Higgs 1968:619.
6 Kislev 1992:87-93; Bar-Yosef and Meadows 1995:66; Bar-Yosef 1998a:167,168-9; Anderson 1998:147; Willcox 1999:478-80,489-94; Colledge 2001:146-151.
And rice in the Chinese Neolithic as well (Pei 1998; Zhao 1998).
7 MacNeish 1992:10.
8 Harris 1989:19.
9 Hole 1984.
10 McCorriston and Hole 1991:49.
11 Hayden 1995a:277-8.
12 MacNeish 1992:320.

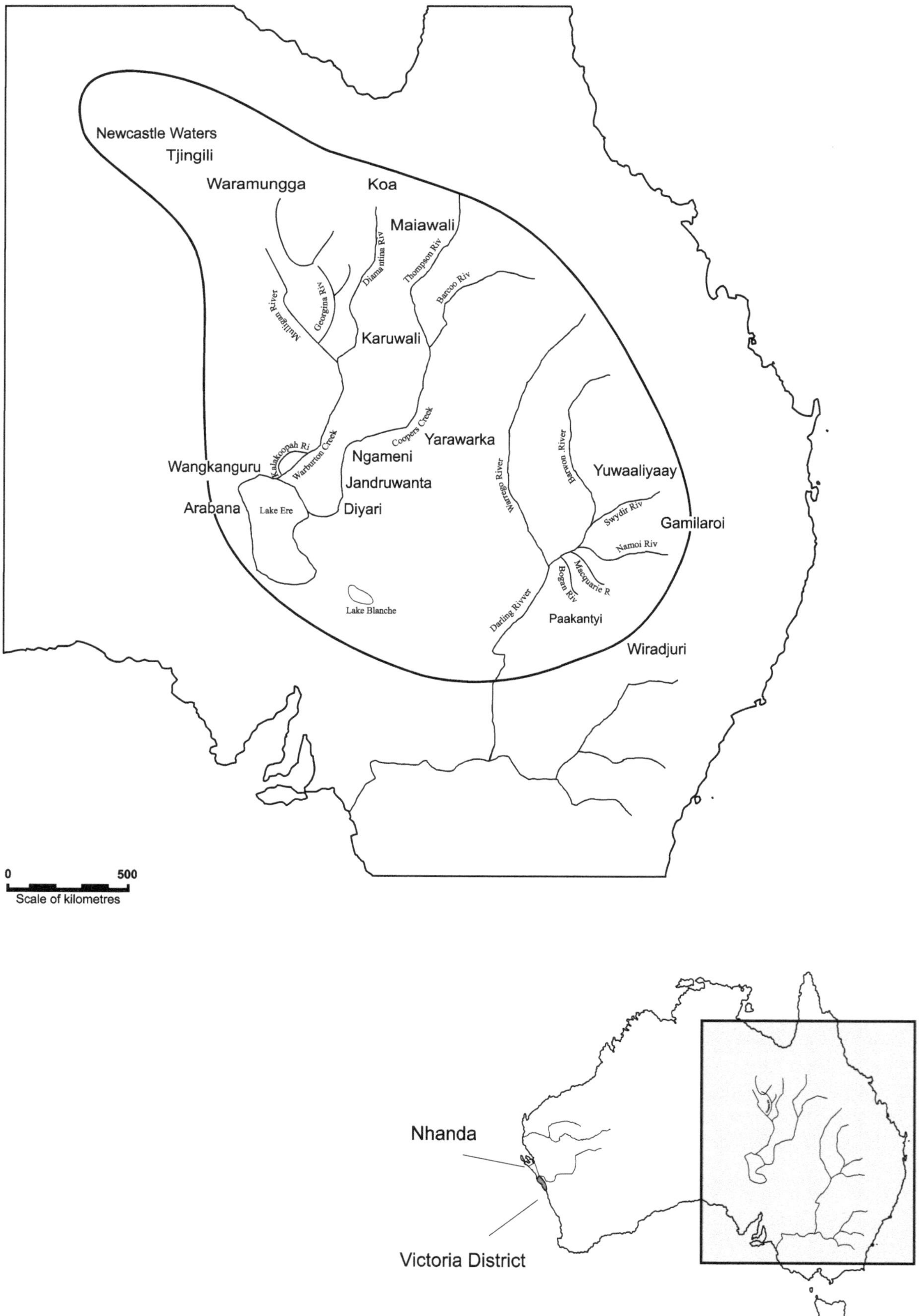


*Figure 8: The Darling River and the "Corners" Region*

Schemata such as these are typical examples of efforts to identify predictive common factors and variables. Their applicability is questionable, however, because they are often based on studies of societies that ultimately developed seed plant subsistence economies and because they often rely upon broad or open-ended generalisations and judgments. Nevertheless, there are some consistencies evident in terms of specific food sources, these being:

1. availability of domesticable plant species,
2. highly seasonal availability of such species,
3. systematic harvesting and processing of these species,
4. systematic preservation and storage of foods produced,
5. availability of other rich food sources.

Seemingly correlated with these factors are sociodemographic and socioeconomic features such as:

1. a higher degree of sedentism than is the norm in hunter-gatherer societies,
2. higher population densities in relative terms,
3. regional economic integration (e.g. regional trade in food).

The Natufians of south west Asia have frequently been nominated as the first incipient agricultural society.[13] In this part of the Middle East, particularly in the Levant, the western part of the "Fertile Crescent", groups of hunter-gatherers formed what is known as the Kebaran culture, which flourished when the climate began to rapidly improve following the Last Ice Age. Between 18,000 and 14,500 years ago the first signs of sedentism, storage and specialised seed grinding appeared, in the latter part of this period, the Geometric Kebaran.[14] This was followed by the Natufian, where seed harvesting and processing became a significant part of subsistence activities. Permanent settlements started to emerge at this time, based upon a "mixed economy" of collecting, harvesting, fishing and hunting.[15] Initially it appears wild crops of the ancestral wheats, einkorn (*Triticum monococcum*) and emmer (*Triticum dicoccum*), as well as barley (*Hordeum vulgare*), were grown at settlements such as Mureybit (Syria), Netiv Hagdud (Israel) and Abu Hureyra (Syria), around 11-12,000 years ago.[16] This marked the commencement of the Neolithic, with the Pre-Pottery Neolithic A (PPNA: 11,500 – 10,800 cal. BP) quickly followed by Pre-Pottery Neolithic B (PPNB: 10,800 – 9100 cal. BP). The first signs of domesticated cereals appear at PPNA Jericho (Israel), Iraq ed-Dubb (Jordan) and Tell Aswad (Syria), dated to 10,800 – 9800 years ago, as well as at a number of other hamlets and villages in the region, such as Gilgal, Hatoula and PPNB Nahal Oren and PPNB Ras Sharma.[17] Coinciding with these developments was the domestication of sheep (Syria, Turkey: 10,700 – 10,200 cal. BP),[18] and goats (Iraq – 11,000 cal. BP),[19] dogs having probably been domesticated two or three thousand years previously.[20] By the latter part of the PPNB, from around 9800 cal. BP, a mixed farming economy had evolved, although hunting, fishing and wild food procurement (such as nuts and lentils) still provided a substantial proportion of the food supply.[21]

While the *Paakantyi* example, in this context, merits closer consideration it is also necessary to look beyond their country and view it in a broader regional context. There are two reasons for this. The first is suggested by the evidence. Not only was large-scale harvesting taking place outside of *Paakantyi* territory but other parallel or relevant developments were also evident in adjacent regions. Secondly, there has been some debate regarding a proposition put forward by Jack Harlan that there were "Centres" and "Non-Centres" involved in agricultural origins. This distinction, as the terms suggest, contrasts plant domestication taking place in a small definable geographic area with larger geographical areas in which domestication occurs in multiple locations.[22] Harlan's concept is not universally accepted and others (e.g. MacNeish[23]) assign these terms differently to some of the examples of pristine agricultural development around the world. Nevertheless the concept serves as a useful reference point, and developments in these parts of Australia may serve as a "test" of the applicability of these concepts. Consequently we will look beyond the *Paakantyi* and the Darling River to a contiguous area encompassing north western New South Wales, south western and western Queensland, north eastern South Australia, and south eastern, eastern and central Northern Territory. This is a region where, in broad terms, the "Corners" of those states meet (Fig. 8), but it is also co-extensive with much of the Lake Eyre Basin and the drainage basin of the middle and upper Darling River.

It was within this region, on the lower-central Darling

[13] Higgs 1968:619; Redman 1978:71; McCorriston and Hole 1991:49.

[14] Bar-Yosef and Belfer-Cohen 1989:484-5; McCorriston and Hole 1991: 58; Bar-Yosef 1998a; Bar-Yosef 2002:380; Radiocarbon CONTEXT Database.

[15] Wright 1978:203; Flannery 1986:13-14; Bar-Yosef and Belfer-Cohen 1989:468,473; Henry 1991:365; McCorriston and Hole 1991:46,49; Smith 1995a:77,79; Cauvin 2002:15,19-20; Edwards and Hardy-Smith 2004.

[16] Flannery 1986:15; Hillman and Davies 1990:159,200-1,206; Bar Yosef et al. 1991:420; Bar-Yosef and Meadows 1995:66; Smith 1995a:68-72; Colledge 2001:150; Cauvin 2002:xvii-xviii Chronological Table.

[17] Kenyon 1957:53,71;1960-83:3:2,6,8,19; Noy, Legge and Higgs 1973; Noy, Schuldenrein and Tchernov 1980; Henry 1989:225-6; Hillman and Davies 1990:166; Bar-Yosef et al. 1991; Smith 1995a:69; Colledge 2001:150; Yasuda 2003b:22,27,Figs 4,9; Radiocarbon CONTEXT Database.
Bellwood (2005:52) claims there are "signs" of domesticated rye at Abu Hureyra I dating from 12,700 BP.
Kislev et al. (2006) also claim cereal cropping may have been preceded by the planting of figs (11,400 - 11,200 BP) at Gilgal.

[18] Smith 1995a:56; Yasuda 2003b:22,27,Figs 4,9; Radiocarbon CONTEXT Database.

[19] Smith 1995a:56. Yasuda 2003b:22,27, Figs 4,9; Radiocarbon CONTEXT Database.

[20] Bar-Yosef and Meadows 1995:57

[21] Contenson 1983:61; McCorriston and Hole 1991:51; Bar-Yosef et al. 1991:417-23.

[22] Harlan 1971; 1992b:48-53; 1995:55.

[23] MacNeish 1992.

River, that explorer Major Thomas Livingstone Mitchell first noted in an account from 1835 that,

"*In the neighbourhood of our camp the grass had been pulled, to a great extent and piled in hay-ricks, so that the aspect of the desert was softened into the agreeable semblance of a hay field ... we found the ricks or hay cocks extended for miles .... All the grass was of one kind, a new species of 'Panicum' related to 'P. effusum' ... not a spike of it was left in the ground.*"[24]

Shortly after Mitchell observed that the "grass was beautifully green beneath the heaps, and full of seeds."[25] Mitchell also commented on the nature of the habitations in this region, comments that were to be echoed by others on numerous occasions in the following years. As he continued his journey he wrote, "we found a native village, in which the huts were of a very strong and permanent construction."[26] One of the huts was "unusually capacious ... capable of containing twelve or fifteen persons."[27] Elsewhere on this part of the Darling, Mitchell observed "permanent huts on both banks of the river" and places where "roads appeared in all directions," converging at a particular point.[28]

Another noted explorer, Charles Sturt, when investigating the desert country 250 kilometres further north in 1845, in north west New South Wales, observed, "there was a quantity of the Grass spread out on the sloping bank of the creek to dry, or ripen in the sun."[29] In the same vicinity he later recorded,

"*Just as we halted ... we observed some grassy plains, spreading out like a boundless stubble, the grass being of the kind from which the natives collect seed for subsistence at this time of the year.*"[30]

Sturt, like Mitchell, had similarly noted "large heaps that had been threshed by the natives were piled up like haycocks" in this particular area,[31] as did one of his men, Brock:

"*here it is quite like a harvest field ..... In every hollow we found the remains of the natives' labour in the shape of straw, from which they had beaten out the seed.*"[32]

Brock also indicated that when Sturt had returned to their base camp he reportedly told his men that he had seen "seed growing in prodigious quantity."[33]

[24] Mitchell 1839:1:237-8.
[25] Mitchell 1839:1:291.
[26] Mitchell 1839:1:262.
[27] Mitchell 1839:1:262
[28] Mitchell 1839:1:225,240
[29] Sturt 1849:1:285.
[30] Sturt 1849:1:294 [Evelyn Ck.].
[31] Sturt 1849:1:285 [Frome Ck.].
[32] Brock 1975:133 [Evelyn Ck. 13/3/1845]/9418
[33] Brock 1975:189.

"The grass," as Sturt and Mitchell called it, was principally *Panicum decompositum*, known to Aboriginal populations in plant or seed form as *cooly*, *tindil*, *katoora*, *ya(r)mmara* or *parpar* and barley-grass, native millet or umbrella grass to Europeans. It is a close relative of common millet, *Panicum miliaceum*.[34] This exploitation of *Panicum* and other seed species was more than just a local phenomenon however. In 1846, in the vicinity of the Macquarie Marshes, 300 kilometres to the north west of his position in 1835, Mitchell noted that *cooly* was harvested there by the women, in "great quantities."[35] A little later, on the Narran River, in the upper Darling region, Mitchell again reported,

"*Dry heaps of this grass, that had been pulled expressly for the purpose of gathering the seed, lay along our course for many miles.*"[36]

He later expressed the view that these heaps "formed the natives harvest field."[37]

Heaps "which had been threshed for seed" were also "often seen" by yet another explorer, Lindsay, on the lower Finke River, south eastern Northern Territory, in the 1880s.[38] Similarly, on upper Coopers Creek in south west Queensland, explorer A. C. Gregory had observed in 1864 that the *Maiawali* or *Koa* people,

"*reap a Panicum grass. Fields of 1,000 acres* [400 hectares] *are there met with growing this cereal. The natives cut it down by means of stone knives, cutting down the stalk half way, beat out the seed leaving the straw which is often met with in large heaps; they winnow by tossing seed and husk in the air, the wind carrying away the husks.*"[39]

Would-be explorer Daniel Bunce also reported "very large heaps" in south west Queensland in 1847 "that had evidently been cut by a sharp instrument,"[40] as did Conrick in 1874 on Farrars Creek in south west Queensland.[41] Mr Poole, one of Sturt's party on his 1845 expedition, while reconnoitring in the north west of New South Wales chanced upon,

"*a tribe of about 60 in number engaged in collecting grass seeds. They behaved very civilly ... were all living on the seeds of a kind of rice (Panicum effusum) which grew abundantly on the flooded lands near the creeks. At this season of the year* [summer] *the natives live principally on seeds both of Acacia and grass ...* "[42]

[34] Maiden 1898:49.
[35] Mitchell 1848:110.
[36] Mitchell 1848:90. Mitchell (1848:98) again observed heaps of *P. decompositum*, as well as *P. oleracea* further up the Narran River.
[37] Mitchell 1848:61.
[38] Lindsay 1890:5.
[39] A.C. Gregory 1887:132; P. L. Watson 1983:35.
Wells, who had been surveying in this area between 1883 and 1886, commented that "the rolling downs present the appearance of vast wheat-fields" (1930:17).
[40] Bunce 1857:169.
[41] Conrick n.d:47.
[42] Browne in Allen 1972:77; 1974:314

The actual process of harvesting, threshing and winnowing of *yarmmara* as carried out by the *Yuwaaliyaay* was described in some detail by Kate Langloh Parker in 1905. This was on the Narran River, and involved the same people observed by Mitchell in 1846.

"*When the doonburr, or seed, was thick on the yarmmara, or barley grass, the tribes gathered this grass in quantities ... Each fresh supply of yarmmara, as it was brought in by the harvesters, was put in this* [brush-fenced] *yard. When enough was gathered, the brush-yard was thrown on one side and fire set to the grass, which was in full ear yet green. While the fire was burning, the blacks kept turning the grass with sticks all the time to knock the seeds out. When this was done, and the fire burnt out, they gathered up the seed into a big opposum-skin rug, and carried it to the camp. There the next day they made a round hole like a bucket, and a square hole close to it. These they filled with grass seed. One man trampled on the seed in the square hole to thresh it out with his feet; another man had a boonal, or stick, about a yard* [90 cm] *long ... and nearly a foot* [30 cm] *broad; with this he worked the grass in the round hole, and as he worked the husks flew away ... ... When the grain was sufficiently clean, it was put away in skin bags to be used as required ...* "[43]

Reports of intensive utilisation of "grass" species abound in the Corners Region.[44] But, these were not the only plants being heavily utilised there. Acacia seeds, other seed plants such as *munyeroo* [*bowa, toong-ara, thukouro, kunaurra*, Purslane: *Portulaca oleracea*[45]], the sporocarps of *ngardu* [*Marsilea drummondii*] and the corms of *yaua* [*Cyperus bulbosus*] were significant in the subsistence of different groups in the region. For example, on Coopers Creek in the north east corner of South Australia, A. W. Howitt recalled that in 1862,

"*we found, near Coopers Creek, a large number of people living almost exclusively on portulac seeds.*"[46]

Which, according to another source, was obtained,

"*by pulling up the plants and throwing them into heaps, which, after a few days, they turn over and an abundant supply is found to have fallen out and can be easily gathered up*"[47]

Shortly after, Howitt's party encountered another group in the same vicinity, at Lake Lipson. Here he observed:

"*These, like the natives at Dampurnoo, seemed to be laying in a stock of bower* [Portulaca seed]. *Harry found one 'plant' near our last night's camp. The seed is wrapped up in grass and coated with mud and the parcel would probably hold a bushel and a half* [approx. 50 kg]."[48]

John Davis, a member of one of the expeditions searching for the lost explorers Burke and Wills in 1861, remarked that near Lake Coogiecooginna in the Strzelecki Desert, *ngardu* "is procured in almost any quantities in the flooded flats by sweeping it up into heaps."[49] Surveyor Lewis made a similar observation in 1875 about the "nardoo flats" of the Mulligan River in south west Queensland "extending northward as far as the eye could reach."[50] Echoing this, Howitt had noticed that in "many places, miles of the clay flats are thickly sprinkled with the dry [*ngardu*] seed."[51]

Whereas seeds formed a significant part of subsistence in some areas of the Corners Region, this was only one element in a well-developed economy. Hunting and fishing, for example, was carried out on a systematic basis through the use of nets, the "possession of every tribe."[52] Some of these nets of were of considerable dimensions, with specialist net-makers taking up to three years to manufacture them.[53] The *Yuwaaliyaay*, for instance, had nets up to 1.5 m high and 270 m long which they regularly employed to catch emus.[54] And this is just one of many reports of nets of similar size being in use in this region.[55] Such nets were often utilised in a widely reported technique for trapping large numbers of ducks. This began by alarming the ducks so that they flew along a river, creek or billabong. As they approached the nets a hawk's call was imitated and a piece of bark thrown into the air, causing the ducks to dip in flight, and straight

---

43 Parker 1905:118.
Coghlan (1898a:48) also describes this method as does Roth (1897:91), though in the latter case "star grass" [*Eleusine aegypta*] was being threshed.

44 e.g. Newland 1890:22; Coghlan 1898a:48; Spencer and Gillen 1899: 7; Newland 1920-1:13; Brock 1975:111,133,194.

45 Smyth 1878:1:226; Curr 1886:2:78,193; Maiden 1889:53.
*Munyeroo* is actually the name of the plant, not the seeds. It also seems to have been applied to a close related species, known as *Claytonia/ Calandrinia balonensis*, usually called *parakeelya*.

46 Howitt in Allen 1972:77 (17-21 April 1862).
Extensive growth of munyeroo was also reported by Mitchell (1848:98)on the Darling, Beckler in north west New South Wales (Spencer 1918-19:13), W.J. Wills (1861:23-5) along the "Barcoo" [Upper Coopers Creek], by Wells in 1883 or 1886 (1930:16) around Poeppel Corner, Norton (1907:68) west of Dubbo in 1907, and by Duncan-Kemp (1934:73-4), "sandy flats and sandhills are covered", on the lower Diamantina.

47 Palmer 1884b:103.

48 Howitt in Allen 1972:78 (27 April 1862); Howitt in Smyth 1878:2: 302

49 Davis 1863:117. See also p. 207; McKinlay 1861:11,14,22,41.

50 Lewis 1875b:20.
Lewis made other comments of a similar nature regarding this area, such as, "nardoo flats, many of them resembling dry lakes," (1875b:20). Hodgkinson (1877:216-7), Bedford (1887:104) and Coghlan (1898a:48) also report extensive use of *ngardu* in this area (Georgina and Mulligan Rivers).

51 Howitt in Smyth 1878:2:302.

52 Mitchell 1839:1:305.

53 Duncan-Kemp 1961:231.

54 Parker 1905:108

55 e.g. Sturt 1833:1:89-90 [80 m - Darling River],92,106; Mitchell 1839:1:305; 1848:104; Landsborough 1862:101 - Barcoo; Gason 1874:34 [18 m]; Cox 1880:635 - *Kalkadoon* [*Kalkadunga*] and "Mithure"; Jung 1884:109 [Paroo River]; Morton in Curr 1886:2:158 [north west New South Wales]; Bennett 1927a:410 [*Dalleburra*]; Duncan-Kemp 1934:121. See also Satterthwait 1987.

into the nets.[56] Reports of nets being used for fishing also abound in this region (see Figure 5c), and include the lower Darling to Coopers Creek, Bogan River to the Mulligan, Eyre Creek, the Barcoo to the Diamantina and Kallakoopah Creek to the Ashburton Range just south of Newcastle Waters, even surprisingly on Margaret Creek on the west side of Lake Eyre South.[57]

Fish traps, weirs and dams were another common feature as well in the Corners Region (see Figure 5b). At least three types of fish traps, the "fence"[58], "log"[59] and "stone wall"[60] type were constructed, to be found on the Balonne, Barcoo, Barwon, Bogan, Bulloo, Burke, Darling, Diamantina, Finke, Georgina, Gwydir, Hamilton, Hugh, Namoi, Narran, and Thompson Rivers, Farrars, Kyabra, Lindsay, Morney, Nebine, Scott's, Warburton and Whitula Creeks, the Macquarie Marshes and the Chillen Chillenba Channels.[61] The "mother" of all these was to be found at Brewarrina, however. Situated on the Barwon River, an upper branch of the Darling, where three tribal boundaries intersected,[62] the traps consisted of a labyrinth of "pens" or "yards" with walls about 1 m high. The whole system extended for up to 300 m, "yards" or groups of "yards" being owned, operated and maintained by particular groups. During summer these tribes peaceably settled at the Brewarrina traps [Ngunnhu], catching "immense quantities" of fish, with small groups continuing to live there throughout the year.[63] The degree to which procurement methods such as these were relied upon led explorers and their compatriots to often comment that the spear, usually the implement of choice for hunting elsewhere, was rarely to be seen in this region.[64]

Manipulation of the environment in this region to enhance the regularity and yield in food supplies, by "investing" in "capital works", went beyond the construction of fish traps however. It would appear dams of considerable dimensions were also being constructed. One mentioned earlier, on the Bulloo floodplain, seems to have had an earthen wall 100 m long which was 2 m high in the centre, with a base 6 m wide, and requiring 180 cu. m of fill.[65] Similar dams appear to have been constructed on Coopers, Farrars and Whitula Creeks, and the Diamantina River, where they were known as *mokhani*.[66] These were distinctly different from the numerous small water storage dams found in the Great Victoria and Simpson Deserts, known to the *Karuwali* as *nilpoonyah*,[67] presumably one of the adaptations that allowed the colonisation of those "Barrier Deserts".[68] While the *mokhani* certainly stored water, their principal function appears to have been to "hold live fish supplies."[69] As mentioned previously, it has also been suggested that this type of dam intentionally increased the area flooded, thus promoting the growth of seed plants,[70] a simple form of irrigation akin to the *Iliaura* and *Wanji* practice of blocking creeks referred to earlier.

Other forms of specialisation had developed in this region for catching fish, apart from nets, traps and dams. Fish hooks, first invented in the Natufian,[71] were used by the *Ngemba*, *Karuwali* and at least some of the groups along the Darling River,[72] and possibly on part of Coopers Creek.[73] More unusual was a specialised method for catching fish described by Duncan-Kemp:

[56] e.g. Teulon in Curr 1886:2:191; Newland 1890:22; Tully 1896:23; Roth 1897:98-99; Parker 1905:106, etc. Another method observed in 1861, where the ducks were patiently shepherded towards waiting nets, is described by Beckler (1993:127-8). A similarly ingenious method for catching water hens on land is also described by Wright (1861-2:81), Beckler (1993:128) and Becker (1979:124). Bush turkeys were another commonly trapped species using a lure of a grasshopper on a long stick with a noose (Roth 1897:98). An almost identical method was reported in south west Victoria by Robinson in 1841 (Clark 1998-2000:4:147-8,175[Fig.4.26],196,200), Griffiths (1845:145) and Smyth (1878:1:192)

[57] e.g. Mitchell 1839:1:100,305; 1848:113,367; Sturt 1849:1:106-7; Gregory 1858:4; Howitt 1861:11; King 1861:4; McKinlay 1861:21,46; Brahe 1861-2:89; Landsborough 1862:101; Davis 1863:113,203, 219-20; Stuart 1864:62-3; Gason 1874:34 [18 m x 1 m]; Lewis 1875b:14,26,42; Hic et Ubique 1977:13; Hodgkinson 1877:217,521; Howitt in Smyth 1878:2:305; Andrews 1879:85; Sanger 1883:1224; MacGillivray in Curr 1886:2:341; Scrivener in Curr 1886:2:183; Teulon in Curr 1886:2:191; Cornish and Salmon in Curr 1886:2:24; Palmer in Curr 1886:2:331; Heaghney et al. in Curr 1886:2:377; Fraser in Curr 1886:2:378; MacGlashan in Curr 1886:3:19,22-23; Ridley et al. in Curr 1886:3:306; Wells 1894:518-9; Gason 1895:169,172; Parker 1897:124; Roth 1897:94; Coghlan 1898b:48; McClellan 1901:258; Meston 1901:63; Parker 1905:107,109; Eylmann 1908:375,382; Horne and Aiston 1924:62-3; Bennett 1927a:410; Duncan-Kemp 1934: 112,121-2,258 [15 m x 2.4-4.5m]; Johnston 1943:298; Lamond 1950:169; Tindale 1951:257; Duncan-Kemp 1952:126; Durack 1959:94,96; Duncan-Kemp 1961:41,118; Duncan-Kemp 1968:3, 124; Kirwan and Breen 1981:288-93; Clarke 2003:150 [54 metres]; Tolcher 2003:14.

[58] Known as "*chaboona*" by the Karuwali (Duncan-Kemp 1968:275).

[59] Known as "*barameda*" by the Karuwali (Duncan-Kemp 1952:237).

[60] Called "*Thillung-gurra*" or "*Thylungra*" (Durack 1959:94).

[61] Sturt 1833:1:41,72; Mitchell 1839:1:93; 1848:103,113,121; McKinlay 1861:69; Wills 1861:17; Wright 1861-2:85; Davis 1863: 290; Stuart 1864:252-3; Lewis 1875b:13,26; Fraser in Curr 1886: 2:378; Ridley et al. in Curr 1886:3:306; Roth 1897:95; Worsnop 1897:100,102; Parker 1905:109-10; Norton 1907:102; Banfield 1909: 53-4; White 1914:143; North 1916-17:126,128; Horne and Aiston 1924:64; Duncan-Kemp 1952:105,142,186,237; Durack 1959:94,96; Duncan-Kemp 1968:124,274-5; McCarthy 1970:59; Dargin 1976:45; McKellar 1984:47,67; Heritage Concepts 2004:23,25-6.

[62] The *Ngemba*, *Yuwaaliyaay* and *Weilwan*. The eastern limit of the *Baranbinja* was also only a few kilometres to the west of Brewarrina (Tindale 1974:Australia North East Map).

[63] Lang 1865:19-20; Rankin 1898:202; Mathews 1903; Parker 1905: 109- 10; Dargin 1976:32,37.
Flood 1993:257 claims the traps originally extended for 1.6 kilometres.

[64] e.g. Sturt 1849:2:138-9; Davis 1863:204; Hic et Ubique 1877:13; Curr 1886:2:169,192; Allen 1974:312.

[65] Rowland and Rowlands 1969.

[66] Fraser in Curr 1886:2:378; Duncan-Kemp 1968:46,274-5.
Parker (1897:48) provides what may be another reference to these dams being used by the *Yuwaaliyaay*. See also Kimber 1984:18.

[67] Duncan-Kemp 1968:180,360. Kimber (1984:19; 1996:48) gives a description obtained from Walter Smith of the construction methods used to make what appear to be *nilpoonyah*.

[68] Rowlands and Rowlands 1969:132; Kimber 1984:19.

[69] Duncan-Kemp 1968:46. See also Fraser in Curr 1886:2:378; Kimber 1984:19 (Birdsville and Carravilla Station)
Duncan-Kemp (1968:274-5) also appears to indicate that the more complex fish trapping systems had a storage function as well.

[70] Tindale 1977:347; Jones 1979:143; Kimber 1984:19.

[71] Childe 1934:52,62; Hurum 1977:19.
There is some evidence that fish hooks may have been invented earlier in Moravia (Hurum 1977:19).

[72] Dunbar 1943-4:176; Duncan-Kemp 1961:211-2; 1968:124,275,319; Hardy 1969:12; Gerritsen 2001a.
Other evidence indicates that fish hooks may have been used further north on Whitula Creek and Cloncurry River as well (Fraser in Curr 1886:2:378; Palmer in Curr 1886:2:334)

[73] Lamond 1950:169; Farwell 1971:67.

*"An aboriginal youth stood motionless at the damp sand's edge watching his two dogs fish the stretch of quiet water. Up and down the shallows ran the tame dingo pups, nosing the water with bright-eyed, prick-eared interest. One pup swam into the outer shallows, edging fish in towards the banks, while the second pup plunged its black-tipped nose deep and with a quick upward twist threw a large perch on the sand where the native gathered it in a reed carry-all. All afternoon the dogs had engaged in co-operative fishing - and these pups were taken a few months previously from a wild-dog litter."*[74]

With fish a major food source in areas where there were significant water sources it is not surprising Mitchell described the *Paakantyi* of the mid- to lower Darling River as "Icthyophagi" ["fish eaters"], or elsewhere simply as "fishing tribes".[75] Correspondingly, the lakes and creeks of the Strzelecki Desert, "seem full of fish", "abounds with fish", have "any quantity of fish" according to Davis and McKinlay,[76] comments echoed by many other travellers in the region around this time.[77] In north east South Australia, on Warburton Creek (the southern extension of the Diamantina River), the *Ngameni* caught "thousands of fish" in their nets,[78] leading Surveyor Lewis to note that:

*"This is also the first place I ever saw natives return and live in the same wurlies* [shelters] *which are very large and well built; the camps may be considered more as villages than anything else; their principal food is fish (we saw large quantities in their possession)."*[79]

While fish and seeds were staples, in many locations other significant food sources contributed to local subsistence. In this vein Mitchell reported "heaps" of fresh-water mussel shells on the Darling, concluding that "this shell-fish is the daily food."[80] He passed another midden field on the upper Barcoo in 1846, describing the mussels there as being "of extraordinary size" and in such quantities that they "resemble snow covering the ground."[81] Along the Mulligan River in south west Queensland, Lewis observed that "the banks are covered with millions of shells, showing heavy feeding on the part of the natives."[82] Coghlan was another of a number of individuals who made similar observations, in his case in reference to the nearby Georgina River.[83] At Brewarrina middens extended "for many miles on both sides of the river ... an accumulation of shells which reach a depth of many feet".[84] Further evidence suggesting "large prehistoric populations"[85] has been provided by recent archaeological surveys carried out in the north east of South Australia. There "unusually large accumulations of kitchen refuse", including midden fields up to 10,000 sq. m in extent, were reported at Lake Toontoowarannie in the Strzelecki Desert, as well as around other lakes and permanent waters on the edge of this Desert, at Innamincka and along Coopers Creek.[86]

Gum seems have been another major provision in favoured locations in the Corners region, with Sturt reportedly finding "a number of bark troughs filled with the gum of the mimosa, and vast quantities of gum made into cakes upon the ground" on a tributary of the Castlereagh River in 1829.[87] Similarly Mitchell observed in 1846 that "great quantities" of a fruit, *Cucumis pubescens* were consumed in the vicinity of the confluence of the Narran and Balonne Rivers.[88]

It should not be assumed from the foregoing, however, that all the major food sources in the Corners Region were necessarily staples that were acquired through intensive harvesting. Some significant supplies were secured through the employment of specialised procurement methods. The *Yuwaaliyaay*, and others in south west Queensland, for example, used an emu calling "cornet" to assist in the capture of emus.[89] Another specialised form of food procurement involved pelican husbandry on the lakes in the Coopers Creek/Strzelecki Desert region. On one of these lakes, Lake Lady Blanche, McKinlay claimed to have seen "tens of thousands of pelicans."[90] With this method the *Diyari* herded the hatchlings into a yard where their parents continued to feed them. Before they were ready to fly the *Diyari* then simply killed and cooked the young birds. As the pelicans bred constantly the cycle was repeated and an ongoing food supply ensured, as long as the surface waters lasted.[91] Women hunting emus with bolas, described by Duncan-Kemp, was one other unusual form of specialised procurement.[92]

---

[74] Duncan-Kemp 1961:59.
[75] Mitchell 1839:1:268,270,272,274.
[76] Davis 1863:90,186. See also pp.39,184,187,207,215,219-20,245; McKinlay 1861:34. See also 21,31,35,41,46.
[77] See also Sturt 1833:1:106; Wills 1861:28; Howitt in Smyth 1878:2:305; Andrews 1879:85.
[78] Lewis 1875b:26. See also 1875a:3.
[79] Lewis 1875a:3.
[80] Mitchell 1839:1:306
[81] Mitchell 1848:321.
[82] Lewis 1875b:20. He called it Gerty Creek though it is now known as the Mulligan River.
[83] Coghlan 1898a:48. Davis (1863:245) also commented that "plenty of mussels" were evidently being consumed at the waterholes in south west Queensland. See also: Landsborough (1862:39,94,96,99-100) and Duncan-Kemp(1934:48) in respect to the upper Georgina, Thompson and Barcoo Rivers, Eyre Creek, along with their tributaries; Stuart (1864:113,122,294) regarding such occurrences on the west side of Lake Eyre and at Newcastle Waters.
[84] Matthews 1977:63.
[85] South Australian National Parks and Wildlife Service 1988:12.
[86] South Australian National Parks and Wildlife Service 1988:12; Williams 1988:59; Puckridge 1992:20.
[87] Sturt 1833:1:118-9.
"A Lady" (1860:217) also refers to the consumption of abundant quantities of wattle [Acacia mollissima] gum in area on the upper Gwydir River, just east of the eastern limit of the Corners Region.
[88] Mitchell 1848:110.
[89] e.g. Tully 1896:23; Roth 1897:97; Coghlan 1898b:48; Parker 1897:109; 1905:106-7,143; Mathews 1977:11; Kerwin and Breen 1981:297.
[90] McKinlay 1861:38.
Similarly Abbott (1881:7-8) talks of "millions" of water birds on Narran Lake and the Narran River and Duncan-Kemp (1934:79) of the "hooves of stock ... stained yellow" from the abundance of wild birds' eggs on the lower Diamantina/Goyders Lagoon area.
[91] Kerwin and Breen 1981:295-7; See also Harney 1961:45.
[92] Duncan-Kemp 1952:27-8; 1968:27.

In addition to the exploitation of high yield food source a vast array of other minor or transitory foods were utilised in the Corners Region. These included honey, tubers, bulbs, fruits, berries, various reptiles and birds, many sorts of eggs, caterpillars, grubs, moths, grasshoppers, ant larvae, frogs, shrimps and other crustaceans, tortoises, a range of small and large mammals (euros and kangaroos), and dingo pups. Which of these was favoured depended, by and large, on their local and seasonal availability, and the availability of other foods.[93]

Overall there was enormous seasonal and geographical variation in food sources in this portion of Australia, with patchy but rich resources, uncoordinated drainage patterns, particularly in the *kirrenderri* "Channel Country", and highly variable and unpredictable rainfall. Consequently some groups would depend on just two seed species as staples while others would exploit 20 such species as part of a broad-spectrum strategy. *Panicum* seeds may have been a single staple, providing 30% of sustenance, whereas elsewhere a range of seeds may have constituted 80% of the diet, or fish, the mainstay for some, were only plentiful for others when "freshes" occurred in their locality.[94]

Attention has been drawn on a number of occasions to the high fertility of the Channel Country of south west Queensland and north east South Australia by a range of scholars. Consequently this area has been described at various times as "a vast floodplain,"[95] or "the largest natural irrigation area in Australia,"[96] possessing innumerable permanent and semi-permanent lakes, waterholes and extensive "floodout" zones up to 50 kilometres across, with the ecotone between the drainage lines and the sandplains being identified as the most productive areas.[97] But water was a critical factor in this heterogeneous environment leading to what can only be described as a "pulsation" pattern of exploitation. Many groups would live, often in settlements with permanent shelters, adjacent to or in close proximity to permanent watercourses, lagoons or other water sources until rains fell, or a flood came down the local creek or river following rains, perhaps hundreds of kilometres away. They would then rapidly disperse to take advantage of the bounty produced as the arid countryside sprang to life, only to retreat again to the more secure supplies of food and water as the country dried out.[98] A number of observers attested to this mode of existence.[99] As a consequence the net natural Net Primary Productivity (NPP) of each area was quite variable, depending on the availability of dependable food resources, the reliability water sources and the local environment's natural response when moisture was plentiful. Another critical variable relates to how their environment was exploited by the local population, what technologies and techniques were brought to bear to produce a return, to provide for their subsistence and possibly generate surpluses. A detailed inventory of local food sources and the technologies and techniques for exploiting them, if this were possible, might provide some indication of the relative "richness" of local populations in the context of their highly diverse and varied environments. In essence this would involve quantifying the inputs and solving the complex equation of a production function. Not only is this virtually impossible with limited information, it may not be necessary if we consider the degree of sedentism, population concentration and population density as concrete manifestations of each area or locality's "richness" and the techniques developed to exploit it.

Accounts of what were ostensibly more permanent shelters and "settlements" have already been mentioned in passing. It will be recalled Mitchell had seen, "a native village, in which the huts were of a very strong and permanent construction" on the lower-central Darling, with one large enough to accommodate 12-15 people. And Lewis similarly commenting that the *Ngameni* "return and live in the same wurlies which are very large and well built; the camps may be considered more as villages than anything else." These are not isolated narratives however. On the upper-central Darling River early in 1829 Sturt reported:

*"Early in the day we passed a group of seventy huts, capable of holding twelve to fifteen men each. They appeared to be permanent habitations, and all of them fronted the same point of the compass."*[100]

Such a village would have had a population of 800 to 1000 residents, a town really. Unfortunately, at the time of Sturt's arrival a smallpox epidemic appears to have been in progress, which he ascertained was "sweeping off" those who stayed, "in great numbers".[101] Consequently he encountered far fewer people than indicated by the size of the settlement. But some 16 years later, 400 kilometres west north west of there, close to the New South Wales/South Australian border, Sturt again struck upon another village.

In his detailed description of this village Sturt noted:

*"They were made of strong boughs fixed in a circle in the ground, so as to meet in a common centre; on these there*

93 e.g. Sturt 1849:2:140; McKinlay 1861:32,35,38,41; Davis 1863: 186,207,245; Lewis 1875b:20; Hodgkinson 1877:215,217; Howitt in Smyth 1878:2:302; Andrews 1879; Heaghney et al. in Curr 1886:2: 377; Lindsay 1890:5; Coghlan 1898b:48; Spencer and Gillen 1899: 7,22; Roth 1904:25; Parker 1905:112-119; Duncan-Kemp 1934,1952, 1961,1968; Allen 1972,1974; Jones 1979; O'Connell, Latz and Barnett 1983; Tolcher 2003.

94 Jones 1975:23; Smith 1986:31; Flood 1999:265.

95 Gregory 1906:66.

96 Towner 1954-6:72.

97 Jones 1979:3; Watson 1983:41; Williams 1988:53; Veth, Hamm and Lampert 1990:51.

98 Allen 1972,1974:311; Jones 1979; Smith 1986.

99 Sturt 1839:1:298,2:140; McKinlay 1861:48-51; Davis 1863:202-3; Andrews 1879:85; Newland 1890:22; Duncan-Kemp 1934; Brock 1975: 168,194

100 Sturt 1833:1:89 (5/2/1829)

101 Sturt 1833:1:93.

*Figure 9: "Native Village in the Northern Interior"*
*(Sturt 1849:1:254)*

*was ... ... a thick seam of grass and leaves and over this again a compact coating of clay. They were from eight to ten feet [2.4 - 3.0 m] in diameter, and about four and a half feet [1.35 m] high, the opening into them not being large than to allow a man to creep in. These huts also faced north-west, and each one had a smaller one attached to it as shown in the sketch ...*"[102]

A drawing based on Sturt's sketch of that village appears below:

Reports of "villages" and clusters of "permanent", "well-built", "substantial" "clay-covered" or "multi-room" dwellings are in fact quite common in the region under consideration.[103] On the middle Barcoo River in 1846 Mitchell had observed "huts of a very numerous tribe ... large hut in the centre ... the whole being substantial construction"

[102] Sturt 1849:1:254.

[103] For example:
Gwydir River – Mitchell 1839:1:76-7 - groups of "three or four", "semi-circular or circular, the roof conical, and from one side a flat roof stood like a portico, supported by two sticks" near Mt Arrowsmith - Brock 1975:100-1 - a "group" of "extremely well constructed" huts
Georgina River - Duncan-Kemp 1968:189 - "thirty humpies, dwellings of the spoonbill tribe [totem group] who lived permanently on this waterhole"
*Diyari*/*Wongkanguru* - Horne and Aiston 1924:18,153 - [*poongas*] "Some are very well built"; "the old blacks stay in permanent camps near water"Strzelecki Desert - Sturt 1849:1:387 - "a village consisting of nineteen huts ... two or three of large size, to each of which two smaller ones were attached, opening into its main apartment" Between Gwydir and Namoi Rivers – Mitchell 1839:1:121 - [7 huts] "one large hut in the centre ... the whole being substantial construction"
*Ngurawola* - Reuther 1981:VII,226/1411 - "the people do not change over their wurleys, even when somebody has died"
Inland New South Wales - Fraser 1892:49 - "huts ... in shape of a round dishcover, 6 or 7 feet high, and will hold a family of twelve persons; a portion may be partitioned off as a separate room ... These substantial huts ..."
Boulia/North West Central Queensland – Roth 1897:106;1910:63-4 -"these structures [construction described 1897:106] ... entail no inconsiderable amount of time, toil, and patience in their making"
Macumba River - Andrews 1879:85 - "They make the best wurleys I have seen anywhere - all covered in securely, and having a hole for the exit of the smoke as well as the entrance hole ... covered all over with grass, rushes, roots, earth etc. and are quite dry."
Lake Hodgkinson - Hodgkinson 1877:215 - "constructing a framework of timber ... plastered with clay ... a circular hut about 9 feet in diameter, 5 feet high"
Gum Creek, near Mt Arrowsmith - Sturt 1849:1:233-4 - "native huts ... differently constructed ... a thick coating of red clay ... of considerable size"
Punthamara - Kyabra Creek - Durack 1959:94 - "gunyahs of mud and bark"
*Maiawali* - Diamantina River - Hill 1901:25 – "habitations ... dome shaped plastered up with mud ... eight or nine of these are built in a circle, all close to one another"
Upper Barcoo – Mitchell 1848:319 - "some large huts ... of a more substantial construction, and also on a better plan"
Mulligan River
– Lewis 1875b:20 - "what might be called a native village on the east side [of waterhole on Mulligan River]"
- Kelly 1968:563 – framework of huts still standing after 60 years or more on Mulligan River. Largest group contained 16 huts.
Mytton's Lagoon - Hodgkinson 1877:211 - "large deserted camp lay on each side" - Marked as "Native Village" on Map See also Wills 1861:21.
Marshall River, north Simpson Desert - Barclay 1916:126 – "quite a village of wurlies on the bank"Coopers Creek - Hic et Ubique 1877:13 - dwellings "shape of a bee-hive ... round and plastered with mud ... neatly made ... with arched doorway and sunken floor ... permanent residences"

*Figure 10: Permanent Shelter from North Eastern South Australia (Elkin 1970:opp. p. 139)*

Well beaten paths, and large permanent huts".[104] Lindsay claimed to have encountered village-like settlements around Poeppel Corner in 1886, including one residence "large enough to accommodate thirty or forty natives."[105] Another "explorer", Conrick, had in fact chanced upon just such a structure, "90 ft [27 m] in circumference ...used for holding corroborees" in the same area in 1874,[106] while Goyder reported a similar, though uninhabited, structure in some sort of settlement south west of Lake Blanche in the south of the Corners Region in 1857. By his account:

"*They* [the "wurleys" in the settlement there] *are constructed in a similar manner to those described by Captain Sturt, and are very warm and comfortable, the largest capable of holding thirty to forty people, quite round, from three to four feet high, and entered by a semicircular opening*"[107]

A cluster of 16 huts were seen by Brock in north west New South Wales and another 32 on the Paroo River in north west New South Wales, by Gow.[108] "Large numbers" of more substantial dwellings were noted in the Coopers Creek, Lake Hope area of the north east of South Australia by A. W. Howitt in the 1860s.[109] Alice Duncan-Kemp, in the early 1900s, referred to "a cluster of sixty grass-and-sand built shelters" she had seen near a place called Pharmaleechie Channel, which collectively accommodated "two hundred or more people".[110] On the Diamantina River in 1875 Lewis passed by a "native camp ... over forty gunyahs capable of accommodating six or seven people each," constituting a settlement in excess of 240 people.[111] Permanent dwellings were, if historical observations are a guide, common in the Corners Region. A photo of a typical example from north east South Australia appears below:

Along the Diamantina River F.H. Wells described such huts in detail:

"*The wurleys at a permanent camp are dome shaped circular erections built of logs, cane grass and mud. In the cold weather fires are kept burning constantly inside. During the summer months the natives keep inside their wurleys in the daytime, thus avoiding the intense heat, but sleep outside at night.*"[112]

and Sturt, in depicting the abodes occupied by the "interior natives," used similar terms. They:

"*... were made of strong boughs with a thick coating of clay over leaves and grass. They were entirely impervious to wind and rain, and were really comfortable, being evidently erections of a permanent kind to which the inhabitants frequently returned.*"[113]

While most dwellings appear to have been similar in form and construction a few seem to have been much larger. It will be recalled Mitchell had commented on an "unusually capacious [hut] capable of containing twelve or fifteen persons." He drew a plan of this structure, a drawing based on this appears below:

Given that there is evidence of concentrations of people in settlements with permanent habitations it is, nevertheless, difficult to talk in terms of population density in this region, although attempts have certainly been made to estimate population densities.[114] While these estimates appear to indicate overall population density was higher than might otherwise be expected, they are effectively only an average. For a region as highly variable as this, such estimates are in a sense meaningless, with large areas that are eerily

[104] Mitchell 1848:324.
[105] Lindsay 1890:4.
[106] Conrick n.d.:37.
Conrick founded Nappa Merrie Station on Cooper's Creek and was actually reconnoitring rather than "exploring".
[107] Goyder 1857:4.
[108] Gow 1861:69; Brock 1975:71 (25/11/1844).
[109] Howitt in Smyth 1878:2:302.
[110] Duncan-Kemp 1952:211.
[111] Lewis 1875b:23.
[112] Wells 1894:517.
[113] Sturt 1849:2:139.
[114] Jones 1979:137-8.

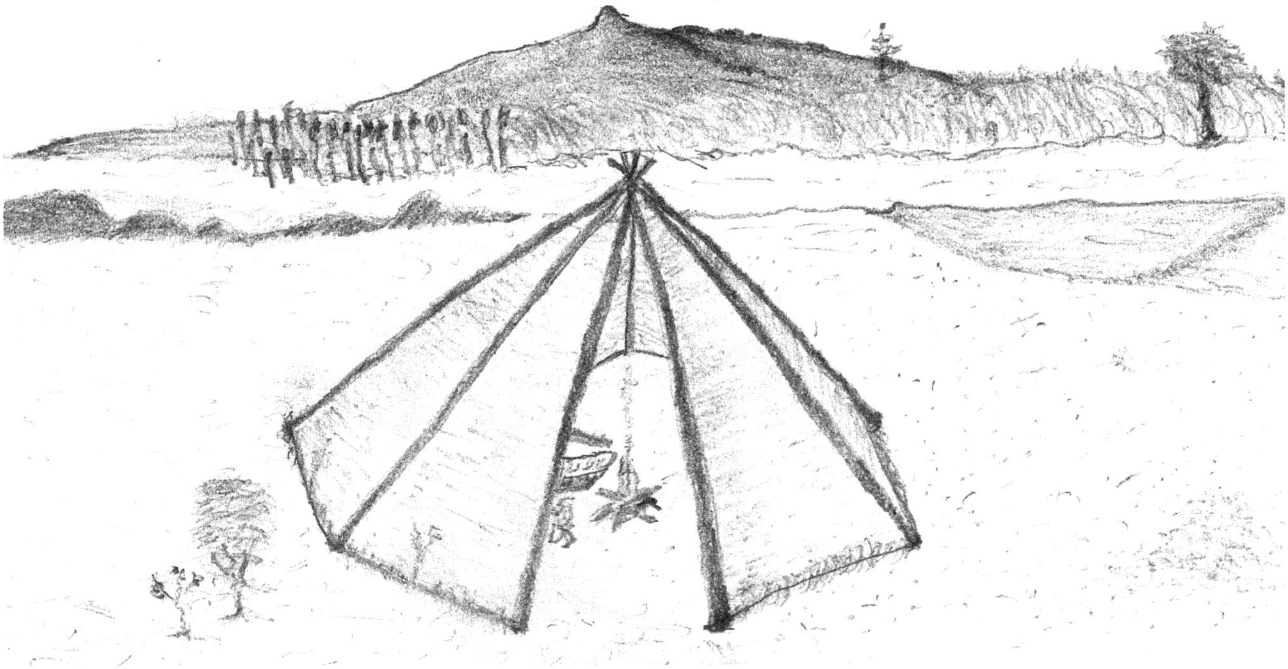

*Figure 11: Larger Structure on the Darling River*
*(based on Mitchell 1839:1:262)*

reminiscent of the surface of the Moon or Mars, being completely uninhabited and uninhabitable. Conversely there were areas which were more favourable and appear to have been quite populous. Research elsewhere has purportedly shown that population densities associated with permanent watercourses in Australia, such as the Murray-Darling, were 20-40 times those found in drier areas not so endowed.[115] It is no surprise then that the relevant archaeological evidence, while extremely limited in extent, does appear to indicate that higher population densities were a feature of parts of the Corner Region, usually associated with the claypans, smaller lakes, waterholes, mound springs and bordering dunes along creeks and rivers in the northern and eastern Lake Eyre Basin, as well as along the Darling.[116] This also appears to have been a relatively recent phenomenon.[117]

At Dalhousie Springs, for example, an almost continuous chain of sites extended for 3 kilometres "mostly on the crest and western slopes" of the local dune system.[118] Similarly, following an archaeological survey along the Paroo River in north west New South Wales in the early 1930s, K. M. Cobb expressed his "astonishment at the vast number of old camps" there.[119] Confirmation of this was provided in another survey in 1944, by C. C. Towle, who noted that "flakes, core-like implements and mill stones were lying everywhere in great abundance" on the Paroo.[120] In corresponding fashion another, more recent, archaeological survey reported "flaked stone" covering an area of 3 square kilometres on the banks of a large waterhole, Lake Idamea, in south west Queensland.[121] Yet another recent survey, in the Lake Eyre South region, documented "vast numbers" of surface sites, some with "millions" of artefacts.[122] Adding to this sense of the intensity of occupation, policeman cum ethnographer Aiston reported in 1921 that Derwent Creek, a tributary of Warburton Creek near Cowarie in north eastern South Australia, "was practically one big workshop for about six miles [9.6 kms]"[123]

Although historical ethnographic sources, the main source of data, are too erratic in frequency and quality to provide reliable estimate of actual population densities, nonetheless impressionistic commentaries, observations and estimates do provide some indication of the nature of occupational densities in particular localities. Sturt's encounter in 1829 with a "town" on the central Darling with a population of possibly up to 1000 residents has already been noted. In a like vein Mitchell, on the upper central Darling in 1835,

115 Birdsell 1953.
116 Cobb 1932:85; Lampert 1988; Williams 1988:59; Veth, Hamm and Lampert 1990:48-9; Holdaway, Fanning and Witter 2000:56; Pardoe 2003:45-50.
117 Lampert 1985:60; 1988:73; Veth, Hamm and Lampert 1990:48; Pardoe 2003:46; Hughes and Hiscock 2005:9.
118 Lampert 1985:60; 1989:10.
119 Cobb 1932:85. See also n327.
120 Towle 1944:73. See also Mitchell 1949:195 and n327.
121 Davidson 1983:30. See also n327.
122 Hughes and Hiscock 2005:12.
123 Aiston 1921:5.
See also: Cooper (1941:3) re Frome River site south east of Lake Eyre South; Mitchell 1949:185 re Lower Coopers Creek; Mulvaney and Kamminga 1999:316-7 re Rodinga Range, Simpson Desert.

thought "the buzz of population gave to the banks at this place the cheerful character of a village in a populous country."[124] Near Innamincka, on Coopers Creek in the north east of South Australia, Sturt observed that "it was evident we were advancing into a well peopled country."[125] Similarly on Morphett Creek in north west New South Wales he surmised that "the natives ... were at this place in considerable numbers."[126] In the end he nevertheless judged that the population on Coopers Creek was the most "numerous".[127] Such comments appear frequently in the accounts of early explorers - "numerous", "large tribe", "legion", "vast number", "large numbers", "immense numbers", "population dense", "big mob" - being typical examples.[128]

Most Australians would find such comments rather surprising given that much of this area is usually seen as being desolate and inhospitable. They even features in some well-known colloquial expressions, such as "back o' Bourke" or the "never-never", forbidding places where no person in their right mind would go. However, the comments cited regarding population densities were not isolated pronouncements. Mary Gilmore, whose works and their veracity were discussed in an earlier chapter, claimed, citing her uncles as the source, that gatherings in excess of 5000 people occurred at the Brewarrina fish traps.[129] This is a figure that is difficult to credit. Large gatherings were transitory events, where a number of different groups met to conduct ceremonies, engage in corroborees, arrange marriages and trysts, barter and socialise. Although gatherings of up to 2500 people are known,[130] there are no other recorded instances in Australia of a gathering of the Indigenous inhabitants in traditional circumstances as large as this. Nevertheless, Gilmore's claim may not be too far fetched as another observer claims to have seen an "immense corroboree" in this area, and 500 people still attended the last ceremonial gathering at Brewarrina in 1885, despite massive depopulation in that part of the region.[131] Furthermore, to the north west of Brewarrina, in south west Queensland, Duncan-Kemp claimed that, prior to the "coming of the white man" there were 3000 *Karuwali* people living just at Dangeri alone, a lagoon on Farrar's Creek.[132] About "one thousand aborigines," according to missionary William Ridley, occupied the county between Surat on the Balonne River and Walgett on the Barwon in 1855,[133] while A. L. P. Cameron claims to have seen gatherings of 800-1000 people in south west Queensland in 1868.[134] Despite 50% mortality from the 1830 smallpox epidemic Scrivener stated that the population on the Paroo was still 500 in 1863.[135] Kopperamanna, in the north east of South Australia, was reportedly the "headquarters" for 1000 or more people in the 1870s when the mission became established, another 700 lived at Lake Killalpaninna in 1867, and it was in this neighbourhood that Police Trooper Gason said he had seen 1000 people performing the *Mindarie* corroboree.[136] An early settler at Coongie Lakes, part of the Coopers lakes, referring to Aboriginal women collecting seeds and roots on the flats, observed that they were "as thick as sheep grazing."[137] On Warburton Creek, just north of Lake Eyre, there were, in 1876, "six hundred to eight hundred souls."[138]

Reports of such large groups arise in the most unexpected places. Nearby, on Kallakoopah Creek, an anabranch of Warburton Creek running through the Simpson Desert, Lewis stumbled across a "camp" of 350 people early in 1875.[139] Gatherings of 500 people also reportedly took place in the northern Simpson Desert.[140] In the course of his explorations in the Coopers Creek area, Sturt recalled, as his party came over the brow of a sandhill that:

"*... on gaining the summit we were hailed with a deafening shout by 3 or 400 natives, who had assembled on the flat below.*"[141]

Davis and McKinlay, two members of one of the Burke and Wills rescue expeditions in 1861-2, as they passed through this part of the north east of South Australia, Coopers Creek and the Strzelecki Desert, also mention hundreds of people on many of the lakes and creeks. Specific numbers reported by Davis included 2-300 people at Lake Coogiecooginna, another 3-400 on a nearby creek, 150 at Lake Lady Blanche, 200 at Lake Hope,[142] 4-500 on Hayward's Creek, and 2-300 at Lake Cann-boog-o-nannie [Lake Jeannie].[143] Seemingly the "Aborigines seemed to pour out of every 'nook and

124 Mitchell 1839:1:225.
125 Sturt 1849:2:61.
126 Sturt 1849:2:109.
127 Sturt 1849:2:135.
128 Examples are too numerous to list but see McKinlay 1861:15-51; Davis 1863:95-215; Conrick n.d.:31,37,39-40,44,47 and Fry 1997:9 for many instances.
129 Gilmore 1934:192.
130 See Dawson 1881:3 (south west Victoria) and Chapter 6:110-1,n93 (Hutt River).
An attack by "no less than one thousand blacks" near Roma in south east Queensland in 1847 (Telfer n.d.:26 in Morwood 1984:328) indicates the presence of a population of 3290 to 4340, if it assumed the attackers were all adult males and there was a typical population structure. See Gerritsen 2000:19 for an explanation of this methodology.
131 Wyndham 1889:41; Dargin 1976:61.
Norton (1907:102) also claimed that "very large numbers" assembled at Brewarrina in the early Contact Period.
132 Duncan-Kemp 1961:190; 1968:26.
133 Ridley 1864:444.
134 Cameron 1885:344.
135 Scrivener in Curr 1886:2:182.
136 Gregory 1906:71; Watson 1998:43; Proeve and Proeve 1952:75-7; Gason 1895: 174.
137 Farwell 1971:159.
Basedow (1925:xiv) claimed that "many thousands" had lived in the north east of South Australia.
138 Paull in Curr 1886:2:18.
139 Lewis 1875b:14.
140 Kimber 1984:14 [W. Smith informant].
141 Sturt 1849:2:75.
142 Apparently there were 6-700 people at Lake Hope when first encountered by Samuel Joseph Stuckey (Farwell 1971:141).
143 Davis 1863:95,120,173,183,204,215; McKinlay rarely gave specific figures though he frequently referred to "a large number", "great numbers", "a vast number" "groups of fifties and hundreds, and often many more" (McKinlay 1861:15,23,31,35,36-8,41,43,51) By his account (1861:46) there were "considerably over three hundred" at Haywards Creek and 2-300 at Cariderro [Browne] Creek (1861:37). Ross (1871:98) also encountered "Natives ... by the hundreds" in this area.

corner' where there was water,"[144] with McKinlay noting, for example, at Tooma-thoo-ganie Creek in the Strzelecki Desert that their "habitations are very numerous on the creek, so they must be pretty strong in numbers here."[145] An almost identical observation was made at one point by A. C. Gregory when passing down Coopers Creek three years earlier (1858), "judging from the number of encampments seen, at least a thousand must visit the banks of the creek."[146]

Indications of localised population concentrations, increasing circumscription and sedentism, other than just those provided by raw numbers, permanent habitations and constancy of settlement, were of course in evidence in the Corners Region. Wide, well-beaten paths, like those noted by Grey in the Victoria District of Western Australia, or "roads" as Mitchell described them on the Darling, seem to correlate with these factors. And once more the pattern appears to be repeated in the region, Sturt, for example, speaks of paths being like "well trodden roads"[147] on the upper Darling. As he approached Innamincka on Coppers Creek he again noticed that "the paths became wider and wider as we advanced. They were now as broad as a footpath in England by a roadside, and were well trodden."[148] Along Coopers Creek a little further on he reported "having ridden along a broad native path the whole of the distance [8 kms]."[149] Later they "crossed the [Coopers] creek, and traversed a large sandy plain intersected by numerous native paths, that had now become as wide as an ordinary gravel walk."[150] Corresponding accounts emanating from south west Queensland include Hodgkinson recording "native tracks everywhere" near Bilpa Morea claypan and "well beaten" paths along the Diamantina.[151] Yet another explorer, Kennedy, likewise met with a "well-beaten native path" in 1847 on the lower Culgoa River, and travelled along another "for two hours running" the following day.[152] Numerous other accounts from the Corners Region mention such paths, including some in which "loose stones" and other "natural obstacles" had been removed and in "some places piled in large heaps".[153]

"Burial grounds", "graveyards" or "cemeteries" are another social feature usually associated with higher levels of sedentism or a significant degree of circumscription. They reputedly first appeared in the Natufian.[154] Determining what qualifies as a cemetery or burial ground is, however, not always straightforward. In his analysis of burial grounds in south east Australia, Colin Pardoe contended that they have four defining characteristics. These are, that there are multiple burials, that the burials are close together ("contiguity"), that there is some means applied to delineate the burial area ("boundedness") and that the burials have not arisen from temporary but repeated camping in the vicinity ("exclusivity").[155] The last characteristic is difficult to determine in the absence of site specific archaeological investigations and fine-grained historical ethnographic studies, though we have, in the preceding pages, indications that some populations in the region had fixed settlements. Certainly there are localities in the region, according to Pardoe, that could justifiably be described as cemeteries or burial grounds, particularly on the upper Darling.[156] One of these was seen by Mitchell in 1835 on the Bogan River at a place called "Milmeridien", a drawing of which appears below.

According to Sturt, in January 1829 "Mr Hume," a member of his expedition, "passed a native burial ground containing eight graves. The earth was piled up in a conical shape," near the junction of the Macquarie and Castlereagh Rivers.[157] Immediately to the west of the traditional owners of these lands, the *Weilwan*, were the *Ngemba*, who also had burial grounds.[158] One informant, Honery, asserted that the *Weilwan* burial grounds contained "hundreds of graves,"[159] with *Gamilaroi* cemeteries also reportedly of similar extent.[160] Other traditional burial grounds existed on the Coopers Creek floodplain at places such as Nockatunga on the Wilson River, a tributary to Coopers Creek in south west Queensland, as well as at Haddon Rig, also in south west Queensland, at a location 65 kilometres west of Tibooburra (*Wadikali*) in north west New South Wales, and on the upper Bogan and Paterson Rivers.[161] There are many other examples reported.[162] Close to the Ashburton Range,

144 Davis 1863:172.
145 McKinlay 1861:51.
146 Gregory 1858:10.
Jung, camped on Coopers Creek in 1865, reported "hundreds" of people arriving there from the north and north west (Nobbs 1992: 136).
147 Sturt 1833:1:86
148 Sturt 1849:2:61.
149 Sturt 1849:2:65.
150 Sturt 1849:2:67.
John Conrick, while residing at Napa Merrie Station on Coopers Creek (SW Queensland, close to NE South Australian border) in 1871 also spoke of "plain beaten roads" in the area (Tolcher 2003:23)
151 Hodgkinson 1877:214.
152 Kennedy 1852:267.
153 Gregory 1884b:207; Also 1858:7 (Coopers Creek).
Other examples can be found in Howitt 1861:10 (Innamincka); Wills 1861:17 (Georgina/Diamantina River),27 (Coopers Creek); Stuart 1864:298-9 (north of Newcastle Waters); Howitt 1907:27 (tributary to Strzelecki Creek); Dow 1938a:32 (Mulga Creek, north east of Broken Hill).
Dow (1938a:31) records numerous stone mounds near Broken Hill including one [Wilson River] that was 1.35m high.
154 Henry 1989:39,49; McCorriston and Hole 1991:50.
155 Pardoe 1988:1-2.
Littleton 2002 has questioned the exclusivity criteria, showing that the reality may have been more complicated than envisaged in Pardoe's schema.
156 Pardoe 1995:705.
157 Sturt 1833:1:51.
158 Dunbar 1943-4:146.
159 Ridley 1878:254.
160 Ridley 1875:159.
161 Puckridge 1992:20; McKellar 1984:62; Gresser 1964:131; Black 1943:437; Smyth 1878:1:99; Fraser 1892:36.
162 For example:
Maranoa – central Queensland (McLellan 1901:258; Meston 1901:63).
Western New South Wales – "usually associated with campsites" (McCarthy 1970:21).
Western NSW – Naualko – Lower Paroo (Freeman 1902:539)
Upper Barcoo, central Queensland (Lumholtz 1889:181).
Central Queensland (Lumholtz 1889:278).
Bogan River (*Yuwaaliyaay*)(Parker 1905:25,85-6,92,101; Heritage Concepts 2004:26).

*Figure 12: Burying Ground at Milmeridien*
*(Mitchell 1839:Plate 20 opp.1:321)*

just south of Newcastle Waters in the centre of the Northern Territory, drover Ashwin also encountered a "burying-ground", of platform burials in 1871, where "there were hundreds of skeletons, skulls and bones."[163] These had in fact been noted 11 years earlier by the first British explorer to successfully cross the continent, John McDouall Stuart, who had come across extensive platform burials in the same area, leading him to conclude that the local populace "must be very numerous."[164] Burial grounds were also found in proximity to the central Darling. Paleontologist Robert Etheridge, for example, observed what he described as an "extensive burial ground", with six burials in an area "thirty yards square" [729 sq. m] here in 1903.[165]

The type of burial was probably similar to ones seen by Mitchell almost 70 years earlier, when he encountered a "burial ground" on this part of the Darling. On a hill overlooking the "native village" mentioned earlier he remarked on:

"*three large tombs of the natives, of an oval shape, and about twelve feet* [3.6m] *in the greater axis. Each stood in the centre of an artificial hollow, the mound, or tomb in the middle being about five feet* [1.5m] *high; and on each of these were piled numerous withered branches and limbs of trees, no inappropriate emblem of mortality.*"[166]

As with the burial ground at "Milmeridien", Mitchell sketched these tombs.

It would appear from the sketch that a considerable amount of effort had been invested in the construction of these graves. More elaborate burials involving greater care and effort are another indication of a greater degree of sedentism, as well as greater sociopolitical complexity.[167] This relationship is not a simple or direct one, however. The most dramatic example of such "status burials" can be found in the well-known pyramids of Egypt. There the political, and religious, importance of the ruler found its expression in the mobilisation of a large labour force drawn from an already sedentary population. Often, in societies where the

Marsden Cave, central Queensland (Flood 1993:148).
Gunnedah, central northern New South Wales (Idriess 1953:ix,xi, xv).
Bulgeraga Creek, northern New South Wales (Mathews 1896:46).
*Murawari* (J. Matthews 1977:33).
Nogoa River Catchment – 10 crypts containing 36 individuals dated 889 – 342 BP (L'Oste-Brown, Godwin and Morwood 2002:45).

163 Ashwin 1932:62.

164 Stuart 1864:220 (27/6/1860).

165 Etheridge 1916:6.
Etheridge was Curator of the Australian Museum and went on to found its Ethnology Department in 1906.

166 Mitchell 1839:1:262. The accompanying map of his explorations shows a "Burial Ground" at this location.

167 Tainter 1978:125-7.

*Figure 13: Tombs of a Tribe*
*(Mitchell 1839:Plate 16 opp.1:262)*

first signs of increasing sedentism and status differentiation are becoming evident, the graves are a little more elaborate and a little larger than usual for some individuals, with valuable or symbolic grave goods, such as necklaces of rare materials, being buried with them. A couple of possible examples have been identified in Australia.[168] Burials where it appears extra labour was invested are commonly reported from the Corners Region, usually in the form shown in Mitchell's tombs, though often larger. When exploring in south west Queensland and north east South Australia in 1876, W. O. Hodgkinson came across what he termed "remarkably large" graves near "Mytton's Lagoon" and "Lake Hodgkinson".[169] One of these he described in some detail:

"*The graves of the natives, of which we passed several, are of astonishing size, one just completed being a collection of sand, boughs, and logs, neatly formed into an oval 18 or 20 feet* [5.4 - 6.0 m] *by a width of 10 to 12 feet* [3.0-3.6 m], *and rising to a summit 3 to 4 feet* [0.9 - 1.2 m] *above the level of the ground, forming the shape of an egg half buried, with its extremities of equal size. Around this erection is a wide well-cut trench about 3 feet* [0.9 m] *and 2 feet 6 inches* [0.75 m] *deep.*"[170]

An even larger example was seen by Stuart in 1860 on Hamilton Creek, north west of Lake Eyre, 1.35 m high and 18 to 21 m across.[171] That these interments may have been status burials was suggested by A. W. Howitt, who had seen *Diyari* and *Jandruwanta* examples on Coopers Creek and around Lake Hope:

"*The graves are just such as spoken of by Captain Sturt - large mounds of sand, covered with logs and brush. I believe them the graves of great men (Pinnarou), or of persons slain in war. They are not, as I think, numerous enough for the burial of the commonality.*"[172]

Howitt's reference to Sturt relates to a mound grave seen by Sturt just north of Coopers Creek, that one being "twenty-three feet [6.9 m] long, and fourteen [4.2 m] broad."[173] These mound graves approach the dimensions of the small barrows of the European Neolithic. A drawing of the grave Sturt saw, along with a photo taken of another at Innamincka on Coopers Creek, appear below.

A *Ngurawola* burial ground where this type of grave

[168] e.g. Grave 108, Roonka IIIa (Pretty 1977:305); Burial in Mound FM/1, south west Victoria (Gerritsen 2000:61)
[169] Hodgkinson 1877:211,215.
"Myttons Lagoon" is possibly Mooradonka Waterhole on Spring Creek, a tributary of the Diamantina River. "Lake Hodgkinson" appears to have been Lake Knappanbarra.

[170] Hodgkinson 1877:215.
[171] Stuart 1864:144.
[172] Howitt in Smyth 1878:2:306.
[173] Sturt 1849:2:57.

*Figure 14a: Native Grave*
*(Sturt 1849:2:57)*

*Figure 14b: Jandruwanta Grave, Innamincka*
*(Basedow 1925:Plate 25 Fig.2)*

*Figure 14c: Graves on Arrabury Station*
*(original photo held by SA Museum)*

predominated can be found on Arrabury Station in south west Queensland. A photograph of that burial ground is shown below.

Examples such as these are only a sample of reports of multiple and/or more elaborate burials from the region.[174] As a final illustration of their extent and variety mention ought to be made of the unusual group of 11 square stone mounds, the largest 2.7 m by 1.5 m, 0.45 m high, marking the site of another small burial ground at Poolamacca in the lower part of north west New South Wales.[175] The significance of this feature in its local context will be discussed later.

Yet another instructive sign of sedentary communities who may be about to begin practising agriculture, if not doing so already, is storage. Archaeologically based arguments and evidence have traditionally pinpointed the advent of storage, through the presence of storage structures, typically clay- or stone-lined pits for grain or other seed foods, as a key innovation that is relatively detectable.[176] In Australia it is common to find explorers, colonists and 19th century commentators disparaging local Aboriginal populations, claiming they had no thought for the future, they never bothered to preserve surpluses and so forth. But storage is difficult if groups are mobile, do not have food sources that are amenable to storage and have not developed suitable preservation techniques. It has been suggested that one of the critical innovations in pre-agricultural groups in south west Asia was the parching of grains, allowing surpluses to be preserved and therefore stored.[177] The role of storage is usually seen as a means of evening out resource fluctuations so that temporary surpluses or momentary abundances are converted into a form that can be consumed over a longer period.[178]

The Australian evidence suggests there are three types of food "storage" - caching, stockpiling and storage proper, or what O'Shea terms "direct storage".[179] Caching usually involves small stores of food, preserved in some manner and left in a protected location, such as in a tree, on the roof of a shelter or buried in the sand, for an emergency or in time of need.[180] This form is most evident in highly seasonal or unpredictable environments such as the harshest desert areas. Thus we see examples such as the *Kukatja* and *Pintubi* of the northern part of the Great Sandy Desert in Western Australia collecting acacia seed pods and eucalypt seeds and covering them with spinifex, consuming them in lean period in the last few months of the year.[181] The

[174] e.g. Smyth 1878:1:119 (Diyari); Birtles 1937:153; Elkin 1937: Plate II,Fig.A.

[175] Black 1950:32,36, Plate 188,Fig. 3.

[176] Rafferty 1985:115,131; Henry 1989:39; Hillman and Davies 1990: 198; Price and Gebauer 1995:6; Wills 1995:227; Varien 1999:12, 33,39.

[177] Jones 1979:147-8.

[178] Testart 1982:524; Ingold 1983:554; Ellen 1988:129; Cashdan 1989: 33.

[179] O'Shea 1981:169.

[180] Gould 1969:265.
In the course of this research well in excess of 70 examples of caching, stockpiling and storage undertaken by Indigenous inhabitants in traditional circumstances in Australia have been identified.

[181] Cane 1984:80; 1989:104,109.
According to Cane enough was collected to help them through the lean period during the hot season and the erratic monsoonal "wet" that followed.

*Ngaluma* of the north west coast of Western Australia did likewise, covering small quantities of spinifex seeds (*Triodia* sp.) with bark as an emergency supply.[182] Desert fruits, like "bush tomatoes" (*Solanum* spp.) and quandongs (*Santalum aciminatum*) were sun- or fire-dried in small quantities and either carried about or left in a protected spot by Western Desert and Central Australian groups such as the *Iliaura*, *Luritja*, *Mardu*, *Pintubi* and *Walbiri*.[183] Stockpiling was another quite common occurrence, usually in preparation for ceremonies, corroborees or other sorts of multi-group meetings. Traditionally the *Wik-Mungkan* stockpiled yams (*Dioscorea* sp.), "nonda plums" (*Parinarum nonda*) and water lily tubers (*Nymphaea* spp.) for such purposes.[184] The *Watjandi* of the lower Murchison River used to prepare for the "Caa-ro Ceremony" by laying down yams (*Dioscorea hastifolia*) and "all kinds of foods."[185] When the *Wongkanguru*, to the north east of Lake Eyre in Central Australia, planned a corroboree stocks of *ngardu*, *munyeroo*, *wadroo* [vegetable foods] as well as cooked snakes, "rats" and rabbits were prepared, and then buried in the sand, kept in *pirrhas* [long wooden bowls - coolamons], placed on the roof of a hut or on a specially built platform.[186]

Storage proper in Australia was also a lot more common than is generally acknowledged. As well as seeds and fruits, nuts, gum, tubers of various sorts, eggs, meat, fish, fish oil and even mussels were stored.[187] One of the few areas where storage proper appears to have had a significant role in the local subsistence regime was in the Corners Region. Already, the practice of storing up to 50 kg of *munyeroo* seeds in grass and mud containers, reported by Howitt, has been mentioned in passing. Sturt too, it will be recalled, stumbled on the preparation of "vast quantities" of gum cakes near the Castlereagh River in 1829. Charles Coxen, in 1836, while exploring in the "interior", probably the western part of *Gamilaroi* territory, intersected by the Barwon and Namoi Rivers 100 kilometres to the east north east of the Castlereagh, was another to note the storage of large quantities of gum, and much else:

182 Withnell 1901:23.

183 Carnegie 1898:230-1; Finlayson 1943:84; Gould 1969:261; O'Connell and Hawkes 1981:101; Walsh 1990:35; Tonkinson 1991:31,37,50.
The *Mardu* also stored large quantities of *Cyperus bulbosus* tubers and "seeds" (Walsh 1990:35).

184 Thomson 1939b:216.
See also Walsh 1990:35.

185 Oldfield 1865:230.

186 Horne and Aiston 1924:7,37

187 e.g. *Tecticornea arborea* seeds by *Wadjari* (Tindale 1974:145); "Wattle" seeds by the *Narrinyeri* (Berndt, Berndt and Stanton 1993:109); *Parinarum nonda* [Nonda plums], Cape York (Thomson 1939b:216); cycad nuts in trenches up to 6m long, *Karawa*, Arnhem Land and north Queensland (Harvey 1945:191; McCarthy 1962:1:35); bunya nuts by the *Gubbi Gubbi* of the Blackall Ranges in south east Queensland (McCarthy 1962:1:35); wattle gum by *Narrinyeri* (Berndt, Berndt and Stanton 1993:109); *Nymphaea* spp. tubers, Cape York (Thomson 1939b:216); "yams ... in great stacks", East Alligator River NT 1872 (Lewis 1922:131); turtle and birds eggs, *Karawa* and *Tiwi* (Harvey 1945:191-2); kangaroo meat, Gascoyne area (Scott 1972:95), *Narrinyeri* (Berndt, Berndt and Stanton 1993:79); fish, *Narrinyeri* (Harvey 1943:109; Berndt, Berndt and Stanton 1993:79,110-11); fish, Moreton Bay (Hall 1982:86); fish, west central Northern Territory (McCarthy 1962:1:35); cod and stingray oil, *Narrinyeri* (Berndt, Berndt and Stanton 1993:115) *Velesunio ambiguus* mussels, Lake Victoria, New South Wales (Simpson and Blackwood 1973)

"*I found a considerable store of grass-seed, gum from the Mimosa, and other stores, carefully packed up in bags made from the skin of the kangaroo, and covered over with pieces of bark, so as to keep them properly dry. The weight of the bags containing the grass seed and gum was about 100 lbs* [45 kg]*; the seeds had been carefully dried after being collected from the small grasses of the plains.*"[188]

Others describe this method of storage, seemingly involving seeds of the grasses discussed earlier and *munyeroo*. For example, on the Darling, Bonney recounted that:

"[skin] *bags are very useful .. to hold a store of grass seed after a good harvest of it. These seeds are very small, some resembling fine gunpowder.*"[189]

It will also be recalled that Kate Langloh Parker had observed this method of storage when describing the harvesting, threshing and winnowing of seed grains, "When the grain was sufficiently clean, it was put away in skin bags to be used as required."[190] And a store of about 50 kg of seeds, similarly in bags, was chanced upon by an Overland Telegraph worker on the Finke River in 1870.[191] A "pioneer" on the upper Darling, Newland, indicated that "quite large quantities [of grass seed] were stored for future consumption in various articles in aboriginal domestic use,"[192] as well as mentioning another preservation method for *Panicum* seeds:

"*They also make the ground up 'parper' into balls, dry them and preserve them for an indefinite time.*"[193]

Whereas the *Diyari* employed clay covered seed-grain pits,[194] according to Alice Duncan-Kemp in south west Queensland ground seed "powder was stowed away in small dilly bags of grass until required", in this instance referring to a number of different seeds or sporocarps, *ngardu*, *kooni* [*munyeroo*], and *karabadi* [seeds of the coolabah tree].[195] Although the *Karuwali* made seed cakes for caching as emergency supplies and for food when "on the road,"[196] they, and other groups in the region, preserved and stored a wide range of foods as well, often in the form of "flour". Dried fish, fish meal and fish flour are the most frequently mentioned, by Duncan-Kemp, Reuther, Parker, Johnston and Berndt for example, in reference to the *Karuwali*,

188 Coxen 1866:4. Also in Smyth 1878:1:143n.
Coxen goes on to say:
"Such instances of forethought are doubtless rare, and, I believe, are only to be found beyond the influence of civilisation." (1866:4)

189 Bonney 1881 in Allan 1972:78; Worsnop (1897:81-2) also refers to an informant who came across "about 1 cwt." [51kg] of "bags of seeds" on the Finke River in 1870.

190 Parker 1905:118.

191 Worsnop 1897:81-2.

192 Newland 1920-1:13.

193 Newland 1920-1:13.

194 Reuther 1981:VI,17/982[4]; VI,17/27; VI,17/152[5]; VI,17/209; Spencer 1918-19:12.

195 Duncan-Kemp1934:261.

196 Duncan-Kemp 1934:56; 1952:111.

*Yuwaaliyaay* and the *Diyari*.[197] Even caterpillars, witchetty grubs, grasshoppers, and meat such as liver, underwent this process, the dried and powdered protein being mixed with flour made from *papar*.[198] Such commodities, including *ngardu*, were apparently traded and bartered as well.[199] Various roots, according to Newland, were pulped and "kneaded into large balls and kept for future use" by the *Paakantyi*.[200] On the Georgina River of south west Queensland the currant-like fruit of the *koonkaberry* bush produced an abundant harvest, with the local people "eating, gathering, drying and storing all they could."[201] Salting or smoke-drying of geese was another preservation technique apparently practised by the *Karuwali*.[202] While policeman Aiston, along with his collaborator Horne, claimed the *Wongkanguru*, on the north side of Lake Eyre, "have no idea of saving for an emergency"[203], they nevertheless describe stockpiling for a corroboree[204] and mention the discovery of "three boxes containing munyeroo" at a deserted camp and, under a piece of bark, "another lot of munyeroo."[205] Moreover, in the same area F. W. Andrews, the Collecting Naturalist of Lewis's 1874-5 Lake Eyre Expedition, said in reference to the dried fruits of "pig's face" (*Mesembryanthemum* sp.), "they gather [it] in large quantities, and store by until wanted or as long as it will keep."[206] More recent evidence, relying on oral traditions, appears to confirm this feature of traditional subsistence in the region. Tindale, in the 1970s, mentions storage of mulga and grass seeds in caves and hollow trees, as did O'Connell *et al.* and Kimber in the 1980s, for periods up to 18 months.[207] The corms of a staple, *Cyperus bulbosus*, known by a variety of names in arid and semi-arid areas, such as *yelka*, *yalka*, *yaua*, and *nalgoo*, or in English vernacular as "bush onions", were also reportedly cached and stored in Central Australia.[208]

Possibly the most striking example of storage on a significant scale originates with the drover A. C. Ashwin. As one of the early colonial intruders in the very north of the Corners Region, on the south western edge of the Barkly Tableland in the Northern Territory, Ashwin, sometime in 1871, came across a *Tjingili* settlement of "50 small mia-mias [shelters]" enclosed by a fence approximately 180 m across. Within this settlement there was a "large mia-mia, about 7 feet [2.1 m] high in the middle and about 16 feet [4.8 m] diameter." Inside this shelter were 17 "large wooden dishes four or five feet [1.2-1.5 m] long filled with grass seed as large as rice." By Ashwin's reckoning these dishes, each "a foot deep [30 cm]" and "covered with paper-bark", contained "about a ton [1 tonne] of seed."[209] While there is indirect corroboration and some validation of Ashwin's claim,[210] nothing on this scale was specifically reported in other parts of the Corners Region. Coxen did, however, mention "bags" of 45 kg and Andrews, Sturt and Newland talked in terms of "large quantities" or "vast quantities."

In this regard Sturt's comment in regard to the "Native Village in the Interior", shown in Figure 9 have added significance. It will be recalled that Sturt had noted that these "huts also faced north-west, and each one had a smaller one attached to it as shown in the sketch." Later, he commented in regard to this arrangement that:

"*it appeared to be a singular but universal custom to erect a smaller hut at no great distance from the larger ones, but we were unable to detect for what purpose they were made, unless it was to deposit their seeds; as they were too small even for children to inhabit.*"[211]

These small huts are evident in Figure 9. Brock, another member of the expedition, also mentions that among a group of "extremely well constructed" huts near Mt. Arrowsmith, in north west New South Wales, were "several smaller ones for the children."[212] But it does not appear that separate shelters were erected for children in traditional Indigenous life in Australia, if the historical ethnographic literature is any guide.[213] So, if Sturt's conjecture was correct and these smaller structures were used for storage, then they would certainly qualify as storage

---

197 e.g. Stirling and Waite 1918-21:133; Duncan-Kemp 1934:147; 1961: 118; 1968:208; Reuther 1981:IV,66/3051,145/3712,205/1243; Parker 1897:26; Johnston 1943:297; Berndt 1953:194,196.
Duck eggs may also have been stored for weeks in south west Queensland (Marshall 1937:59).

198 Howitt 1904:799; Roheim 1945:49-51; Duncan-Kemp 1961:213; 1968: 208; Reuther 1981:IV 254/1637,265/1731,X10.
The *Tjingili* dried shredded kangaroo meat for "future use" without making it into meal (McCarthy 1962:1:35).

199 Mathews 1909a:497-8; 1909b:7; Duncan-Kemp 1934:147,209; 1961:118; Goddard 1939: 160.

200 Newland 1890:22.

201 Duncan-Kemp 1968:186.

202 Duncan-Kemp 1961:192,193.

203 Horne and Aiston 1924:7.

204 Horne and Aiston 1924:7,37.

205 Horne and Aiston 1924:33.

206 Andrews 1879:83-4.

207 Tindale 1974:104; 1977:346; O'Connell, Latz and Barnett 1983:93; Kimber 1984:19. See also Devitt 1992:45.
Eylmann (1908:289) mentions seeing people in Central Australia, possibly in the vicinity of Lake Eyre, in possession of 1 hectolitre (approximately 100 kg) of *Eucalyptus microthera* seeds.

208 O'Connell, Latz and Barnett 1983:86; Aboriginal People of Central Australia 1995:54; Latz 1995:158; Isaacs 1996:23.
Reuther (1981:VI,17/883,8) indicates in-ground storage with patches being reserved.

209 Ashwin 1932:64.
Ashwin further reported finding 6 similar "dishes" containing "rice seed" stored in the trees some distance to the north, at "Frew's Ironstone Pond" [Milner's Lagoon] (1932:62).

210 Eylmann (1908:369,371,373,Plate 24:Fig.3), along with Spencer and Gillen (1912:2:380) confirm that the *Tjingili* and *Waramanga* and other groups in Central Australia possessed large containers (up to 120 cm long) which were used for storage. Heaghney (in Curr 1886:2:375) describes a *Bidia* or *Kulumali* (south west Queensland) "coolamon" 75 cm long, 20 cm wide and 20 cm deep, McCourt (1975: 140) obtained a "coolamon" of the dimensions described by Ashwin (130 cm long, 37 cm wide, 30 cm deep) from the Coopers Creek area while Palmer (1884a:325) refers to a type from north central Queensland 120 cm long.
Calculations of the estimated volumes and weight (based on assumed density of 50% that of water) of the seed contained in each of these "dishes" suggest that Ashwin's estimate was reasonably accurate.

211 Sturt 1849:2:139.

212 Brock 1975:101 (30/12/1844).

213 It is a remote possibility these smaller structures were dog kennels. James Dawson (1881:89) reported that in western Victoria their "domesticated" dingos "would not sleep or take shelter under the

facilities.[214] The *Diyari*'s clay covered grain pits and the *dudther* bark pile structures employed by the *Yuwaaliyaay* to keep excess witchetty grubs,[215] probably also qualify as storage facilities. From this, and the other evidence, it would seem that overall a range of storage methods were used in the Corners Region. Unfortunately, because the particular forms seem to have been made of perishable materials, unlike those in the Levant, they are unlikely to be detected archaeologically. Nevertheless, the scale of storage in the Corners Region in some instances was quite significant when compared to practices in the Natufian and early Neolithic. Storage pits and bins in south west Asia were not common initially in the Neolithic, it being suggested baskets may have been the favoured mode of storage.[216] Whatever the case, they would probably have only contained amounts equivalent to that contained in a few skin bags.

As can be seen from this discussion, with the extent of seed utilisation, the intensive utilisation of other food sources, often employing elaborate methods involving considerable labour investment, reports of large settlements of permanent dwellings, with indications of a greater degree of sedentism and higher population densities than usual in Aboriginal groups, along with at times extensive storage, the proposition that the peoples of the Corners Region were incipient or proto-agriculturalists certainly has strong support. But there is evidence indicating that their economy may have been more than that. In Chapter 3 various instances of planting were compiled, and one of particular relevance here was Thomas's unverified claim of the supposed cultivation of *munyeroo*. Given that this was a major food source in some parts of the Corners Region, if true, it points to the possibility of agriculture.[217] While Thomas's evidence is questionable, this is not the only case of sowing and planting in the Corners Region, we need look no further than the examples detailed in Chapter 3 for evidence of that. It should be realised, however, that that list did not include all instances of planting, or necessarily indicate the extent of sowing and planting. For example, it will be recalled Walter Smith, Kimber's informant, mentioned broadcast seeding in Central Australia of *ngardu* and grass seeds such as *P. decompositum*. According to Smith, a fully initiated man of part *Arabana* descent, the people here would:

"*chuck a bit there. Not much, you know. Wouldn't be a handful ... a little bit, spread it you see - one seed there, one seed there, ... chuck a little bit of dirt on, not too much though. And soon as the first rain comes ... it will grow then.*"[218]

In fact Smith named a number of groups he knew to have been involved in these practices, as Kimber's map shows:

The information provided by Smith is highly significant in this context. Nevertheless, without some form of corroboration it is difficult to place complete confidence in that information. Certainly other species seem to have traditionally been planted in this region, as outlined in Chapter 3, including two species of *Ipomoea* in the desert regions, as well as "native tobacco" (*Nicotiana* spp.) and possibly others. Yet many who have studied Indigenous cultures in this region categorically deny that planting took place. Horne and Aiston, in reference to the *Wongkanguru*/*Wangganguru*, observed that the "growth of food from seed was quite beyond them ... no idea existed of planting a seed with the hope of a harvest."[219] Yet they themselves describe two examples of broadcast seeding. Regarding the first instance they state:

"*the seed of the 'wirra' [Acacia salicina] was also thrown broadcast, but this seed had first to be crushed, and therefore rendered impossible of fertilisation. These ceremonies are generally performed after a shower of rain has fallen or at any rate seems imminent.*"[220]

However, the "crushing" of the seed, which involved hammering each seed between two miniature hammerstones at the location they wanted new plants to grow,[221] rather than preventing germination, as Horne and Aiston presumed, is in fact the recommended method of ensuring rapid germination of these seeds. It is called "scarification", and with acacia seeds scarification "is required to breach the hard seed coating and allow germination."[222] Logically one would do such a thing when moisture was available, such as that provided by "a shower of rain," to ensure rapid germination. The principal attraction of the *wirra* was its leaves, which were burnt and the ashes used to neutralise the alkaloids in the narcotic *pitjuri* before it was chewed.

roof of their master, a separate place was generally erected for them". However, there are no observational records of this (Gerritsen 2000), and it would appear to have been the universal custom elsewhere in Australia, particularly when cold, that tame dingos slept with their owners, for mutual warmth (See for example Lumholtz 1889:179; Jones 1970:268; Olsen 1985:88). There is however a modern example of brush shade shelters being provided for dogs by the *Alyawarra* (Memmot 2007:136, Plate 11)

[214] Allen (1968:45-6) first suggested this.
An oral account by informant George Dutton, given to linguist and ethnographer Luise Hercus, provides some support for this contention. Dutton indicated "grass-seed flour" was stored in a "small humpy [shelter]" and that a snake was placed inside to keep away vermin. He was not however sure if this was a traditional practice (Memmot 2007:136).

[215] Stow [K.L. Parker] 1930:70-1.

[216] Henry 1989:211-3; Bar-Yosef and Belfer-Cohen 1992:34,38; Bar-Yosef 1998a:170.

[217] It may well have been at least incidentally planted, the plants being gathered and piled in a heap till the seeds dropped out. Given the fineness of the seeds it is quite likely some escaped in the process at the location this took place and sprouted when suitable conditions arose. In Central Australia stone rings were constructed to contain the *munyeroo* plants in these situations, indicating that there was some awareness of seeds escaping. See Latz 1995:249-50.

[218] Kimber 1984:16.
Smith lived all his life in central Australia, was fully initiated, as stated, and had an extensive knowledge of traditional customs. For a full account of his life and activities readers can refer to Kimber's biography of Smith, *Man From Arltunga: Walter Smith, Australian Bushman*.

[219] Horne and Aiston 1924:8.

[220] Horne and Aiston 1924:8

[221] Horne and Aiston 1924:133-4.

[222] Wrigley 1988:387; K. Johnson and Burchett 1996:93.

*Figure 15: Map of Groups Involved in Broadcast Seeding in Central Australia*
*(Kimber 1984:Fig. 1,13)*

In addition to this it appears *yaua, Cyperus bulbosus*,[223] a staple food in Central and eastern Central Australia,[224] was also broadcast seeded. There is little ambiguity in Horne and Aiston's statement in this regard:

"*The ceremony of making sure of plentiful growth* [of *yaua*] *was very similar to the 'wirra' bush ceremony, only in this case the seed was not ground*"[225]

The accompanying song, correctly translated, ended with the words, *charrilla charrilla koppara nunta*, "plant plant root grow."[226]

Like Horne and Aiston, Duncan-Kemp claimed that the *Karuwali* did not plant seeds, "they did not believe it was necessary to cultivate food ..."[227] But earlier had described the "Katoora Ceremony" in the following terms:

"*From their woven dilly bags the gins sprinkled seed food over the ground ..... Katoora or barley-grass seed lay in little hillocks, already swelling and creeping to repeated applications of water which the gins poured on them to make 'wunjee all the same walkabout (grass to grow)'* "[228]

Qualified support for the claim that *katoora* was being planted can be found in Mary Gilmore's evidence, discussed previously. Although her evidence was called into question, and related to planting carried out by the *Wiradjuri*, who lived to the east of the Corners Region, she too describes "small pinches" of grass seeds being sown.[229]

An added dimension to Duncan-Kemp's account was the fact that the *katoora* seed [presumably *Panicum decompositum*] was obtained from the north and north west. It was:

"*gathered from the coarse barley-grass of downs country was obtained by barter from the Kalkadoon, Goa, and Pitta Pitta tribes of the Winton and Boulia district. The Myorlis of Davenport country passed it down in reed or grass wrappings*"[230]

This was the same area where Gregory had reported the harvesting of thousand acre [400 ha] expanses.

Coincidentally, in describing broadcast seeding in Central Australia, Smith reported that these seeds were also obtained from the north, from the *Waramanga* in the direction of the Barkly Tableland. They were transported in special bags called *apwa*, made of emu feathers and kangaroo sinews, each containing up to 23 kg of seed.[231] The *Waramanga*, coincidentally, lay just south of the *Tjingili*, the people whom drover Ashwin had reported as having a tonne of seed stored. And this is not the only instance of seeds being traded over long distances, another contemporary oral account describing the traditional "Bookartoo" expeditions includes "grass seed in bags" as one the trading items carried by the "Bookartoo men" in the course of the expedition from the Lake Eyre region, 500 kilometres south to the Flinders Ranges.[232] Furthermore, the *Paakantyi* to the east were, according to at least one historical source, also trading in seeds,[233] suggesting that this form of trade may have been quite extensive.

Harlan has previously observed that a people's mythology always justifies their predominant subsistence method.[234] By and large this appears to be true. Much of Aboriginal mythology contextualised, where hunting and gathering predominated, that mode of subsistence. Similarly, Demeter of the Greeks, Ceres of the Romans, and Isis of the Egyptians are figures central to mythologies from the Mediterranean and European cultural heritage describing the invention and practice of agriculture. A set of *Wongkanguru* myths collected by German missionary Johannes Reuther at Lake Killalpaninna in the late 19th century are particularly relevant in this regard. These myths refer to a Dreamtime being (*muramura*) called *Markanjankula* who:

"*At first came to Aruwolkanta (waru ngankana), on the Makamba* [Macumba River], *where he found a beautiful level plain. Here he cleared away all weeds and stones, loosened up the soil, sowed some ngardu, and then covered up with earth what he had sown, so that, should a flood come along, the ngardu should come up and the people should have enough to live on and not starve.*"[235]

At a further five points in the legends relating to *Markanjankula* he is described as planting or broadcast seeding *ngardu* or seed species (*wadika*).[236] This was

223 Horne and Aiston labelled it as *Cyperus rotundus* (p. 134) but were incorrect in this identification. *C. rotundus* is not found in this part of Australia (Blake 1942:9-10) and references elsewhere to yaua list it as *C. bulbosus* [*Ngameni*](Johnston and Cleland 1943:152; Latz 1995:158,394)

224 Latz 1995:56,158.
Stirling wrote: in "almost every camp we saw large quantities" and "in many camps it appeared to be the only food that was being used" (Stirling 1896:60).

225 Horne and Aiston 1924:134.

226 Aiston 1930:49.
Aiston pointed out that his collaborator Horne had slightly mistranslated the song in their 1924 account.
While the broadcasting of the seeds was in the context of an increase ceremony other increase ceremonies for *yaua* did not employ the seeds (See Duncan-Kemp 1934:150; Mountford 1976:293) indicating planting was the intent.

227 Duncan-Kemp 1968:207.

228 Duncan-Kemp 1934:146-7.
This is completely unlike a grass seed increase ceremony recorded by Mountford (1976:298) where again the seeds themselves were not involved.

229 Gilmore 1934:221.

230 Duncan-Kemp 1934:147.

231 Kimber 1984:16.

232 McBryde 1987:259-61.
Trade in seeds and fruits was also reportedly undertaken by the *Mardu* people of the north central Western Desert (Walsh 1990:35).

233 Mathews (1909a:497-8). He erroneously called it *nardoo*.

234 Harlan 1992b:31-5.

235 Reuther 1981:X,17/p. 26.

236 Reuther 1981:VII,218/1352;VII,226/1415;VII,328/2271;X,17/p. 26;X,17/p. 27.

to ensure "they continually had food".[237] There are also numerous references to storage, including the digging of "a seed-grain pit".[238]

The planting or sowing of seeds seems to have involved three different strategies. With *Markanjankula*, seeds were sown in a locale subject to flooding. This is a form of decrue planting. Smith alternatively indicates seeds were planted in a suitable spot so that as the "first rain comes ..... it will grow then."[239] Thirdly, seeds were broadcast, according to Horne and Aiston, when "a shower of rain has fallen or at any rate seems imminent."

Elements of Smith's narrative indicate that planting was quite deliberate, "we got him growing, that grass," that the planting was translocational, in locations outside the natural distribution, "different sort of country," and that particular locations were preferred, "good camp ... or might be a good soak or something, good sandhill."[240] If read in conjunction with the *Markanjankula* account, along with Horne and Aiston's information that the *wirra* was planted on the sandhills and *yaua* on the watercourses,[241] it could be concluded that planting, principally of *yaua*, *ngardu* and *Panicum* grass species, was on the sandhills, floodout zones, claypans, around springs and perhaps at the smaller lakes where more permanent settlements were established. Not surprisingly, these are the very types of locations where Bar-Yosef and Meadows claim cultivation first commenced in the Near East, "in well-watered areas such as lakeshores and alluvial fans or land near perennial springs."[242]

Numerous reports of large areas of grass seeds being harvested in the Corners Region have already been cited, although with little indication of the types of locations where this was taking place. Whether this was a circumstance where the harvesters were simply taking advantage of a more extensive natural distribution, or one where they were reaping the results of extensive planting, is a difficult question to answer. Whatever the case, the fact remains that in many of the instances where the location of such plants was specified, these locations corresponded to those where it has been deduced planting may have been occurring. For example, recall Poole's description of his encounter with 60 people in north west New South Wales, where he had observed that they were harvesting "seeds of a kind of rice (*Panicum effusum*) which grew abundantly on the flooded lands near the creeks." On Fromes Creek, in the very corner of north west New South Wales, Sturt mentioned another encounter in 1845 where "there was alluvial soil mixed with sand, and they had an abundance of grass upon them, the seed having been collected by the natives for food"[243]

Mitchell, and Newland, commented in a similar vein, that "grass is only found on the banks of the [Darling] river ... enriched by the floods, it is every where productive of grass"[244] Riding "over flooded lands of somewhat sandy soil," around Evelyn Creek/*Wanjiwalku* territory Sturt, as before, noticed that they were "covered with different kinds of grass, of which large heaps that had been threshed by the natives were piled up like haycocks."[245] Exploring on the Warrego River in south west Queensland in October 1847, E. B. Kennedy wrote of "luxuriant pasture" and that the "grass has the appearance of a wheat crop so thick and high it has grown."[246]

Earlier, attention was drawn to the "nardoo flats" in such places as the Coopers Creek the lakes of the Strzelecki Desert and Mulligan River, exemplified by Howitt's observation that in "many places, miles of the clay flats are thickly sprinkled with the dry [*ngardu*] seed." This observation is indeed a curious one because, according to Reuther, "it dries on the bush in pods, and thereby remains preserved,"[247] an observation supported by other observers and expert opinion.[248] Moreover, there is ample corroboration of such associations and the extent of the phenomenon. Echoing this McKinlay, in passing "a large flooded flat" along Coopers Creek, averred that it must have been "recently been under water, with a great abundance of clover [*ngardu*[249]] and grass as far as the eye could see."[250] Near Lake Kadhi-Baerri he also noted "plenty of clover and grass the whole distance travelled, about eighteen miles [28.8 kms]."[251] In his posthumously published account Wills had recorded on 27 May 1861:

*"I passed some blacks on a flat collecting nardoo seed. Never saw such abundance of the seed before. The ground was quite black with it."*[252]

Many similar comments were made by the explorers criss-

237 Reuther 1981:X,17/p. 27.
238 Reuther 1981:VI,17/27;VI,17/152[5];VI,17/209;VI,17/883,8;VI,17/982[4];VI,17/1141[1].
239 Kimber 1984:16.
240 Kimber 1984:16.
241 Horne and Aiston 1924:133-4.
242 Bar-Yosef and Meadows 1995:49.
243 Sturt 1849:1:284-5.

244 Mitchell 1839:1:302; Newland 1890:22.
245 Sturt 1849:1:294.
246 Kennedy 1852:260.
247 Reuther 1981:II,247/1122.
248 e.g. Spencer 1918-19:170-1; Basedow 1925:150.
This is confirmed by expert opinion (Pers. Comm. Murray Fagg, Centre for Plant Biodiversity Research, Australian National Botanic Gardens - 5,8 April 2002).
249 McKinlay uses the term "clover." The ngardu plant bears a passing resemblance to clover and other explorers in the area at the time use the term although they were clearly referring to ngardu. John King, the sole survivor of the ill-fated attempt by Burke and Wills to cross Australia recalled :
"I saw a plant growing which I took to be clover, and, on looking closer, saw the seed, and called out that I had found the *nardoo*." (King 1861:3)
Wright, leader of one of the support parties for the Burke and Wills expedition was even more specific :
"there was an abundance of marsilice a plant creeping close to the ground, with leaves not unlike clover, and bearing a seed largely used by the natives as food." (Wright 1861-2:82)
250 McKinlay 1861:15.
251 McKinlay 1861:11.
252 Wills 1861:29.

crossing the region at this time.[253] Perhaps one of the more revealing of these came from McKinlay who concluded:

"*The whole country looks as if it had been carefully ploughed, harrowed and finally rolled, the farmer having omitted the seed.*"[254]

This is quite possibly the result of extensive harvesting of "hectares" of *yaua* which leaves the area like "a ploughed field".[255] Of course, as the preceding evidence of planting indicates, these "farmers" certainly may not have omitted the seeds. It is not surprising, then, that Robert Etheridge should claim in 1894, when describing an implement thought to be a hoe, that the belief Aboriginal Australians were "devoid of any knowledge of husbandry" was a "mistake", one simply based on prejudice.[256]

***Incipient or Actual Agriculture?***

Returning again to the issues raised in the beginning of this chapter, regarding the nature of developments in the Corners Region, it will be recalled that the common features of an incipient agricultural society were adduced to be:

1. availability of domesticable plant species,
2. highly seasonal availability of such species,
3. systematic harvesting and processing of these species,
4. systematic preservation and storage of foods produced,
5. availability of other rich food sources.

and that these appeared to be correlated sociodemographic and socioeconomic factors such as:

1. higher levels of sedentism,
2. relatively higher population densities
3. regional economic integration (e.g. regional trade in food)

Given the evidence presented thus far, a case could be made that it was not just incipient agriculture emerging in the Corners Region but something more. But rather than rush to judgement at this point a summary of the evidence pertaining to the ecological factors and food procurement and preservation practices, and further analysis of the sociodemographic and socioeconomic, will be undertaken for the sake of completeness, and to enable a fair comparison to be made. So, in specifically addressing these features and factors, it would seem from the evidence presented so far that undoubtedly a number of plant species were heavily exploited in the Corners Region, *Panicum* grasses, *ngardu*, *yaua* and *munyeroo* being the principal ones. These were indeed systematically harvested and processed. The question of their domesticability cannot be directly determined but it would seem that most plants become "domesticated" if they are being repeatedly planted and harvested on a significant scale. By taking control, through planting, of a plant's life-cycle a new selection regime based on human preferences is established which usually, though not always, produces genetic and morphological changes in the subject species. And as has been shown, it would appear that *Panicum*, *yaua* and *ngardu* were being planted, the only issues being, to what extent and whether this would lead to the domestication of these plants, a matter which will be addressed in due course.

The evidence that has been presented also shows in seasonal terms the region was highly variable, reliant mainly upon unpredictable summer rains and floods. Furthermore, there are indications that storage was an extensive, consistent and integral feature of the food procurement strategies of populations in at least some parts of the Corners Region. Significant quantities were set aside in some instances, and storage facilities, such as skin bags, sun-dried clay containers and large wooden dishes (*pirrhas*/coolamons), were utilised, with grain pits and purposive structures, as described by Ashwin and Sturt, quite possibly being employed as well. There can be little question that other rich food resources were present and heavily exploited. Fish, waterbirds, molluscs, gum and smaller marsupials were some of the other foods present in very large quantities and procured in enormous amounts, often employing equipment and facilities such as nets and fish traps produced through considerable labour investments. Furthermore, indications of unexpectedly large populations, concentrated along rivers and creeks, and floodout zones, waterholes and lagoons, may well fall within the rubric of higher population densities.

Other lines of evidence put forward earlier showed that there was trade in food, *ngardu*, *Panicum* seeds and fish flour being examples, in the Corners Region. Paralleling the exchange in food was an extensive trade in another consumable, *pitjuri*, the plant product containing nicotine related compounds made from the dried stems and leaves of a bush (*Duboisia hopwoodii*) found in south west Queensland. Production was centred on the Mulligan-Georgina Rivers district but it was a highly significant element in transactions in central Australia, it being described by Duncan-Kemp as the "gold standard" of the region.[257] The *pitjuri* trade extended across almost the whole width of the continent, from the rivers flowing into the Gulf of Carpentaria to Lake Alexandrina, the outflow of the Murray River on the south coast.[258] If ethnographic

253 e.g. Mitchell 1848:60-1; McKinlay 1861:22,32-3,36; Brahe 1861:89; King 1861:4; Wills 1861:16,29; Wright 1861-2:81-2,88.
Wills also appears to indicate there were extensive areas of *munyeroo* growing along the Barcoo (Wills 1861:23-5).
254 McKinlay 1861:50.
Interestingly Mitchell (1848:274) came across patches of up to "two acres" which "resembled ground broken up by the hoe" on the Belyando River in 1846, but he was unable to ascertain for "what purpose" this was done.
255 Latz 1995:158.
256 Etheridge 1894:110.
257 Duncan-Kemp 1968:185.
258 Bedford 1887:110; Roth 1897:100; Berndt, Berndt and Stanton 1993: 117; Tonkinson 1993:xxiv.

accounts are taken as a guide the total distributional area where *pitjuri* was bartered was in the order of 1.4 million square kilometres.[259] Obviously, large amounts were being harvested,[260] but its real significance can be seen in the fact that an annual trading cycle, beginning in March of each year, was reputedly triggered by the *pitjuri* harvest. Trading parties came from hundreds of kilometres away, taking back hundreds of kilograms of *pitjuri*.[261] Though information is scanty, individual men and women were seen at times carrying from 13.6 to 31.8 kg of *pitjuri*.[262] At least 12 ceremonial exchange centres of varying importance have been identified in the region – Carrandotta, Roxborough Downs, Boulia, Carlo, Bedourie, Annandale [Kati-thari], Peecheringa, Goyders Lagoon [Goondaritchina], Birdsville, Innamincka, Kopperamanna, and Anna Creek.[263] Such was the demand for *pitjuri* that it was commonly reported "local Blacks will give anything they possess" to obtain this commodity.[264] On one occasion up to 500 people were apparently seen waiting at Goyders Lagoon for the *pitjuri* to arrive.[265] Thus the *pitjuri* trade, if viewed in conjunction with the evidence of exchanges involving food items, points unequivocally to some degree of economic articulation in this region.

The final factor under consideration, the degree of sedentism manifested in eastern central Australia has an important bearing not only on the question of whether proto-agriculture was evident in this part of Australia, but the presence of rudimentary agriculture there as well. To arrive at any conclusion regarding this requires a closer examination of that question and some related issues. This begins with the most commonly employed framework classifying forms of mobility and sedentism, developed over 30 years ago by Murdock and his collaborators, which posited five categories: fully nomadic, semi-nomadic, semi-sedentary, sedentary but impermanent, and sedentary and permanent.[266] Elsewhere I have argued that this and other such schema are completely inadequate in describing situations where sedentism is evolving and where there are subtle but important distinctions in the degree of sedentism.[267] Instead, building on studies of the *Nhanda* of Western Australia and the traditional Aboriginal culture in south west Victoria,[268] as well as the research presented here, the concepts of seasonal, multi-seasonal, seasonally mobile, sequential and partial sedentism will be employed as an alternative to Murdock *et al.*'s "semi-nomadic" and "semi-sedentary" categories.

Seasonal sedentism, as the name implies, involves populations remaining in the same location for a single season or substantial part of a single season. It is a common adaptation, often found in otherwise quite mobile hunter-gatherer groups who live(d) in climatic zones where there is one harsh season each year. In Australia this type of sedentism was seen on the west coast of Tasmania, noted for its cold, wild winter weather, where the *Needwonne* lived in winter camps composed of "commodious ... weather-proof", "semi-circular dome" shaped dwellings "from 10 to 12 feet in diameter and eight feet in height."[269] At the other end of the climatic scale the *Kukatja* of the Great Sandy Desert in north western Australia retreated to a waterhole or reliable water source for a period in the latter part of the dry season, food resources having been conserved in that area, to await the rains.[270]

Multi-seasonal sedentism can be seen as an obvious extension of seasonal sedentism, except that the sedentary period extends over successive seasons. This seemingly arises not from climatic exigencies but because the relevant groups either have rich, often multiple, local food resources and/or possess the technical capability enabling higher production and extractive efficiency of existing natural resources. Late Natufian and early Neolithic settlements in south west Asia are believed to have exhibited this form of sedentism, implying that multi-season sedentism may be an indicator of the commencement of dedicated food production strategies.[271]

The "seasonally mobile" category describes groups who are almost fully sedentary but who may became mobile for a particular season or limited period of time. "Sequential

[259] Calculation based on Hodgkinson 1877:214; Cox 1880:635; Bedford 1887:110; Roth 1897:51,100,134-5; 1901:31; Bruce 1902:83-4; Eylmann 1908:305-6; Bennett 1927a:410; Aiston 1937:376; Pearson 1949:203; Duncan-Kemp 1961:191; 1968:184-5; Watson 1983:29-30,34 7; Berndt, Berndt and Stanton 1993:117; Tonkinson 1993:xxiv.

[260] Duncan-Kemp (1968:185) stated that "weeks were spent in gathering the leaves," resulting in the accumulation of "great coolamons full."

[261] Howitt 1891:76; Watson 1983:39.

[262] Howitt 1891:76/10878 (70lbs); Duncan-Kemp 1961:191 (30lbs); Gason in Watson 1983:31 (70lbs)

[263] Hodgkinson 1877:214; Roth 1897:135; Howitt 1904:714-5; Horne and Aiston 1924:20,34,81-2,104; Duncan-Kemp 1934:208-10,214; Aiston 1937: 373-4; Harney 1950:42; Burchill 1960:36-7; Watson 1983:36- 7; McBryde 1987:260-1; Tibbett 2002.
Listing is from north to south.
Annandale is Annandale Homestead in south west Queensland (25° 23' S 138° 38' E) but may be Annandale Carrawallia Waterhole (24° 57' S 138° 43' E).
Anna Creek is the Anna Creek 80 kilometres west of Lake Eyre.

[264] Roth 1901:31. See also Bedford 1887:111; Bancroft 1879:7; Bruce 1902:84

[265] Aiston 1937:374.

[266] Murdock 1967:51; Murdock and Wilson 1972; Murdock and Provost 1973.

[267] Gerritsen 2000:37. See also Harris 1968:611-33.
Memmot's (2007:13,150) recent attempt to account for more permanent/sedentary settlements in Australia is also inadequate in my view as it misses a number of key examples and does not provide a considered theoretical analysis.

[268] Gerritsen 2000; 2002.

[269] Robinson 1971:144 (5 April 1830).

[270] Cane 1984:62; Veth 1987:104-5.
The same pattern is described See also Tonkinson (1991:40-1) in relation to the *Mardu* from the same region.
Similarly people in low-lying parts of Arnhem Land lived in stilted huts for five months of the year during the monsoonal season (McCarthy 1962:42).

[271] Noy, Schuldenrein and Tchernov 1980:69,82; Bar-Yosef et al. 1991: 423; Lieberman 1993:600,607,609; Habu 1996:38; Kuijt and Mahasneh 1998:160; Pringle 1998:1449; Cauvin 2002:2,15,17,19-20; Edwards and Hardy-Smith 2004:257-8.
Similarly Early Formative settlements in the Tehuacan Valley in Mexico are reported to have been occupied on a multi-seasonal basis (Stark 1981:354).

sedentism" applies to groups who move a few times each year to locations where they remain sedentary. This form of sedentism is most apparent where territories are highly circumscribed, with the inhabitants following an "embedded" round, moving at particular times of the year to fixed settlement sites.[272] In respect of the last type listed, "partial sedentism", in some instances it would appear a proportion, but not all, of the members of a particular group may be sedentary, with the rest of the group still engaging in some form of mobile subsistence activity. For example, women, children and the elderly may be permanently based at a village-like settlement while the men go away for a time on hunting expeditions or suchlike. This can sometimes be found in groups who undertake what has been called "logistic foraging".[273]

Having proposed this classificatory system to describe degrees of sedentism, what form, or forms, of sedentism were there in evidence in the Corners Region? Given that this is a large and highly varied region it would be unrealistic to expect that the sedentism of different groups was of a uniform character. Another difficulty relates to the heavy reliance on historical ethnographic evidence. In many instances this basically exploits an observational "window" where relatively intact traditional lives and settlements were being only briefly or incidentally described before they were profoundly and irreversibly altered by the enormous impact arising from European occupation. Howitt, for example, in describing the extent of these impacts, claimed that the *Paakantyi* population alone declined from 3000 to just 80 by 1884 as a consequence of the British colonists' incursions into their territory.[274] But significant European influences even appear to have preceded their actual physical presence on the "frontier".[275] Smallpox, in particular, had a devastating effect, especially in larger settlements and areas with higher population densities. Sturt, on the Darling in 1829, as mentioned before, noted "a violent cutaneous disease" raging in the district where he had encountered the apparently deserted village of 70 huts. This disease was "sweeping them off in great numbers" and as a consequence "the population had thinned."[276] Other early explorers and informants made similar comments,[277] and in referring to one of these epidemics (possibly the one witnessed by Sturt) pastoralist Bonney was informed by a *Paakantyi* local that one-third of the population had died at the time with the rest fleeing, leaving "the dead unburied."[278]

A further complication relates to a common custom when a death occurred. Some commentators asserted that the whole camp was abandoned if someone died, this custom being specifically ascribed by Gason to the *Diyari* (who also seemed to have destroyed the dwelling concerned), and by Bonney and Newland to the *Paakantyi*.[279] This could possibly explain a proportion of the unoccupied assemblages of seemingly permanent shelters reported by various witnesses. Sanitation may have also been a limiting factor in more sedentary groups. Although the *Wongkanguru* were quite fastidious according to Horne and Aiston,[280] when it became too much of a problem they simply shifted camp.[281] Such things need to be kept in mind, especially when explorers report finding a deserted settlement consisting of ostensibly permanent dwellings - is the absence of people due to some catastrophe or a sudden death, had the residents simply relocated to pursue a different subsistence activity, had they gone elsewhere to trade or to participate in ceremonies? Whatever the case, and in spite of these limitations and uncertainties, it is remarkable how often seemingly permanent dwellings were reported in the Corners Region. Already we have seen the comment by Lewis that the "village" he encountered on Warburton Creek was "the first place I ever saw natives return and live in the same wurlies, which are very large and well built." Typically Sturt, with an eye for detail, described them as being:

"*made of strong boughs with a thick coating of clay over leaves and grass. They were entirely impervious to wind and rain, and were really comfortable, being evidently erections of a permanent kind to which the inhabitants frequently returned.*[282]

Almost identical terms were used by a range of other amateur ethnographers - "beehives of sticks ... covered with tufts of grass or weeds, and finally earth or sand"[283], "dome-shaped circular erections built of logs, cane grass and mud"[284], "round with a dome shaped roof ... built of sticks, reeds and cane grass interwoven and plastered up with mud"[285], "huts ... in the shape of a round dish-cover, 6 or 7 feet high and will hold a family of twelve persons"[286],

---

272 Kelly 1992:57; Gerritsen 2000:40.

273 Binford 1980:10,17; Gerritsen 2000:40.

274 Howitt 1908:85.
Newland (1926:88-9) also notes a catastrophic decline in the *Paakantyi* population in the Post-Contact Period.

275 Lourandos 1997:324.

276 Sturt 1833:93,105.

277 Mitchell 1839:1:307. See also Kimber 1988; 1996:93-4.
Gregory (1858:10) made the curious observation on Coopers Creek in 1858 that "the numbers of burial mounds which have been constructed about two years ago greatly exceeded the proportion of deaths which could possibly have occurred in an ordinary season of mortality."
He ascribed this to drought but an epidemic could have been an equally plausible cause.

278 Bonney 1884:123-4.
Parker (1905:39) too speaks of a tradition of what appears to have been smallpox (called *dunnerh-dunnerh*) which "decimated" the *Yuwaaliyaay*, so that, "terror-stricken," they too "did not stay to bury their dead".

279 Gason 1895:170; See also Smyth 1878:1:119 and Howitt 1891:89; Bonney 1884:125; Newland 1926:105. See also Schulze (1891:231) (Lower Southern Arrente).

280 Systematic refuse removal appears to be a Neolithic trait (Hardy- Smith and Edwards 2004).

281 Horne and Aiston 1924:18.

282 Sturt 1849:2:139.

283 Howitt in Smyth 1878:2:302.

284 Wells 1894:517.

285 Hill 1901:25.

286 Fraser 1892:49.

"large and well built"[287] are just a few of the offerings.[288] A number of these commentators stated that these were "winter" habitations, or used when the weather was "cold",[289] evidently implying the usual occupants were more mobile in summer when temporary shelters and windbreaks were often used.[290] Others maintained they were also used to "keep ... the mosquitoes out"[291] or "used during the rainy season."[292] However Sturt, it will be recalled, in referring to the "interior natives", noted their "erections of a permanent kind to which the inhabitants frequently returned." Lewis concurred with Sturt, asserting that the *Ngameni*, the eastern neighbours of the *Wongkanguru*, "return and live in the same wurlies."[293] Balancing this is the assertion by Hill on the central Diamantina that the *Maiawali*, despite the fact that they possessed "dome shaped" dwellings "plastered up with mud," were constantly on the move, "rarely staying in one place."[294] Though in general terms the *Wongkanguru* reputedly had "standing camps at all the waters in their country,"[295] and in summer used a large common brush shelter called a *wilpie* during the daytime, the more permanent dome shaped abodes, the *poonga*, were still the principal residence.[296]

As seen above, in some instances these seemingly permanent dome or cupola-shaped dwellings were described as "winter" habitations. This appears to stem from an assumption on the part of some as they passed by (e.g. Stuart)[297] that such structures could only be built to protect the inhabitants from cold or inclement weather. But the evidence indicates they were also occupied in the "rainy season" or when mosquitoes were troublesome, both characteristic features of the summer in these parts of Australia.[298] Furthermore, it just raises the question of why people in this region built these types of dwellings when those in much colder or wetter parts of Australia did not? And some writers, with the opportunity for extended observations, provided an alternative view. F. H. Wells, in alluding to "wurleys at a permanent camp," in south west Queensland noticed that in "the cold weather fires are kept constantly burning inside. During the summer months the natives keep inside their wurleys in the daytime, thus avoiding the intense heat, but sleep outside at night."[299] Similarly, Duncan-Kemp, also referring to these type of residences in south west Queensland, spoke of them being "snug and weatherproof, they are built to withstand gales in summer and to keep out chill winds of winter ... the entrance ... in winter is closed by a specially fitted bark door."[300] She, furthermore, mentions in passing "thirty humpies, dwellings of the spoonbill tribe who lived permanently on this waterhole."[301]

Commentary by the missionary Reuther supports this notion, that some groups at least were in constant occupation of their settlements. He spoke of the tribe "named the Nguraworla because the people do not change over their wurleys, even when someone has died in [their] hut."[302] Here he was referring to the *Ngurawola* of the Strzelecki Desert, the people who Davis and McKinlay constantly reported in the hundreds around their lakes and waterholes. But the *Diyari* to the south of them clearly did not stay in the same site on an ongoing basis. Both King, the only survivor from Burke and Wills' disastrous attempt to cross the continent, and Wills before he died, lived for a time with, or in close proximity to, the *Diyari*. In their accounts they mention a number of occasions in which the *Diyari* they associated with relocated.[303]

Yet another residential pattern, indicative of partial sedentism, appears to have been documented in the Corners Region. At Brewarrina, site of the extensive fish trap system discussed earlier, two different authorities claimed that a proportion of the population were in constant residence there. According to Palmer there were "generally a few blacks camped on either side of the river in small camps ... to watch their tribal interest and repair the walls [of the traps] when requisite," and R. H. Mathews also commented that, "some straggling members of the tribe located about the fishing grounds all the year round."[304] Other examples of partial sedentism were in evidence among the *Wongkanguru* too, Horne and Aiston maintaining that "all

287 Lewis 1875a:3.
288 See Tolcher 2003:14-5 for a detailed description of the materials used and the method of construction.
289 Stuart 1864:82,144; Howitt in Smyth 1878:2:302; Machattie in Curr 1886:2:366; Heaghney in Curr 1886:2;375; Roth 1897:106.
290 Stuart 1864:144; Howitt in Smyth 1878:2:302; Andrews 1879:85;
291 Heaghney in Curr 1886:2:375; Also Machattie in Curr 1886:2:366.
292 Sturt 1849:1:254.
There seems to have been some confusion about when these substantial structures were occupied. Roth (1910:63-4) states there were two types, one used when cold, the other when the weather was wet. But it is not clear why two different types of the same type of structure were constructed, particularly as the area referred to received rain in both summer and winter (Memmot 2007:140).
293 Lewis 1875a:3.
294 Hill 1901:24-5
Hill was quite possibly confusing the traditional settlement pattern with the post-contact pattern. Aboriginal groups were often forced to become more mobile following occupation of the lands as key resources (e.g. waterholes) were taken over and they were increasingly marginalised. In these situation they usually reverted to more temporary shelters, for obvious reasons (Gerritsen 2000:3-6,20-22,36).
295 Horne and Aiston 1924:54.
296 Horne and Aiston 1924:17-8,20.
297 Stuart 1864:82,144.
298 Mosquitoes - Spencer and Gillen 1912:1:31; Dr Rolf Gerritsen (Pers. Comm.).
Wimberley (1899:70), in describing the stone and clay structures built by the Yauraworka, implies frequent or sustained residence, "They hold out the rain well, and are very warm in winter"

299 Wells 1894:517.
300 Duncan-Kemp 1952:9.
301 Duncan-Kemp 1968:189.
302 Reuther 1981:VII,226/1411. Although not explicitly stated Horne and Aiston (1924:152-3) appear to indicate that the *Wongkanguru* did not abandon camp upon the death of one of their number but simply destroyed their wurley, the wood being placed on the gravemound so that the spirit of the departed could use it to build themselves a new *poonga* and make a fire.
The *Diyari* abandoned the camp of the deceased but did not, according to Berndt and Vogelsang, destroy the structures (Berndt and Vogelsang 1939:169)
303 King 1861:5; Wills 1861:29.
304 Palmer 1893:1:97; Mathews 1903:153. See also Dargin 1976:32.

the old blacks stay in permanent camps near water."[305] Later Aiston pointed to a further instance where "the oldest man [*pinnaru*] of each moiety usually stayed at the home place," guarding their sacred *moora* totem objects.[306] Mitchell made a comparable judgement regarding partial sedentism, noting that "some of the tribes on the Darling are not migratory, but remain; in part at least, the gins [women] and children possibly, at some particular portion of the river."[307]

Undoubtedly the historical ethnographic evidence suggests that various forms and degrees of sedentism were in evidence in parts of the Corners Region. Elsewhere, cross-cultural studies, as well as archaeological evidence, have been used to establish that particular groups exhibited a high degree, if not full, sedentism. Correlates or indicators of sedentism that have been proposed as a result of these studies have included the presence at one site of the remains of flora and fauna from different seasons of the year, the presence of commensals (species such as mice or house wrens normally found in or around permanent, sedentary habitations), settlement size, a long and intensive occupation history, permanency of dwellings, significant amounts of labour invested in the construction of dwellings, the morphology of dwellings, the presence of partitioning or multiple rooms in the dwellings, separation distance between dwellings being of smaller dimensions than the dwellings themselves, artefact density, debris distribution, organised refuse disposal, specialised lithic assemblages, expedient core technology, thick cultural deposits, pottery, heavy artefacts, storage facilities, communal structures (such as kivas and mounds), special purpose communal extractive facilities (such as fish traps), cemeteries, and bone strength and form of the long bones diaphyses in skeletal material.[308] As so few in-depth archaeological investigations have been carried out in the Corners Region in respect to these issues, little of this type of evidence can be drawn upon to verify any specific propositions. Nevertheless, the ethnographic evidence is capable of shedding light on at least some of these factors.

What could be construed as sizeable settlements were reported by a number of the European interlopers. Sturt's village of 70 seemingly permanent huts capable of housing up to 1000 people is one example, as was his "Village in the Interior". Lewis, it will be remembered, had seen a "village"-like settlement near Kallakoopah Creek of 350 souls, while Duncan-Kemp also mentioned "thirty humpies, dwellings of the spoonbill tribe who lived permanently on this waterhole" in one instance and, "a cluster of sixty grass-and-sand ... dwellings [which] housed two hundred or more people" in another. By way of comparison, Pre-Pottery Neolithic B (PPNB) villages in the Levant, covered areas ranging from 0.5 to 12 hectares, having populations calculated to be from 100 to 500.[309] PPNB Jericho for example, described as a "large village", covered 2.5 hectares, equating to a population of 230.[310] Yet in most of these early Neolithic villages full sedentism is only assumed,[311] otherwise it is believed to have been multi-seasonal, "most of the year" or "during all seasons of the year."[312]

Dwelling permanency and the amount of labour invested in their construction are interrelated factors in terms of determining the degree of sedentism. As Rafferty put it, the "expectation that sedentary people would build more substantial houses than people who are non-sedentary seems to be more or less correct both archaeologically and ethnographically,"[313] and Kelly, "Cross-cultural studies demonstrated that investing labour in houses is related to low (or no) residential mobility."[314] While there is no measure given in such studies of what constitutes "more substantial," or what is sufficient labour investment to constitute a sedentary habitation, clearly there is some sort of correlation. In general terms, more mobile Aboriginal groups in other parts of Australia built temporary structures, with the work usually being carried out by women, who completed the task in the space of a half to one hour.[315] Contrast this with one of the more detailed descriptions of the building of the seemingly permanent dome-shaped dwellings, called *unnakudyi* by the *Pitta-Pitta*. This involved digging an elliptical pit 45 cm deep, constructing a framework of branches, wedging in wet grass and then coating it all with mud that hardened. A ring of mud 30 cm thick was then constructed around the doorway.[316] According to Roth, the witness in this instance, it entailed "no inconsiderable amount of time, toil, and patience in their making."[317] Much the same could have been said of

[305] Horne and Aiston 1924:153.
Ryan (1969:37) describes a settlement pattern by the *Yantrawantra* [sic] and others that appears to have been sequentially sedentary, though possibly describing a post-contact phenomena, but also mentions (1969:66) "a very old man ... who lived on the river [Pituri Creek] in a mud hut", indicating partial sedentism.
[306] Aiston 1935:5.
[307] Mitchell 1839:1:294.
Jones (1979:133-4), in one of the few studies of the traditional Aboriginal culture in the Cooper Basin, also arrived at the conclusion that a proportion of population was camped permanently on the river frontage.
[308] Flannery 1972; Clemens 1979:72-3; Andresen et al. 1981:33-4; Stark 1981:357; Rafferty 1985:115,129-31,135-6; Parry and Kelly 1987:285,297,300-2; Keeley 1988; Pardoe 1988:1-2; Henry 1989:38-9; Kent 1989:2; Bar-Yosef et al. 1991:423; Tchernov 1991:315; Kelly 1992:56-7; Sassaman 1992:250,256; Larsen and Ruff 1994:28; Habu 1996; Haaland 1997:376; Adams and Bohrer 1998; Bar-Yosef 1998a:168; 1998b:1-2; Lieberman 1998; Rocek 1998:207-9,211; Russo 1998; Valla 1998; Varien 1999:6-12,33,39,62-88; Gallivan 2002; Edwards and Hardy-Smith 2004.
For a critique of some of these methodologies see Edwards 1989.

[309] Bar-Yosef and Meadows 1995:76; Bar-Yosef 1998a:168.
Calculations based on formula developed by Hassan (1981:67).
[310] Bar-Yosef and Belfer-Cohen 1992:34; Bar-Yosef and Meadows 1995:76; Bar-Yosef 1998a:168,170.
Calculation based on Hassan's formula.
[311] Moore 1983:109; Bar-Yosef and Meadows 1995:62; Habu 1996:39; Bellwood 2005: 23.
[312] Bar-Yosef et al. 1991:423; Lieberman 1993:607,609; Lieberman 1998:76.
[313] Rafferty 1985:129.
[314] Kelly 1992:57.
[315] No systematic studies of the building of shelters and the division of labour involved have been undertaken in Australia although many examples can, of course, be identified (e.g. Schulze 1891:230; Fraser 1892:49; Ferguson 1987:128).
[316] Roth 1897:106.
[317] Roth 1910b:64.

the *Yauraworka* residences in the very north east of South Australia. These were constructed of - "flat slabs and stones ... The supporting structure ... particularly strong to carry the weight of the stones. The crevices are filled with clay to render them watertight, and earth is banked up both inside and outside."[318] Furthermore, the fact that the *Wongkanguru*, who possessed "very well built" structures of the *unnakudyi* type, had male hereditary specialist hut builders provides additional support for the notion that there was a qualitatively different character to the habitations used by some of the Corners Region populations.[319] Reinforcing this proposition are accounts of what appear to have been quite large domiciles in the area, such as the one encountered by Goyder south west of Lake Blanche in 1857, which he stated "was constructed in a similar manner to those described by Captain Sturt, ... very warm and comfortable, the largest capable of holding thirty to forty people" It takes little imagination to realise the degree of labour necessary to build a structure of this form and capacity.

The morphology of habitations and the presence of structural differentiation within structures, in the form of partitioning or multiple rooms, has been linked to higher degrees of sedentism. In the case of the former factor, initially residences of sedentary peoples are usually "rounded" or "circular", the dome shaped dwellings so frequently noted in the Corners Region being typical examples, but become "rectangular" as time goes by.[320] This is evident in the Levant where circular structures of the PPNA were replaced by rectangular ones in the PPNB.[321] However, based on available information, none of the abodes recorded in the Corners Region approached a rectangular morphology.[322] Regarding internal structural differentiation, it would seem the situation was a little different. Again most PPNA dwellings were single room structures, though some had common walls and the first partitioning appeared at that time, becoming common in the PPNB.[323] Considering the Corners Region it would appear that at least some of the "more permanent structures" capable of holding "a family of twelve persons" incorporated, if Fraser is to be believed, "a portion that may be partitioned off as a separate room."[324] What may have been an even more elaborate type of construction was seen by Sturt at a deserted village, probably in the Strzelecki Desert. There were, he said, "two or three huts in the village of large size, to each of which two smaller ones were attached, opening into its main apartment."[325]

Other indicators of sedentism, such as storage facilities, communal extractive facilities like fish traps and dams, and cemeteries have already been discussed. Once more the evidence points toward a high level of sedentism in at least some parts of the Corners Region. In addition, the much larger dwellings housing from 30 to 40 people, seen by Lindsay, Conrick and Goyder, because they obviously were something other than a family residence, could be considered to fall under the rubric of communal structures.

Expedient core technology, the practice of flaking from cores to suit the needs of the moment rather than relying on "curated", specialised stone tools, cannot be used as a specific diagnostic criterion for sedentism, or the commencement of conscious food production, but only as a general indicator.[326] The only detailed archaeological study of large lithic assemblages in the region, at Stud Creek in north west New South Wales,[327] could, however, be interpreted in terms of the systematic employment of expedient core technology, providing further confirmation of the proposition. Heavy artefacts, another general indicator of sedentism, can encompass a range of implement types. Rock mortars and pestles, as well as large grindstones, characteristic features of early Neolithic sites in south west Asia, are typical examples.[328] In Australia grindstones were fairly common, with some degree of specialisation, a particular type, for example, being used for grass seeds, another for *ngardu*.[329] Large grindstones, some up to 90 cm long and weighing 54 kg were usually left at locations

[318] Basedow 1925:103.
Such structures appear to have been first reported by Wimberley in 1899, described as "having a wooden frame covered with stones and about a foot through" (Wimberley 1899:70). See also Kenyon (1930-1:72) for further details. This is also supported by oral tradition (Tolcher 2003:15).

[319] Horne and Aiston 1924:19-20.

[320] Rafferty 1985:130.
There are exceptions to this. The first Early Formative dwellings in Mesoamerica and Early Neolithic dwellings in north east China were rectangular, although it has been suggested that this was because they were built on sloping ground (Whalen 1988:252-3; Shelach 2000:395-6,398,400,404).

[321] Noy, Schuldenrein and Tchernov 1980:69; Bar-Yosef and Meadows 1995:76; Cauvin 2002:213.

[322] The octagonal structure drawn by Mitchell (see Figure 11) or the "isosceles" form described by Jung (1884:117) could be seen as transitional forms in a trajectory to more rectangular dwellings. Also two rectangular stone dwellings in a group of 17 "stone walled" huts at Marion Downs on the Georgina River have been documented in a heritage survey (Davidson, Burke and Mitchell 1989:17-18,93-95) but there was uncertainty as to whether they were of pre- or post-contact construction.

[323] Kenyon 1957:53,71; Stekelis and Yizraely 1963:5; Kenyon 1981:3: 52,61-2; Henry 1989:226; Bar-Yosef and Meadows 1995:76-7; Cauvin 2002:39-41.

[324] Fraser 1892:49. He also states that in building these "houses," "the men take their share of the work."

[325] Sturt 1849:1:387. Quite possibly the absence of inhabitants was a consequence of recent rains and a pulsation to the sandhills. See 1:339-40.
Jung (1884:118), who spent time in the Corners Region, refers to multi-room dwellings but does not specify where they were found.

[326] Parry and Kelly 1987:285,297,300-2; Torrence 1989; Sassaman 1992: 250,253-4.

[327] Holdaway, Fanning and Witter 2000:51,54-5.
Other sites at Cowarie on Derwent Creek, north eastern South Australia (Aiston 1921:5), on the Paroo River (Tillenbury Station) (Towle 1944), the Coopers Creek floodplain west of Lake Killalpanina (Gregory 1906:76-7), west of the Darling River (Dow 1938b:101), Mt Arrowsmith Creek, 3.2 kilometres south east of Mt Arrowsmith Station, north west New South Wales (Dow 1938c:130) and Lake Idamea (Davidson 1983:30) also possibly exhibit expedient core technology. Judging from these informants comments on the density and variety of artefacts, and other indications (e.g. grinding stones, mounds), it would appear that there had been intensive utilisation of these sites formerly.

[328] Stekelis and Yizraely 1963:8,12; Bar-Yosef 1983:15,19; McCorriston and Hole 1991:49,58; Bar-Yosef 1998a:170.

[329] Pardoe 2003:48.

in which the inhabitants were either permanently based or frequently returned.[330] Their distribution was, not surprisingly, largely contiguous with the *panara* grasslands, of which the Corners Region was a significant part.[331] Consequently, they were employed on the western flanks of the Great Dividing Range (inland New South Wales), throughout north west central and south west Queensland, arid central Australia, the Darling and "other rivers" such as Coopers Creek, but not reaching the Murray River to the south.[332] A particular "pear-shaped" type, found in "western New South Wales, western Queensland and central Australia" [i.e. the Corners Region] has been described as "the finest examples of these artefacts in Australia."[333] Mortars and pestles too were in common use across the continent, with larger ones certainly present in the Corners Region.[334]

In summary it can be seen that particular lines of evidence provide strong indications of a high level of sedentism in parts of the Corners Region. Specifically, these include settlement size as well as permanent habitations incorporating internal partitioning or rooms in some instances and involving considerable labour investment. The presence of storage facilities, cemeteries and occupational specialisation in building shelters provides further support for this conclusion. A range of other indicators, while not diagnostically precise, nevertheless show a high level of congruency with such an inference.

Given this and the preceding body of evidence, what conclusions can be made regarding the degree of sedentism in the Corners Region? Firstly, if the evidence is taken at face value, "sequential sedentism" appears to have been practised by the majority of the *Wongkanguru*, with "standing camps at all the waters." This may explain some instances where "abandoned" settlements were noticed. However, an element of "partial sedentism" is also evident, the old and sick staying at permanent camps. The *Diyari*, too, may have been sequentially sedentary, as well as incorporating partial sedentism, since the custodians of the sacred *moora* stones "usually stayed at the home [keeping] place." Multi-seasonal sedentism was possibly practised by the *Ngameni*, returning to live in the same wurleys, although this is by no means certain. Adjacent to the swamps north of the Namoi River, the north western *Gamilaroi*, Cobb informs us, "formed their camps, lived, hunted, and made their implements while the water lasted, which was generally throughout the winter and early summer, before retiring to the river."[335] They may also have been engaged in multi-season sedentism. Reports of "permanent camps," such as the *Karuwali* living permanently on a particular lagoon, or of people in south west Queensland who occupied their weatherproof wurleys in both summer and winter, point to situations approaching full sedentism. If such people were not fully sedentary they probably only left for brief periods, when rains or floods produced temporary abundances elsewhere, in the "pulsation" pattern described earlier. They were therefore, if not fully sedentary, sedentary to the extent of being only "seasonally mobile".

On balance it would seem reasonable to conclude that an incipient/proto-agriculture culture was clearly manifesting itself in the Corners Region. But it would appear that developments went well beyond that, some groups in the Corners Region were in fact engaged in rudimentary agriculture. Consider here the basic definition of agriculture, as proposed in Chapter 4:

*"propagation and harvesting of plants on a systematic basis, of sufficient extent to provide a significant proportion of the food supply for the people undertaking those activities. It should also involve a strategy to conserve the resultant output through storage or other means. Non-essential complementary features should include activities intended to maintain or enhance productivity (i.e. husbandry), and the expectation that the cultivar will have undergone some morphological change."*

There can be little doubt that plants were being harvested on a systematic basis in parts of the Corners Region, and that the resultant yield did provide a substantial proportion of the food supply for those engaged in that form of food procurement. The drought tolerance and dependability of millets under "poor conditions" made *P. decompositum* particularly suited to the region.[336] Storage, as documented above, was undoubtedly practised to some extent. Husbandry, in the form of ground preparation and basic flood irrigation, may also have been undertaken. There is no definite evidence of the plants in question having undergone morphological changes, the issue having never come under consideration up to this point in time. However, evidence will be presented in the next chapter indicating that plants, in particular *P. decompositum*, may indeed have been in the process of being domesticated. However, the critical question in regard to the status of agriculture in the Corners Region is: to what extent were the subject plants being propagated? Planting, if the evidence presented is accepted, was clearly taking place. Unfortunately direct evidence of its extent is quite limited, although a number of lines of evidence and inferences strongly suggest that it may have been quite extensive in some localities. In the following chapter a case will be made that provides further, strong, circumstantial evidence that in particular areas planting may well have been of significant extent.

If one then accepts at this juncture that agriculture was a feature in parts of the Corners Region it may be possible

[330] McCourt 1975:136; McCarthy 1976:59; Smith 1985:25-6; 1988:95; Gorecki, Grant and O'Connor 1997:141-5.
Idriess (1964:57) claims to have seen grindstones as large as wagon wheels in the Channel Country.
[331] Tindale 1974:99, Fig.31
[332] Smyth 1878:1:214,2:382-3; Roth 1897:104; Newland 1920-1:13-4; Tindale 1974:102; 1977:347; Lampert 1985:60; Lourandos 1997:188- 9.
[333] McCarthy 1976:60.
[334] McCarthy 1976:63,76,85
[335] Cobb 1933:164.
[336] Hancock 1992:193.

to determine if there was a "Centre" or a "Non-Centre". Harlan couched these concepts in terms of the problematic notion of "plant domestication", but the status of the plants being grown is either completely unknown or, as we shall see, possibly only in progress. But other types of evidence may provide pointers at least. Smith's evidence, along with other sources, indicates that planting was quite widespread. A pattern of sowing in single discrete area and radiating from there, as would be expected with a Centre, does not seem evident. The fact that a number of different plant species were involved is also consistent with a Non-Centre. If viewed as a cultural complex - where planting was taking place in conjunction with intensive harvesting, other forms of highly productive food procurement were being widely pursued, storage was being practised, and there was a high level of sedentism, indicated by "more substantial" dwellings and larger settlements, a broad but discontinuous pattern is evident. In fact the term "Oasis Agriculture" may actually be more apt than a "Non-Centre" to describe this pattern.

Examples of this culture complex can be found focussed on northern New South Wales, on the central Diamantina River, the Georgina-Mulligan Rivers area, the central Darling River and north west New South Wales, central Cooper's Creek, the north side of Lake Eyre, particularly around the Macumba River, Warburton and Kallakoopah Creeks, the lakes and floodout zones of the Strzelecki Desert, and, finally, possibly in the vicinity of Newcastle Waters and the Barkly Tableland of central and eastern Northern Territory. If "substantial" dwellings and grave mounds are taken as the principal guide then there appears to have been outliers around the gorges, creeks and floodouts of the Barrier Range of western New South Wales,[337] the Neales River and Hamilton Creek west and north west of Lake Eyre,[338] as well as south west of Lake Blanche.[339] In conclusion, it would appear that the agricultural developments in the Corners Region were more consistent with a Non-Centre than a Centre. But for my final words I turn to Thomas Worsnop. He was one of a number of 19th century compilers of information on Indigenous societies, drawing data from other sources as well as from personal observations and contacts, which he distilled in his book, *The Prehistoric Arts, Manufactures, Works, Weapons etc. of the Aborigines of Australia*, published in 1897. In this Worsnop includes a rather curious statement:

"*Some years ago there was a story current in the colonies that a white captive was held by an interior tribe, whose influence on them was great. He had taught them (so the story ran) to cultivate the nardoo plant and store the seed for future use; and their mills and grinding stones were said to be better made and larger, their huts were reported to be of permanent construction and commodious, and their implements and weapons more efficient ... No narrator of the story could give the locality.*"[340]

Of course, the notion that such developments were the work of a "white captive" is totally fanciful, an attempt, perhaps, to explain the unexpected. Consequently, if the arguments presented here, based as they are on only a proportion of the available evidence, are accepted, then Worsnop's story should not be seen as idiosyncratic but in fact an accurate portrayal of what were the traditional lifeways of the Indigenous inhabitants of parts of the "East Central Australian Non-Centre". However, developments in this part of Australia went beyond being simply an exemplar of early agriculture, as I shall endeavour to show in the next chapter. Indeed, I will argue that, without consciously realising it, the multifarious witnesses present in the region in the 19th and early 20th centuries were coincidentally encountering the "Australian Early Neolithic".

337 Sturt 1849:2:109; Gow 31 May 1861; Brock 1975:71,100-1.
338 Stuart 1864:82,144.
339 Goyder 1857:4
Stuart's journals provide an interesting guide to population "densities" to the south, west and north of Lake Eyre through Central Australia. There are frequent indications of significant numbers of people to the south, south west, west and north west of Lake Eyre. Through Central Australia he then went weeks with almost no contact or indications whatsoever until he reached Newcastle Waters when frequent indications resume. Compare Stuart 1864:10,16,18,77,82,85,88,99,106-7,113,131,143-4,149,154,216-7, 219-220,248-9,265,294,298-9,333,335 to Stuart 1864:168-214,221-230,269-293. Unfortunately the Central Australian area has become a major archetype for traditional arid and semi-arid zone cultures.

340 Worsnop 1897:82.

# Chapter 6.
# An 'Australian Early Neolithic'?

The case put forward in the last chapter, that there had been an agriculturally based culture complex in East Central Australia at the time the region was invaded by the agents of British colonialism, drew heavily upon the observations and experiences of a multiplicity of explorers and other eyewitnesses who had infiltrated the interior of Australia in the course of the 19th and early 20th centuries. Although some of these individuals were perceptive enough to make relatively dispassionate, detailed and informed observations, it is highly unlikely that any of them would have been able to describe what we now call a "neolithic" society. This is not surprising really, as very few individuals had a clear notion at that time what such a society might actually look like. It is only in recent decades that the concept has matured and sufficient evidence accumulated to allow a reasonably comprehensive description. If these observers did rely upon some conceptual framework to interpret what they saw it is likely to have been based on the prevailing paradigm, the idea that cultures went through particular stages, "culture-stages". These "culture-stages" were "savagery", "barbarism" and "civilisation", manifested archaeologically, as cultures progressed through three ages, as "Stone Age", "Bronze Age" and "Iron Age". Each of these "culture-stages" was supposedly a very specific entity, definable in terms of fairly fixed characteristics. A central feature of this schema was the recognition that cultures had changed over time, leading from "lower" to "higher" cultures and "higher civilisation". Cultures were then graded, from "lowest" to "highest". The trajectory followed by any culture in its "progress", as it embraced a "higher" culture, was framed in highly deterministic terms, cultures either moved from one very specific type of culture-stage to another, or not at all. According to some writers of the time certain cultures had even gone backwards, "degenerated" in the terminology then current. Allied to culture-stages concept was an explanation, based on "Social Darwinism", of why different peoples were at different stages. This "theory" was essentially racist, that "higher" stages of culture were reached because of those people's racial superiority, they were "fitter" in survival terms.[1] The corollary was that modern hunter-gatherers were stuck in, or had "degenerated" to, the "lowest" culture stage, "savagery", and having "Stone Age" technology were, therefore, fossil versions of the Paleolithic hunter-gatherers. Adherence to the idea that a certain progression must be followed and that modern hunter-gatherers were examples of "fossil" culture forms is known today as the "progressionist trap".

While the "culture-stages" models was eventually discredited, the three ages model, with modifications such as Paleolithic instead of "Stone Age", and Mesolithic for transitional forms, turned out to be conceptually useful in characterising the archaeological sequences found in Europe and the Near East. But rather different sequences occurred elsewhere, and as the archaeological evidence accumulated globally it was realised that all cultures, including hunter-gatherers, adapt and change through time, and continue to do so. There is no imperative that they must change, however, and there is no prescribed direction in which change takes place. Nevertheless, in a number of instances change did take place in prehistory, leading to cultural complexes such as the Neolithic, where planting and full-scale agriculture, and other modes of food production, ultimately became the predominant economic mode. As alluded to in the first chapter, there is no accepted explanation for why this happened, why it happened in some places and not in others, and why it happened in those places when it did.

The supposed lack of a cultural complex in Australia analogous to the Neolithic is the converse side to these issues. Because it has been assumed that agriculture in any form was absent in traditional Australian Aboriginal culture, the question of a whether a neolithic culture existed in any part of Australia has never been previously considered. Where the issue of the reputed lack of agriculture has come under scrutiny, however, the response by Australian scholars has often been formulated in terms of treating traditional Aboriginal culture in general, or Australia itself, as a "special case", as will become apparent in a later chapter. Consequently many of these scholars have seen it as unnecessary to account for this situation in terms other than some unique characteristic of Australia, or the culture of its Indigenous inhabitants. But these arguments are ill-founded because they are based upon a false assumption, as demonstrated in the preceding chapters. Beginning with the presumption that the case for indigenous Australian agriculture presented in those chapters is valid, I will be endeavour to further demonstrate that there were traditional

[1] Not surprisingly the apex of this cultural development was posited as being that found in nineteenth century industrial Europe, and the finer gradations designating the level of "civilisation" reached by any particular group was, by and large, how alike they were to Europeans in their physical appearance, society and culture.

lifeways and economies followed in parts of Australia which were akin to what we now call the Neolithic.

The Neolithic is a term usually applied to south west Asia as it was between 8000 and 11,000 years ago.[2] Analogs of the Neolithic have also been identified in North America (the Formative), Europe (European Neolithic), China (Chinese Neolithic) and elsewhere. The term itself, and the concept, was first devised in 1865 by Sir John Lubbock.[3] At that time the evolutionary scale previously mentioned, "Stone Age", "Bronze Age" and "Iron Age", had become the dominant paradigm in the developing disciplines of anthropology and archaeology, and in the study of prehistory. Lubbock's intention in devising the term was to create two subdivisions within the "Stone Age" category, "Paleolithic" and "Neolithic". As these subdivisions derived from the "Stone Age" category, their defining features revolved around stone, lithic, technology. According to Lubbock the "Neolithic" stone implements were "more skilfully made, more varied in form, and often polished."[4] Elsewhere he referred to it as the "polished Stone Age; a period characterised by beautiful weapons and instruments made of flint and other kinds of stone."[5]

Lubbock's ideas, while representing a conceptual breakthrough, were limited and quite subjective. It was also realised after a time that fully polished stone artefacts were a later development in that they followed, at least in the Levant, such things as a mixed farming/food producing economy, and the invention of pottery and textiles. Consequently by 1937 V. Gordon Childe was characterising Neolithic cultures as being those based on mixed farming, which engaged in woodworking, the manufacture of pottery and the production of textiles. Instead of polished stone tools and implements, they normally only employed "edge-ground" lithic techniques.[6] Following that, the Neolithic became firmly linked to the notion that this was the period where food or subsistence production economies became established, even though hunting, fishing and wild plant procurement actually continued to be important.[7] Features of the Neolithic now encompassed "permanent dwellings", forming "villages and even towns."[8] Basketry, the making of stone and pottery containers, and spinning and weaving, were also seen as critical indicators.[9] However, pottery and textiles, it turned out, were not developed until the latter Neolithic period in south west Asia, known as the Pre-Pottery Neolithic B (PPNB: 10,800 - 9100 BP).[10] Consequently, although the term is widely used, there is currently no definitional consensus of what specifically constitutes the Neolithic.[11] But this is not particularly helpful if one wishes to establish some basis for comparison. To resolve this impasse one can simply consider all the principal cultural features commonly associated with the neolithic, with particular reference to the early Neolithic, in south west Asia, and the analogous periods elsewhere, which is what I propose to do.

Characteristically the south west Asian subsistence economies in this period initially involved the planting of wild crops of emmer and einkorn wheat, and barley, although even this is not certain,[12] which were harvested while still "green", or at least before they were fully ripe, using hand-held flint sickles.[13] Once gathered these seed were threshed in "cup holes" sunk into stone slabs and then ground on millstones.[14] Some of what was not immediately used was, probably after parching, kept in small stone-lined storage pits, storage bins, larger mud brick silos or possibly baskets.[15]

Dwellings at this time were rudimentary but seemingly permanent. Single-roomed circular or oval structures with dimensions of 3 to 8 m were the norm in PPNA (11,500 – 10,800 cal. BP) settlements, these settlements covering

2 Moore 1995:44-5; Kuijt 2000:6; Mithen et al. 2000:656.
3 Lubbock 1865:3.
4 Lubbock 1865:60.
5 Lubbock 1865:2-3.
6 Childe 1937:85,99-100.
Ground stone technology actually made its first appearance at the end of the Natufian (Cauvin 2002:17).
It would also seem that in the period where edge-ground tools became common in early "Neolithic" contexts (Peterson 2002:28), leading ultimately to polished, formalised tool types such as finely polished axes, expedient lithic technology also features (Torrence 1989:58,64-5).
7 Cole 1967:vii-ix,1; Mellaart 1975:9; Hole 2000; Colledge 2001: 148-150; Cauvin 2002:214-6; Bar-Yosef 2003:51.
8 Cole 1967:61.
9 Cole 1967:61.

10 Bar-Yosef and Meadows 1995:73,79; Moore 1995:39.
In cases elsewhere, such as in China and Jomon Japan, with the earliest known pottery, the making of pottery long preceded agriculturally based food production (Barnes 1993:64-6; Aikens 1995; Yasuda 2003a).
The earliest textiles may be from Jarmo (impressions in clay, 7000BC) and carbonised cloth from Catal Hoyuk (Anatolia, Turkey) dated at 6000 BC, possibly woven from wool or flax (Helbaek 1963; Burnham 1965; Barber 1991: 126-32).
11 Sherratt 1997:274-5.
12 Bar-Yosef 1998a:167,169; Colledge 1998:130; Mithen et al. 2000: 656; Colledge 2001:146-7,150-1; Cauvin 2002:53-9,214,217; Asouti and Fairbairn 2003:1.
Rye may have been cultivated as early as 13,000 BP (Pringle 1998:1449) although, as pointed out earlier, one needs to be cautious in using changes in plant or seed morphology as an indicator of this (i.e. "domestication") because more intensive harvesting may produce such a result simply because of the change in the selection regime applied to that species by humans (See Ladizinsky 1987).
13 Unger-Hamilton 1989; Hillman and Davies 1990; Bar-Yosef and Belfer-Cohen 1991:470; Anderson 1992; Unger-Hamilton 1992:223; Anderson 1998:147-50.
Questions have been raised regarding the assumptions and diagnostic basis for determining the function of the "sickles" (See Colledge 1998:122; Ibanez et al. 1998:137; Kislev, Weiss and Hartmann 2004:2693-2695). Kislev et al. suggest the "sickles" may have been used more like a knife. Perhaps they were used in a manner analogous to the *juan* knife of the Corners Region.
Kislev et al. 2004 also suggest, based on seed morphology, that "ground collecting" was common in the early Neolithic and may have played a part in domestication. They cite some ethnographic examples of ground collecting in support of their case. For reasons too detailed to consider here I would argue that their case is based on dubious assumptions, there are other explanations for the observe characteristics (e.g. hand stripping) and that ground collecting, at least in the Australian examples, and probably the others, was a marginal activity in marginal environments.
14 Bar-Yosef et al. 1991:411; McCorriston and Hole 1991:58; Olszewski 1991:43; Bar-Yosef and Belfer-Cohen 1991:470,475,480, 482,489.
15 Stekelis and Yizraely 1963:11; Flannery 1986:13; Henry 1989:234, 365; Bar-Yosef and Belfer-Cohen 1992:34; Bar-Yosef 1998a:164,170; Kislev et al. 2006:1374

areas ranging from 1000-3000 sq. m.[16] At PPNA Jericho, as was common in the PPNA, "beehive shape" abodes predominated, with some evidence of partitioning of the rooms.[17] Low stone walls 40 cm high served as supports for a framework forming "huts or tents" at Gilgal I, while 80 cm walls "of cobbles coated with mud" and wooden posts and roofing of "some sort of mud mixture" were found at Netiv Hagdud.[18] Even in the PPNB simple wattle and daub habitations, such as those at Tell Aswad, were still to be found, although more rectangular structures had begun to become the norm by then.[19] Individual dwellings would have housed from four or five people at PPNA/PPNB Nahal Oren, or up to nine or ten at an "unusually large" PPNA habitation at Jericho.[20]

The size of settlements increased in the later PPNA and PPNB, the larger ones, such as PPNB Jericho, Mureybit and Tel Aswad, covering 2.5 to 5.0 hectares,[21] are calculated to have populations of 230-330.[22] There is some evidence that a settlement hierarchy was forming with a range of settlement sizes from small seasonal camps to the larger villages described.[23] A preference for alluvial lowland sites adjacent to permanent water sources also appears to have developed.[24] In terms of the degree of sedentism in evidence, it was noted in the last chapter that it is often just assumed that these early Neolithic villages were fully sedentary.[25] Where there has been specific research into this question, using one or more of the methodologies outlined earlier, there are indications that such settlements were occupied at least on a multi-seasonal basis, or "most of the year", possibly "during all seasons of the year."[26]

Apart from the planting and harvesting of crops, which presumably provided at least some part of the early Neolithic peoples' sustenance, hunting, of gazelle and wild boar for example, was still a major pursuit in the quest for food, even into the PPNB. Gathering, of nuts, fruits, berries, tubers, wild grasses and legumes (lentils), also made a substantial dietary contribution.[27] Waterfowl, particularly ducks, were another resource extensively exploited in this period.[28] Fishing, which only assumed some significance in the preceding period, the Natufian, further complemented these food sources.[29] Tame, husbanded animals, specifically ancestral sheep and goats, became incorporated into the Neolithic economy about the same time or shortly after food crops began to assume the primary role in subsistence. These animals were eventually to become fully domesticated,[30] and so by the end of the PPNB the distinctive Levantine mixed economy based on domesticated animals and crops had begun to take shape.[31]

As mentioned earlier, hafted hand-held flint sickles, cup holes sunk in rock slabs and millstones formed the principal seed harvesting and processing equipment. Shallow grinding bowls, stone knives, bladelets, picks and pipe mortars (thought to be for grinding acorns) were also part of the early Neolithic toolkit.[32] Manufactured objects, utensils and constructional features, culminating in pottery, began with clay daubing of dwellings at the beginning of the Neolithic.[33] Plaster, made by firing gypsum at low temperature, first appeared in the Natufian.[34] In the early Neolithic it was used ritually to form human faces on skulls or as human figures, and in the PPNA it began to be used extensively for floors.[35] Clay containers, such as cups, dishes and storage vessels, were also present in the early Neolithic. These were made with little or no tempering, and in some instances seem to have simply been dried in the sun.[36] Baked or fired clay figurines and artefacts made their appearance in the PPNB, the period in which true pottery first developed.[37] Preceding that had been the

[16] Kenyon 1957:71;1960-83:3:2,6,8,19; Noy, Schuldenrein and Tchernov 1980:69; Henry 1989:225-6; Bar-Yosef et al. 1991:408; Kuijt and Mahasneh 1998:155; Peterson 2002:131.

[17] Kenyon 1957:71;1960-83:3:2,6,8,19; Peterson 2002:25.

[18] Noy, Schuldenrein and Tchernov 1980:64-5; Bar-Yosef et al. 1991: 408,411.

[19] Contenson 1983:58; Bar-Yosef and Meadows 1995:76-7; Cauvin 2002: 39; Peterson 2002:26.

[20] Noy, Legge and Higgs 1973:78-9; Kenyon 1960-83:3:42.
These calculations are based on the floor area of those dwellings. The relationship between floor area and the no. of inhabitants is not straight forward. The conversion factor used her (2 sq m per person up to 10 sq m and then one per each additional 7 sq m, which is then corrected upwards by a further 10%)is based on ad hoc calculations by Cook which appear to under-estimate the actual figure (Hassan 1981:73). In order to correct for the underestimate I have applied an arbitrary correction factor of reducing the floor area per person by approximately 20% in both components of the floor area relationship and then inflating the gross figure for the number of individuals by an arbitrary 10%.

[21] Bar-Yosef and Belfer-Cohen 1992:34; Bar-Yosef 1998a:168.
The largest PPNB sites, such as Abu Hureyra in Syria, may have been up to 12 hectares in extent (representing a population of 506)

[22] Calculation based on application of Hassan's (1981:67) formula relating settlement area to settlement population size.
The largest PPN sites were up to 12 hectares and had estimated populations of at least 450 (Peterson 2002:25).

[23] Kuijt and Mahasneh 1998:160.

[24] Bar-Yosef and Meadows 1995:49; Smith 1995a:79; Sherratt 1997:276-7; Belfer-Cohen and Bar-Yosef 2000:30; Brookfield 2002:68-9; Cauvin 2002:63; Bellwood 2005:57;
Jericho is a classic example, situated next to the Ein es Sultan spring (See Ruby 1995:86-90,100). A similar preference appears to have developed in the European Neolithic (Bogucki and Grygiel 1993:410,412).

[25] Moore 1983:109; Bar-Yosef and Meadows 1995:62; Habu 1996:39.

[26] Bar-Yosef et al. 1991:423; Lieberman 1993:607,609; Lieberman 1998:76; Peterson 2002:25.

[27] Wright 1978:203; Contenson 1983:58,61; Henry 1989:19; Bar-Yosef et al. 1991: 417-23; Bar-Yosef and Belfer-Cohen 1992:37; Balter 1998:1443,1445; Bar-Yosef 1998a:171-2; Belfer-Cohen and Bar-Yosef 2000:30; Mithen et al. 2000:660-1; Alvard and Kuznar 2001:305; Peterson 2002:8,24-5,28,30; Asouti and Fairbairn 2003:1-2; Kislev et al. 2006.

[28] Bar-Yosef 1998a:167,172.

[29] Cole 1967:1; Contenson 1983:58,61; Bar-Yosef and Belfer-Cohen 1992:37; Kuhn and Stiner 2001:115.

[30] Bar-Yosef and Meadows 1995:39; Smith 1995a:51-67; Alvard and Kuznar 2001:300.

[31] It must be recognised that there was considerable regional variation in the subsistence economy in south west Asia during this period. See for example Asouti and Fairbairn 2003:1.

[32] Kenyon 1957:56-7; Stekelis and Yizraely 1963:12; Bar-Yosef 1983: 15,18-19; Moore 1983:96; Contenson 1983:61; Bar-Yosef 1998a:170.

[33] Moore 1995:45.

[34] Bar-Yosef 1983:15; Moore 1995:45; Peterson 2002:16.

[35] Moore 1995:45-6.

[36] Moore 1995:46.

[37] Contenson 1983:59; Moore 1995:40,46.

development of "white ware", a variety of containers made from plaster.[38] Mud bricks started becoming a common constructional feature by the PPNB, though wattle or reed and daub persisted in some places, such as at Tell Aswad.[39] Finally, in the later PPNB pottery, initially crumbly, straw tempered, lightly fired wares appeared.[40]

Another key characteristic of the Neolithic was so called long-distance trade, which seems to have commenced in the later Natufian. Shells from the Nile, the Red Sea and even the Atlantic reached the area, presumably through exchange networks, having covered distances of up to 3000 kilometres. Obsidian came 6-700 kilometres from Anatolia (Turkey), greenstone and malachite from Syria, Transjordan and the Arava valley, while heavy basalt objects were bought from the Golan Heights to Mount Carmel, a distance of 100 kilometres.[41]

Looking at the comparable periods in China, their Neolithic, and Mesoamerica, their Formative, there are a number of similarities. In China two contemporaneous Neolithic culture complexes have been identified, one centred on the Yangtze River in central China, another on the Yellow and Wei Rivers in north central China, with possibly a third in the Chifeng region in the north east.[42] Foxtail millet (*Setaria italica*) and broomcorn millet (*Panicum miliaceum*) became the principle food crops in the central north, and rice (*Oryza sativa*) in the south or central regions, these being planted as early as 11,200 years ago, or earlier, in the case of rice,[43] 10,800 BP with the millets. Domesticated foxtail millet is evident, perhaps as early as 8500 - 8300 BP, and "domesticated" forms of rice by 5500 BP.[44] Harvesting implements and seed processing utensils in the millet growing areas included serrated stone and mussel shell sickles, millstones, which had only made their appearance in the preceding millennium, and footed stone querns.[45] In addition rice was dehusked in holes in the ground, with the aid of a wooden "pestle", in the Neolithic rice growing areas of the Yangtze valley.[46]

Little is known about storage arrangements in Neolithic China, although storage pits, often containing carbonised grain, some 2.7 m long and 2.0 m wide, and capable of holding "significant" quantities, have been found at the larger village sites in the Yellow and Wei Rivers region.[47] The earliest, ostensibly permanent, dwellings known from China appeared in the Hupei Basin on the central Yangtze River, and the Yellow and Wei Rivers region in the north. In the Initial, also known as Early, Neolithic[48] these were single room, mostly round structures, 2-3 m across, and built with walls of wattle or posts and coated in clay.[49] Villages at this time incorporated cemeteries and covered areas of one to two hectares, reaching up to eight hectares, representing populations from 145 to 205, up to 400 or more[50]

The millet growing areas in north central China, although probably moister and milder in the Early Neolithic than today, were on the loess plateau dominated by semi-arid steppes subject to unreliable rains, droughts and floods.[51]

38 Kenyon 1983:3:93; Moore 1995:45.

39 Contenson 1983:58; Bar-Yosef and Meadows 1995:76; Moore 1995:45.

40 Bar-Yosef and Meadows 1995:73,79; Moore 1995:39-46; Ruby 1995: 135.

41 Bar-Yosef 1983:22-3; Contenson 1983:58; Bar-Yosef 1998a:165,172: Belfer-Cohen and Bar-Yosef 2000:30; Mithen et al. 2000:656.

42 Shelach 2000; Bellwood 2005:117.
The Neolithic culture complexes referred to here are principally the Pengtoushan type cultures of the Yangtze, Cishan, Peiligang and Houli cultures of the Yellow River, the Lijiacun/Baijia of the Wei River, and the Xinglongwa and Zhaobaogou of north east China.

43 Yan (2003:152) indicates that rice was being cultivated at Yuchanyan in Hunan and Jiangxi as early as 12,000 BP, while Yasuda (2003a:134) and C. Zhang (2003:189-90) claims rice cultivation was also possibly taking place at Yuchanyan as early as 14,000 BP.

44 Zhimin 1989:646-7; Crawford 1992:9,13-15; Smith 1995a:127,130, 133-4,136; Underhill 1997:112-4,120,138; Higham and Lu 1998:870-1; Lu 1998:902,907; Pei 1998; Zhang and Wang 1998; Zhao 1998:886,895; Wang et al. 1999; Shelach 2000:367-8,379; Bellwood 2005:118; Lee et al. 2007.
There does not appear to be a clear consensus about dates or location, with recent arguments and evidence pointing to rice cultivation commencing along the Yangtze River in western or central China from 11,500 - 11,200 BP, perhaps earlier. Another debate centres on the region where rice was domesticated. While the middle Yangtze region (including the upper Huai River) is becoming increasingly favoured, arguments have been advanced for the lower or upper Yangtze, south China and even south Asia or in an overlap zone where millets grew (Bellwood 1995; Blust 1996; Normile 1997:309; Underhill 1997:108,129-35; Crawford and Shen 1998:862-5; Higham and Lu 1998; Pringle 1998:1449; Zhang and Wang 1998; Wang et al. 1999; Lu et al. 2002; Toyama 2003; Bellwood 2005:117-8; Londo et al. 2006). Recent research suggests that rice may have been taken into cultivation and domesticated in two different locations at different times (Londo et al. 2006).
Chinese cabbage (*Brassica campestris*) also appears as a domesticate about this time (Crawford 1992:14-15; Shelach 2000:368).
In regard to the north east Chinese centre (Xinglongwa and Zhaobaogou cultures) there is no evidence available at time of writing as to what was being cultivated, the only plant remains reported so far are wild walnuts (*Juglans manschuria maxim*), although implements thought to be hoes and ploughs have been uncovered (Shelach 2000:379-387). Shelach (2000:379-82) suggests that soybeans may have been cultivated and domesticated here.

45 Ho 1977:432; Chang 1986:91,92,95; Barnes 1993:92; Smith 1995a: 135; Underhill 1997:120-1; Lu 1999:33,61,124-5; Shelach 2000:368.
Spades of wood or bone and stone "hoes" also appear to have been in use, though their exact purpose is uncertain (Lu 1999:114; Yan 1999:133; Pei 2003; Yuan 2003:158-9).

46 Lu 1999:121.

47 Chang 1986:90,94,100; Barnes 1993:92; Hayden 1995a:277; Smith 1995a:134; Underhill 1997:120; Lu 1998:907; Zhang and Wang 1998: 897; Lu 1999:36,88-9; Yan 1999:134; Shelach 2000:367,379.
At Cishan itself it has been calculated the storage pits contained up to 50 tonnes (Shelach 2000:368,377,379,398,401).

48 Some, such as Yan (1999:133), call the later stages of this period the Middle Neolithic

49 Chang 1986:90; Barnes 1993:104; Smith 1995a:130,133-4; Lu 1999: 36-7,88-9; Yan 1999:134; Shelach 2000:396.
By the Middle Neolithic larger habitations, up to 3-5 m in dimension, with square or oblong pit houses, began to become more common (Chang 1986:90; Barnes 1993:104).
Dwellings in north east China, as noted earlier (Chapter 5 n320), were an exception in that the earliest Neolithic structures were rectangular, with no rounded antecedent forms being evident (Shelach 2000:395-6). See n66 below for a discussion of this.

50 Chang 1986:90; Crawford 1992:14; Barnes 1993:104; Smith 1995a: 134; Underhill 1997:120,122; Lu 1999:36,39,88-9; Yan 1999:134; Shelach 2000:398,400; Bellwood 2005:117.
By the Middle Neolithic (6000 - 5000 BP) villages covered areas of 5-10 Ha.

51 Ho 1977:420,425,427; Crawford 1992:13-14; Smith 1995a:135.

Most of the settlements there, in north east China, and in the balmier, wetter rice growing areas of subtropical central China, were located in diverse and distinct ecozones close to water, and situated either on or overlooking alluvial plains and terraces.[52] The degree of sedentism is not certain, although the presence of pottery from the earliest levels, the Initial or Early Neolithic (11,200 - 9000 BP), along with multi-seasonal deer kills, has been seen as an indicator of a high degree of sedentism.[53] By the Early Neolithic, villagers of the Yangshao Predecessor (or Pre-Yangshaonoid) and Pengtoushan type cultures appear to have been "relatively" or "quite" sedentary, although they may have had to shift their more permanent settlements every three years or so because the soils became exhausted.[54]

Unlike the south west Asian Neolithic, pottery, as noted above, appeared in the earliest Neolithic levels in China, and its invention probably preceded the planting of crops and the adoption of village life by several millennia.[55] Domesticated pigs (*Sus scrofa*) and chickens (*Gallus gallus domesticus*) were also a feature in the latter part of this phase of the Chinese Neolithic economy.[56] However, as elsewhere, hunting, fishing, trapping of waterfowl, and collecting of molluscs and wild plant foods, remained an important part of the subsistence base of the settled populations well into the Neolithic.[57] Although many tools were initially only edge-ground (becoming polished later), flaked implements and microliths did not disappear from use completely until the later Neolithic.[58] In summary, sites in China are considered to be Neolithic if they exhibit one or more of the following key traits: the presence of pottery, ground stone tools, sedentism, cultivation or animal husbandry.[59]

The third example of a pristine "neolithic" is the Early Formative of Mesoamerica, also known as the Pre-Classic, which in the semiarid Highlands of Mexico (Valleys of Mexico, Oaxaca and Tehuacan and Balsas River) spanned the period from 3500 BP to 2900 BP.[60] There are indications that planting may have commenced at about the same time here as it did in south west Asia and China, 10,000 years ago, with the tentative identification of domesticated squash (*Cucurbita pepo*) of that age.[61] However staple plants, such as maize and beans, do not appear to have been domesticated until much later, around 6250 - 5500 BP,[62] and systematic planting did not become a significant or dominant part of subsistence until the Late Archaic and Early Formative (from 5000 - 3500 BP - Abejas and Purron Phases).[63] Storage baskets appear in the Early Formative, as well as storage pits up to 1 m across, capable of holding a tonne of maize, became numerous as permanent and semi-permanent hamlets and villages developed during this period.[64]

From the early part of the Early Formative (Ajalpan Phase – 3500 - 2850 BP) hamlets of 50-80 people, and villages with estimated populations in excess of 200, have been found. These types of linear settlements, following the alluvial river bottoms, had grown as large as 650 inhabitants by the end of the Early Formative.[65] The dwellings themselves were initially single roomed with dimensions of 4 m to 6 m, and large enough for 4 or more people. In form they were rectangular by the Early Formative with wattle and daub walls (cane plastered with clay), and grass or palm thatch roofs.[66] The people living in them appear to have been sedentary, or nearly so, and cemeteries in excess of 340 individuals have been noted from the middle of the Early Formative.[67]

---

[52] Ho 1977:427-8; Barnes 1993:103; Smith 1995a:135; Sherratt 1997: 279; Underhill 1997:110-13; Lu 1999:39; Shelach 2000:372-3,400, 404; Pei 2003:180.

[53] Ho 1977:433; Smith 1995a:136; Lu 1999:39,131,133; Shelach 2000: 402-3,405.

[54] Chang 1986:114; Barnes 1993:104; Higham and Lu 1998:872; Lu 1999: 33,89,131.

[55] Chang 1986:90-1,94-6,105; Barnes 1993:66,92; Higham and Lu 1998: 870; Shelach 2000:377; Yasuda 2003a:119-128; C. Zhang 2003:189.

[56] Ho 1977:414-7,466; Chang 1986:93; Barnes 1993:93; Smith 1995a: 94,138; Underhill 1997:122; Shelach 2000:382; Yuan and Flad 2004.

[57] Chang 1986:102,112; Barnes 1993:93-4,100; Smith 1995a:137; Underhill 1997:123,125; Lu 1999:33,36-7,39,60,90,93,115; Yan 1999:135; Shelach 2000:382,385,387; Yasuda 2003a:133.

[58] Chang 1986:81-5,87,93; Barnes 1993:60,64,92; Higham and Lu 1998: 870-1; Lu 1998:906; Zhao 1998:888; Lu 1999:33,36,38-9,60-1,88-9, 125-6; Shelach 2000:377,384-6.
While the Australian "small tool tradition", which included microliths, had sprung up about 5000 years ago, use of microliths disappeared, by and large, by about 1-2000 years ago. (Mulvaney and Kamminga 1999:235)

[59] Underhill 1997:105.

[60] Lamberg-Karlovsky and Sabloff 1995:238; Jordan 1999.
The Early Formative began about 300 years earlier in the Mesoamerican lowlands (Voorhies 1996:17,20).

[61] Markus and Flannery 1996:65; Smith 1997.
The appearance of apparently domesticated squash may be misleading. Firstly, as Ladizinsky (1987) demonstrated, plant morphology can be changed simply by harvesting and without the subject species being planted. This may be the case with *C. pepo* which is highly variable, this species producing a wide range of varieties including squashes, pumpkins, marrows and zucchinis (courgettes). Secondly squashes are "notorious camp followers" that "do well in habitats disturbed by humans" (Markus and Flannery 1996:65), making it highly likely they were only "incidentally" domesticated.

[62] de Tapia 1992:157; Fritz 1995:5; Sherratt 1997:275; Smith 1997: 932; Benz and Long 2000:460; Piperno and Flannery 2001; Smith 2001; Matsuoka et al. 2002.

[63] de Tapia 1992:157; MacNeish 1992:103; Markus and Flannery 1996: 71-2; Jordan 1999; Benz and Long 2000:464; Smith 2001:1325; Bellwood 2005:146,155-6.

[64] Flannery 1972:27,44; Winter 1976:25,27; MacNeish 1986:96; de Tapia 1992:157; Lamberg-Karlovsky and Sabloff 1995:245-6;Markus and Flannery 1996:73; Jordan 1999.

[65] Flannery 1972:44; MacNeish 1972:80; Drennan 1976:350; Flannery 1976a:173; Marcus 1976:80-83; Reynolds 1976:181; Hassan 1981:52, 93; de Tapia 1992:157; Lamberg-Karlovsky and Sabloff 1995:245-7; Marcus and Flannery 1996:71; Jordan 1999; Bellwood 2005:165-6.

[66] Willey, Ekholm and Millon 1964:1:451; Flannery 1972:37-8,44; 1976b:16,19; Whalen 1988:252-3; MacNeish 1992:112; Markus and Flannery 1996:73.
As noted earlier (Chapter 5 n320), the early appearance of rectangular dwellings in the Early Formative in the Highlands of Mesoamerica and in the Initial/Early Neolithic of north east China is unusual. Shelach (2000:398,400) suggests this may have been because in both cases the dwellings were on sloping ground (presumably requiring this structural form). The other possibilities are that there may have been round antecedent structures that have not been identified in the archaeological record as yet, or that there is some other factor operating.

[67] Whalen 1976:77-8; De Tapia 1992:157; Lamberg-Karlovsky and Sabloff 1995:250; Jordan 1999.

Other features of the cultures found in this part of Mesoamerica included grinders and millstones, common in the Late Archaic, which had become specialised ground stone *manos* and *metates*, principally for processing corn, by the Early Formative.[68] Nets and baskets also made their appearance in the Late Archaic, or earlier, as did the first crude pottery (Purron Phase).[69] Nevertheless hunting and gathering of plants continued to provide 60% of the food supply in the early Early Formative, this still being 40% in the Middle Formative.[70] Trade, not surprisingly, was also an attribute associated with this period and included pottery, obsidian, jade, turquoise, "iron pigments", iron ore (for mirrors), mica, mollusc and turtle shells, fish and stingray spines, and sharks teeth.[71] Obsidian appears to have been traded over considerable distances even in the Early Formative, up to 580 kilometres.[72]

Although the characteristics of each of these exemplars are slightly different, and some features appear in the archaeological record in a different sequence in each case (e.g. pottery), nevertheless there are a number of commonalities which allow what constitutes an "Early Neolithic" culture to be formalised. These are:

- Commencement of a food production economy usually based on agriculture or "mixed farming" (i.e. agriculture and animal husbandry).
- The appearance of specialised procurement and/or processing facilities, apparatus, tools or implements.
- The use of ground stone technology, particularly edge-grinding, in the manufacture of stone tools and implements.
- Evidence of conservation of output (preservation and storage as opposed to caching and stockpiling) with a view to providing an ongoing food supply.
- Permanent or near-permanent (i.e. capable of habitation for extended periods of time) dome-shaped or rectangular dwellings with the capacity to house up to 5 people, if not more.
- Permanent settlements, at least on an annual basis, with populations of 50 to 80 initially, growing to 2-300 later.
- Settlement sites located on or adjacent to alluvial lowlands (at the regional level) close to relatively reliable water sources.
- Multi-seasonal sedentism at least.
- The presence of cemeteries in the region.
- Continued reliance on hunting, fishing, capture of waterfowl and the procurement of wild food plants.
- The manufacture of nets and weaving of baskets, mats etc.
- High value or exotic goods being traded over distances of at least 500 kilometres.[73]

Other features, such as pottery and weaving of cloth, appear in the early Neolithic or equivalent period in some, but not all, instances. Extensive use of pottery, for example, is seen a marker for the beginning of the Chinese Neolithic,[74] and the Mesoamerican Formative, but in south west Asia it does not appear until near the end of the early Neolithic, and in fact is often seen as marking the end of the early Neolithic.

In comparing with Australia, it was noted in the beginning of Chapter 5 that authors such as Mulvaney and Diamond were of the view that along the Darling River, part of the "Corners Region" considered in that chapter, "incipient agriculture" was in evidence at the time of "contact". Others, such as Lourandos, have noted what they see as the parallels between the situation as it was in Australia at that time and the Mesolithic of Europe and south west Asia, the North American Archaic and Jomon Japan.[75] Parallels have also been drawn between the harvesting and processing of seeds in the Australian Recent and similar practices in the Mesolithic Natufian culture of south west Asia.[76] Stekelis and Yizraely, however, have pointed out that the lower grindstones from PPNA/PPNB Nahal Oren are identical to those found in Australia.[77] Curiously, Kate Langloh Parker, one of the more sympathetic and insightful observers of the *Yuwaaliyaay* on the Narran River in north west New South Wales, had already commented at the beginning of the 20th century that their grindstones reminded her of those from the British Neolithic.[78] But occasional observations such as these, even if they provide useful indicators, are not helpful in determining if particular lifeways, as they were in some areas of Australia, can justifiably be characterised as something analogous to a "neolithic" culture. A review of the Australian evidence in light of the common features

[68] MacNeish 1986:95-6; Lamberg-Karlovsky and Sabloff 1995:244; Jordan 1999.

[69] MacNeish 1986:96; de Tapia 1992:157; Hoopes 1995:185; Lamberg-Karlovsky and Sabloff 1995:244; Jordan 1999.
The earliest pottery in the Americas did not appear in Mesoamerica however, but in the lower Amazon around 8000 BP (Hoopes 1995:185; Bellwood 2005:158).

[70] MacNeish 1972:80,82; Lamberg-Karlovsky and Sabloff 1995:250; Markus and Flannery 1996:72.
To put this in perspective the *Kukatja* of the Great Sandy Desert obtained only 20% of their food from seeds, although they were plentiful, as a supplemental food source (Cane 1989:111).

[71] Pires-Ferriera 1976a; Pires-Ferreira and Flannery 1976:286.

[72] Pires-Ferreira 1976b:300-304.

[73] There is no information on China concerning in this regard as no research on this question could be found, and such studies may not have been done (Shelach 2000:405-6).

[74] Zhao (1998:895) lists four criteria to delineate the neolithic:
- Pottery
- Ground stone tools
- Settled villages
- Domesticated plants and/or animals.

However two of these are particularly problematic. Firstly pottery may only appear much later in a sequence, such as in south west Asia, and, secondly, the insistence that agricultural plants must be domesticated when, as discussed elsewhere in this book, this is an unsatisfactory criteria for determining whether agriculture was taking place. As discussed elsewhere even in China the evidence indicates Neolithic cultures there grew rice for at least 3000 years before it assumed its final "domesticated" form.

[75] Lourandos 1997:23,333.

[76] Smith 1989b:305; Diamond 1997:310-11.

[77] Stekelis and Yizraely 1963:8.

[78] Parker 1905:119.

distilled from the examination of examples of other "neolithic" cultures can, however, at least provide some basis for determining whether specific Australian examples can be classified as "neolithic".

It will be recalled that in Chapter 4 it was argued that the southern *Nhanda*, and northern *Amangu*, people of the central west coast of Western Australia had been engaged in agriculture. In terms of the criteria above, this means a food production economy was developing. Because yams were being harvested the only specialised harvesting implement required was the *wippa*, the digging stick, a long heavy staff.[79] As there has not been any archaeological excavations relating to the settlements and economy of this district it is difficult to determine if any specialised processing tools were in use. However, a specialised type of grindstone, thought to be for employed in the processing of yams, and a particular type of scraper which may have been used to scrape the skin off the yams, have been found at several occupation sites in the Victoria District and the region to the north.[80] It is believed edge-ground axes were present in the region although none have been reported. The basis for this belief is the existence of a traditional bartering place called *Eyearawa*, which means "edge-ground axe" in languages of other groups immediately to the east, where edge-ground axes have been reported.[81] The inference from this is that edge-ground tools were used in the region but have simply not been collected.[82] With conservation of output, while stockpiling has been reported, there is nothing in the limited historical ethnographic information to hand to suggest dedicated storage practices. But in other parts of the world, yams are commonly left in the ground and harvested as needed. Reports from explorers and pioneers in the 1839-51 period, of innumerable holes where yams had been dug (they were a major hazard when traversing the area, especially to horses and their riders, and were often noted),[83] are consistent with such an output conservation strategy being practiced by the *Nhanda*, simply leaving the yams in the ground until needed.

The size, form and structure of the *Nhanda* habitations, as well as the location and size of their settlements is a topic that has been analysed in greater depth, employing historical ethnographic data and site survey techniques.[84] From this analysis it was concluded that their dwellings were dome-shaped, about 1.8 metres in height and about 3 to 4 m across, formed of timbers 15 cm thick, coated with a layer of clay or turf, and capable of housing at least 10 individuals. On the basis of comments by explorer Grey, that they "were fixed places of residence,"[85] and the degree of labour invested in their construction, it was concluded they were permanent or near-permanent residences.[86] The five settlements that have been identified in the region are presumed to have been permanent as well, on the basis of the permanency of their constituent dwellings and their size. In the two cases where the size in population terms could be determined, these villages had estimated populations of 150-230 and 290 (4 ha) respectively.[87] All, with one exception, were located either on alluvial flats or overlooking alluvial flats.[88]

An assessment of the degree of sedentism in the region points to the *Nhanda* exhibiting at least multi-seasonal sedentism though, being based on limited evidence, this is not absolutely certain.[89] Certainly at least one "burial ground" is known of in the region and there were possibly many more.[90]

As would be expected with the developing agricultural economy here yams, and "gum", provided a significant proportion of the diet. Recall Lt. Helpman's 1849 observation, noted in Chapter 4, that the *Nhanda* "seem to depend entirely upon warran and gum, of which they have great abundance". An analysis of the subsistence base in this region showed that hunting, fishing and utilisation of wild plant foods, apart from "gum", seemed to form only a minor part of their subsistence, particularly in the core area of the Victoria District, with its series of small fertile river valleys.[91] Illustrating this is the comment made by Government Resident Burges in 1851, referred to in Chapter 4, that the *Nhanda* "seem very little addicted to hunting and very few of them are even expert at tracking a Kangaroo".

Very little is known specifically of the *Nhanda* people's material technology, although small nets seemed to have been used by northern *Nhanda* groups, such as the seasonally sedentary *Watjandi*, obtained in trade from Shark Bay, 150 kilometres to the north.[92] Similarly with trade, information is very limited but it is known that Hutt River was a ceremonial exchange centre, where ceremonial gatherings reputedly attended by up to 2000-2500 people

79 Foley 1865:298; Oldfield 1865:277,295.
In most other parts of southern Western Australia this woman's digging stick is known as a *wanna*.
80 Jackson and de Gand 1996:29; Personal Observation 28 April 1998; 27 June 2000.
81 Davidson and McCarthy 1957:424,426; Gerritsen 1994a:176,249,312, 328.
See also Bates 1938:124; 1985:109 for evidence of edge-ground, and perhaps fully ground axes from the north of the state reaching coastal areas in the south.
82 This view is supported by Jackson and de Gand (1996:11,28).
83 See for example Gerritsen 1994a:82-3.
84 Gerritsen 2002.
85 Grey 1841a:2:19.
86 Gerritsen 2002:13-15.
87 Gerritsen 2002:12.
88 Gerritsen 2002:4-11,13-15.
The exception is what appears to have been a hamlet size settlement close to the shoreline at the tip of Peron Peninsula in Shark Bay. In this case there may have been a greater reliance on fishing and marine mammal hunting than yams. See Bowdler 1990; Gerritsen 2001a:25; 2002:15.
89 Gerritsen 2002:16-19,21.
90 Gerritsen 2002:19,26.
91 Gerritsen 2001c:9-10,20.
92 Oldfield 1865:266,274,277.
These nets may well have come from further north (See Gerritsen 2001a:25).

were held.[93] Pearl and baler shells originating from Shark Bay (200 kilometres away), the north west Pilbara coast (650 kilometres) and Broome (1250 kilometres) were traded at Hutt River (*Beedalinyoo*/*Wilya-a*),[94] before (pearl shells) passing on to the south west of Western Australia. Ochre from the Wilgie Mia mine in the Murchison, 400 kilometres north north west, also found its way through the district into the south west, a further 400 kilometres.[95] Considered overall it would appear from the foregoing that the *Nhanda*, particularly those of the Victoria District, exhibited all the characteristics, specifically or proximally, of a neolithic society.

Evidence bearing on the issue of the Australian Early Neolithic in the Corners Region, the East Central Australian Non-Centre (ECAN), is somewhat more extensive, especially as it incorporates some archaeological evidence. The nature of the economic base, the size and form of settlements, population density, the settlement pattern, the degree of sedentism, storage, trade and a range of other matters came under consideration in the preceding chapter. Regarding the question of a neolithic culture in that part of Australia, there is firstly evidence pointing to agriculturally based food production. Propagation of a range of species such as *Panicum* grasses, *ngardu*, *munyeroo*, *wirra*, *yaua*, tubers of the sweet potato family *Ipomoea* and native tobacco species has been documented, the only uncertainty being as to how extensive this was. Direct, indirect and inferential evidence indicated that in some areas at least sowing and broadcast seeding was quite extensive. Given:

- that a number of tribal groups were named by Smith and others as engaging in sowing and seeding
- that some groups were reported to be engaging in sowing or planting of more than one species by more than one source
- that Smith had mentioned a number of locales where seeds were broadcast,
- the fact that the *apwa* seed bags in which the seeds to be broadcast contained up to 23 kg of seeds,
- that there was a trade in such seeds,
- that there was mythological evidence (the legends of *Markanjankula*) of extensive sowing and planting (as well as clearing and weeding),[96]
- that there was evidence of extensive stands and heavy exploitation of the identified species in the very location types where it was deduced sowing of such species may have been taking place,
- that the types of location where it was deduced sowing was taking place matched agriculturally favoured sites in the south west Asian Neolithic and elsewhere,

it is reasonable to conclude that, on the balance of probabilities, agriculturally based food production of some extent was being undertaken at least in some parts of the Corners Region. A little later arguments and evidence will be put forward providing further support for this proposition.

Regarding specialised procurement and processing technologies, no specialised implements have been specifically identified in relation to the harvesting of the species that may have been extensively planted - *Panicum* seeds, *ngardu*, *yaua* and *munyeroo*. Gregory did report, however, that the people harvesting large areas in south west Queensland used "stone knives". These could be the mysterious *juan* knives, large stone knives with a characteristic fur grip, whose distribution is roughly contiguous with areas where there was a heavy reliance on *Panicum* seeds, although flint knives have frequently been reported in the Region as well.[97] Mention was also made in the last chapter of specialised grindstones, such as the "pear-shaped" type, for processing seeds and sporocarps. Some were used for *ngardu*, others for grass seeds, in a process reminiscent of the evolution of the specialised Mesoamerican *metate* grindstone, deriving from an earlier generalised form over the last 5000 years.[98] Furthermore, the Australian seed grinders have use-wear patterns very

---

[93] Edward Cornally, one of Daisy Bates' informants, observed as a youngster a gathering at Hutt River in the 1850s "of over 600" men.(Bates n.d.i:76a quoted in Gerritsen 1994a:225). If the population multiplier ratio, based on Aboriginal population samples (Gerritsen 2000:12), ranging from 3.29 to 4.20 is applied this provides figures of between 1,974 and 2,520. Bates says she was told the participants camped in "the sandy hollow", "near Port Gregory", and "20 miles from Port Gregory". Cornally's stepfather was a Pensioner Guard at the Lynton Convict Station, 6 kilometres from Port Gregory and less than a kilometre from the first Hutt River village. Near here there is permanent water, sandy dunal areas and the bed of the river is sandy whereas Hutt River "20 miles from Port Gregory" lies in mostly rugged, clayey and rocky country north west of Northampton and is usually dry. The Hutt River village and Lynton Convict Station lay 20 miles from Northampton and this is probably what was meant.

[94] *Beedalinyoo* and *Wilya*-a roots mean pearl shell (place of) and conch shell (place of) (Bates n.d.j)

[95] Oldfield 1865:266,269; Bates 1938:124; McCarthy 1938-40:437-8, 191; Gerritsen 1994:222,225,325; Bates 1985:109,162,262,280-2.

[96] The dam, 100 m long and 2 m high, on the Bulloo floodplain, ostensibly created to pen fish and increase the area flooded to promote growth of seed plants may actually have been intended to increase the flooded area for decrue planting. This is perhaps a more logical possibility justifying the labour required to construct it. However, as there nothing specifically known about the origins and use of this structure it is only possible to speculate about its function(s) at this point in time.

[97] Such knives, if they were the ones observed by Gregory, may not have been used exclusively for harvesting as these and other types of knives appear to have had multiple uses including sometimes being used in "duels".
Evans 1872:264,fig. 198; Smyth 1878:1:380; Cox 1880:635; Ridley et al. in Curr 1886:3:305; Teulon in Curr 1886:2:191; Gregory 1887:132; Etheridge 1890:255; 1891:32; Darbishire and Myres 1903: 33,Plate C; Edge-Partington 1903:37; Aiston 1921:5; Bennett 1927a:401,407-8; 1927b:104; Tindale 1957:29; Gresser 1962:53; 1963a:32,36; 1963b:80; Johnson 1963:69; Mulvaney and Joyce 1965; McCourt 1975:119; McCarthy 1976:20,35-6; Morwood 1981: 23, 31,50; Davidson 1983:30,32,fig 3.6; Morwood 1984:352; Knight 1990; Lourandos 1997:292; Mulvaney and Kamminga 1999:244-5.
Pick axes (Etheridge 1890:369,372) also appear in the Corners region but their purpose has not been ascertained.

[98] McCourt 1975:136; McCarthy 1976:59-60; McBryde 1987:271; Smith 1985:25-6; 1986; 1988:95; 1989; Edwards and O'Connell 1995:774; Fullagar and Field 1997; Gorecki, Grant and O'Connor 1997:141-5.

similar to those used in the Early Neolithic in north central China.[99]

Ground stone technology was certainly in evidence throughout the Corners Region, with some axes approaching full polish.[100] Pecking or hammer dressing of axes, to alter the remaining surfaces, a technique commonly associated with the Neolithic of south west Asia and Europe, also occurred throughout the ECAN, though not restricted to it.[101]

As for evidence of preservation and storage, a wide range of examples were provided earlier. Skin bags holding up to 50 kg were commonly reported, usually containing *Panicum* seeds, gum and *munyeroo*. Clay containers were also made to hold such things as *munyeroo*. A great variety of foods, from geese to fish and grasshoppers, were preserved in some way or another, usually in dried form and mixed with seed flour. Storage "structures" such as clay lined pits, bark piles (for witchetty grubs), the clay containers, and quite possibly the small purpose built shelters Sturt and others had encountered in parts of the Corners Region, have also been reported. The scale of storage may have been quite significant if Ashwin's account of up to a tonne of seed is to be given credence.

While it is not possible to be certain that some types of dwellings in the ECAN were permanent, this is a reasonable inference to make. Recalling Lewis, as a prime example, he had noted that the *Ngameni* he had encountered "return and live in the same wurlies which are very large and well built". On the upper-central Darling River, Sturt thought that the large dwellings in the village he had seen there, "appeared to be permanent habitations." Similarly Mitchell deduced that the residences in a *Paakantyi* "village" he saw further downriver were of a "very strong and permanent construction". The very size of some structures that were reported, capable of accommodating 30-40 persons, would seem to indicate they were, in all likelihood, permanent structures. Numerous drawings and photos, as well as descriptions of some of the dwellings seen in the region, show them typically to be dome-shaped. Some of the more common dwellings were easily able to provide for up to 12-15 individuals, a figure mentioned by Sturt, Mitchell, Fraser and Lewis. Others, such as Duncan-Kemp, with more than passing familiarity of the *Karuwali*, said the permanent "gunyahs" she had seen were "capable of accommodating six or seven people each." Multi-room dwellings, a facet only really evident in the PPNB in south west Asia, were also present. As Fraser attested, a portion of some of the larger permanent residences "may be partitioned off as a separate room," and Sturt similarly, that they had "smaller ones attached, opening into its main apartment."[102]

Clear evidence in terms of settlement size, which meets the sixth neolithic criteria of initial settlements of 50 to 80 growing to 2-300, is conveyed through a number of sources referred to in Chapter 5. The town-like village seen by Sturt, for example, had an estimated population of 800 – 1000 people if fully occupied. Other estimates were of "two hundred or more"[103] on Pharmaleechie Channel in south west Queensland, or over 240 people in a settlement Lewis passed on the Diamantina River, also in south west Queensland. And a term often used in accounts of other settlements that were encountered was "village", even if specific figures were not given.[104] There is even evidence suggesting some functional specialisation within settlements, as demonstrated by the store-house in the centre of the village encountered by drover Ashwin. Where specified, these settlements, and other reported concentrations of populations, were found on the banks of rivers, alluvial flats or their fringing sandhills, the margins of lakes, waterholes, lagoons, and around springs.[105] It is just such locations where the richest alluvial soils were likely to be present, along with permanent or semi-permanent water sources.

The degree of sedentism, as another indicator of the Neolithic character of the ECAN, was a matter considered in the preceding chapter. From that it was concluded that some, such as the *Wongkanguru* and *Diyari*, exhibited sequential and partial sedentism. Others, like the groups based at Brewarrina, also appear to have been partially sedentary, with the *Gamilaroi* and *Ngameni* peoples deemed to have been multi-seasonally sedentary. Seasonally mobile if not full sedentism may have been in evidence among groups such as the *Karuwali*. Cemeteries, another marker of sedentism, or at least a high level of circumscription, along with possible status burials, were widely reported in the ECAN. Explorers and other sorts of intruders often drew attention to burial grounds in particular localities, observations supported by the oral traditions of the descendants of the traditional owners. Some of these were said to incorporate hundreds of interments. Single mound graves as apparent status burials 1.35 metres high and 18 to 20 metres across were also in evidence. Significant labour investment is necessary to construct graves such as these, as with the burial ground of stone mounds at Poolamacca in New South Wales, a level of investment that is not normally found in more mobile groups. Overall the pattern of sedentism is quite consistent with what is known of the neolithic analogs elsewhere, these being characterised as

[99] Lu 1999:30.
[100] Fitzpatrick 1914:39; Horne and Aiston 1924:56; Davidson and McCarthy 1957:426; McCarthy 1976:50,52; Tibbett 2002:22-7.
[101] McCarthy 1976:51-2.

[102] Specialisation of function of structures in settlements, specifically the storage structure noted by Sturt and Ashwin, would also appear to indicate permanent and more sedentary settlements.
[103] This statement by Duncan-Kemp is corroborated by comments she made elsewhere of habitations large enough for 6-7 people and having passed a waterhole where there were 30 residences. This corresponds to a population of 180-210.
[104] See pp.46-8 and Chapter 5 n103.
[105] See pp.48-51.

"multi-season", "most of the year", "nearly full", "high degree" as well as "fully sedentary".

It is paradoxical that hunting, fishing and bird catching, as well as wild plant food procurement, is a feature of early neolithic societies, but the transition to specialised farming agriculture or mixed farming is, or was, a slow process. Like the neolithic economies studied earlier, the peoples of the Corners Region certainly made effective use of the natural resources available to them, as was detailed in Chapter 5. But the important point to note is that engaging in hunting, fishing and so forth does not mean that those peoples so engaged were <u>not</u> involved in a neolithic economy. Be that as it may, there nevertheless appears to be a qualitative difference in how these activities were conducted in the ECAN, and some other places in Australia, when compared to more typical Aboriginal societies. Again much greater labour investment, and perhaps new forms of social and labour organisation engendered in cooperative endeavours, appears to have been involved, through the use of such devices as large fishing, hunting and duck nets, fish traps, dams, battues and hunting traps.

Storage also requires another level of social and labour organisation in helping to maximise extractive efficiency in the procurement of plant and other foods. This, in a sense, sets the stage for the transition to agriculture, making that transition possible. Unfortunately many of these types of activities are not necessarily evident in even the best documented archaeological sequence, in south west Asia. Nevertheless, what evidence there is does appear to point to a similar phenomenon commencing in the Natufian, where such things as storage, nets and fish hooks began to form part of the repertoire of subsistence technologies.

While on the subject of nets it is apposite to consider the next neolithic criteria, concerning the manufacture of not only nets but baskets, mats and such like. Here again the evidence is unequivocal. Nets were common, being found, in fact, in just about every part of the Corners Region. Some of the hunting nets were enormous, extending, as we saw in Chapter 5, for up to 270 metres. Finely crafted baskets, carry-alls, bags and other made items were also in evidence, having functional as well as artistic value.[106]

Considering trade, the final criterion, there are well documented indications that it played a significant role in the traditional economy of the Corners Region. Aspects of this factor were considered in the preceding chapter. The Corners Region had in fact a pivotal role in the intra-continental trade network found in Australia, with goods being funnelled through there. The chain of major ceremonial exchange centres indeed formed the main north-south trunk route in Australia.[107] Distances "exotic" trade items travelled easily exceeded the minimum 500 kilometres found in other neolithic entities. *Pitjuri*, for example, was traded over distances of up to 1200 kilometres, baler shells 1500 kilometres and pearl shells 1700 kilometres.[108] Furthermore, the fact commodities such as food and other consumables were an integral part of trade in the ECAN suggests a level of economic articulation not otherwise found in traditional hunter-gatherer societies.

From this, and the other factors considered above, it would seem that a neolithic culture was present in the Corners Region/East Central Australian Non-Centre, at the end of the traditional period. All the criteria appear to have been met, the only indeterminate facet being the extent of agriculturally based food production. And given that there are two examples of neolithic type cultures in Australia, in the core *Nhanda* territory from Hutt River to the Irwin River, and in the Corners Region, it would seem justifiable to conclude that the contention that there was an Early Australian Neolithic is supported. But in order to more clearly illustrate this, compare the descriptions and re-creations of life in the Australian Early Neolithic presented here with an artist's impression of a typical scene in early Neolithic south west Asia. As can be seen, the differences are not great, if at all.

In the following chapter it will be argued that there is another example, in south west Victoria, of an Australian Early Neolithic culture. But before proceeding, three other elements associated with the appearance of neolithic cultures will be considered, pottery, plant domestication and changes in the gender assigned, "sexual", division of labour. All can be seen as general indicators of the trajectory followed by neolithic cultures elsewhere, as well as helping to confirm the conjecture that the two regions in question were indeed neolithic.

The first completely artificial ceramic objects, known as the "Venus figurines", come from Dolni Vestonice in the Czech Republic and are believed to be about 27-30,000 years old.[109] Pottery, however, as baked clay vessels, did not appear until around 16,000 years ago, in Hunan, Guangxi Zhuang and Jiangxi Provinces in China, Nagano Prefecture and Tsugaru Peninsula in Japan, and the Khummi site in far eastern Russia. Other finds from Jomon period sites in Japan, as well as sites in the lower Amur River Basin in eastern Russia, and Hubei Province in north China, date from the same period or perhaps a little later.[110] The reasons why pottery was "invented", what it was used for initially, and why the timing of its appearance in different cultural

[106] Eylmann 1908:373-4,383,Plate 25 Fig.3, Plate 27 Fig.4, Plate 28 Fig.6,Plate 29 Fig.1; Davidson 1933:274-7; Duncan-Kemp 1934: 112,122, 146; Isaacs 1984:104; West 1999.

[107] McCarthy 1938-40; McBryde 1987.

[108] Based on McBryde 1987:263 and evidence (Berndt, Berndt and Stanton 1993:117; Tonkinson 1993:xxiv) that *pitjuri* originating in south west Queensland was reaching Lake Alexandrina in south eastern South Australia.

[109] Akerman and Stanton 1994:14,Map 1; Barnett and Hoopes 1995:2.

[110] Barnes 1993:54-6; Aikens 1995:11,17; Kenrick 1995:4-5,8-14; Higham and Lu 1998:870; Lu 1999:87; Kuzmin and Orlova 2000; Shelach 2000:377; Tsutsumi 2003:242-9; Yasuda 2003a; C. Zhang 2003:189. Yasuda (2003a:121-124) makes claims for the appearance of pottery during the Last Ice Age (23,000-18,000 BP) in southern China and Japan.

*Figure 16: An Early Neolithic Scene*
*(adapted from Leakey 1981:208-9)*

sequences is so variable, are uncertain and may depend on the local context.[111] The most common hypotheses assert pottery was developed for preparing, cooking or detoxifying foods, or for storage.[112] This last factor has been used to account for the relatively "late" development of pottery in south west Asia and the Americas,[113] that these peoples already had other types of storage containers such as baskets, gourds, skins and stone bowls, and that pottery vessels only provided a marginal benefit.[114]

[111] Hoopes 1995; Underhill 1997:115-6.

[112] Aikens 1995:19; Barnett and Hoopes 1995:3; Hoopes 1995:195; Higham and Lu 1998:870; Shelach 2000:377-8; Tsutsumi 2003:249-50; Yan 2003:153-4; Yasuda 2003a:129.

[113] The first pottery in the Americas has been found in the Amazon basin and dated at approximately 8000 BP [7000 bp](Hoopes 1995: 185). It is uncertain at this time whether the pottery found in Mesoamerica arose because of diffusion from elsewhere or as an independent development (Clark and Gosser 1995; Hoopes 1995:185).

[114] Barnett and Hoopes 1995:5.

While pottery originated in a number of places around the globe as part of local cultural sequences, there is not a strong correlation between its initial development and the appearance of sedentism, agriculture or mixed farming (the Jomon are often categorised as "hunter-gatherer-fishers" for example), or other technical developments.[115] So pottery by itself is not an unequivocal indicator of sedentism, though once pottery is used on an extensive basis, the quantities found at settlement sites appear to be a good measure of the degree of sedentism.[116] Consequently, in south west Asia, early "farming and village life" preceded pottery by two millennia, before it appeared almost synchronously across the region.[117] Nevertheless, because of the relative abundance of well-documented material evidence from the region, and the detailed descriptions and contextualisation of the presumed developmental sequence, south west Asia provides a useful comparative case for Australian evidence.

On the available evidence systematic manipulation of "plastic" materials in south west Asia appears to have begun with plaster, made by firing gypsum at low temperature, in the Natufian.[118] Hole claims that this innovation, the "use of pyrotechnology", was an important element in the "profound transformation" that commenced in this period.[119] By the early Neolithic plaster was being used ritually to form human faces on skulls or as human figures.[120] Following this had been the development of "white ware", a variety of containers made from plaster, built up layer by layer, usually in basket moulds. White ware was being produced in large quantities by time pottery made its appearance,[121] with plaster also extensively being used for floors.[122]

Clay daubing of dwellings is another feature claimed to have "marked the inception of the Neolithic",[123] with mud bricks becoming a common constructional feature by the PPNB. However, wattle, or reed, and daub persisted into the PPNB in some places, such as at Tell Aswad.[124] According to current understandings, clay containers, such as cups, vases, dishes and storage vessels, were first made in the early Neolithic as well. These containers, which have turned up at Ganj Dareh (Zagros Mountains, western Iran), on the Euphrates River, Mureybit and Abu Hureyra, were made with little or no tempering, and in some instances seem to have simply been dried in the sun.[125] Baked or fired clay objects (spheres, discs and cylinders) and human figurines first made their appearance at a number of places in the PPNB, continuing to be made "for centuries" more before true pottery developed in the later PPNB.[126] The pottery that appeared then was initially a crumbly, straw tempered and only lightly "fired" ware.[127]

In comparing Australia with the south west Asian sequence, the usage of plaster, usually known as *kopi*, a well documented traditional cultural practice, is one aspect warranting closer scrutiny. *Kopi* was used in most of the Corners Region, as well as extending down to the middle Murray River and adjacent areas,[128] to make a variety of objects such as "widow's caps", grave markers, ceremonial objects such as some "cylcons" (cylindro-conical ceremonial objects) and balls.[129] Widows caps were a type of plaster helmet, some weighing up to 2.7 kg, usually worn by the wife, relatives and friends of the deceased, and like "white ware", built up layer by layer. Just as in south west Asia the plaster was usually made by burning gypsum and then adding water, sand or ashes.[130] While certainly not comparable to "white ware" some of the Australian items were, nevertheless, arguably homologous to those occurring in the earlier Neolithic of south west Asia.

As shown previously, the daubing of dwellings with clayey admixtures was widely practised by the *Nhanda* and in much of the Corners Region. Clay was also employed to cover grain pits in at least part of the region.[131] Again these practices invite comparison with the early Neolithic of south west Asia. But deliberate manipulation and high temperature firing of clay to make pottery does not seem to have occurred in any sort of systematic way in Australia.[132] However, baked and sun- or air-dried objects and containers were certainly produced. Some cylcons, spinning tops and balls for games, and figures, appear to have been made from clay, although whether these were baked or simply dried is uncertain.[133] But baked clay objects were most certainly created, either deliberately or as a by-product of other activities. Along the middle and lower Darling River in the Corners Region, extending down into the lower and central Murray River region and other parts of

[115] Edwards 1989:33-4; Aikens 1995:12-4; Hoopes 1995:195; Wetterstrom 1998:32; Shelach 2000:378.

[116] Edwards 1989:34; Barnett and Hoopes 1995:4; Close 1995:32; Haaland 1997:376.

[117] Moore 1995:39,44.

[118] Bar-Yosef 1983:15; Moore 1995:45; Hole 2000:191; Peterson 2002: 16.

[119] Hole 2000:191.

[120] Moore 1995:45-6.

[121] Kenyon 1983:3:93; Bar-Yosef and Meadows 1995:73,79; Moore 1995: 45.

[122] Moore 1995:45-6.

[123] Moore 1995:45.

[124] Contenson 1983:58; Bar-Yosef and Meadows 1995:76; Moore 1995:45.

[125] Kenrick 1995:6-7; Moore 1995:46.

[126] Contenson 1983:59; Moore 1995:40,46.
Such as Gritille, Ghoraife and Aswad.

[127] Bar-Yosef and Meadows 1995:73,79; Moore 1995:39-46; Ruby 1995: 135.

[128] Davidson 1949:58 (Map 1)

[129] Officer 1901:238-41; Mathews 1908-9:63-4,66-7; Newland 1926:111; Black 1949:105; Davidson 1949:63-4; McCourt 1975:125.

[130] Bar-Yosef and Meadows 1995:73; Moore 1995:45; Officer 1901:238; Mathews 1908-9:63,66.

[131] Reuther 1981:VI,17/982[4]; VI,17/27; VI,17/152[5]; VI,17/209; Spencer 1918-19:12.

[132] Although shards of pottery containers have been found in Arnhem Land (Berndt and Berndt 1947) and elsewhere it does not appear to have been of Australian manufacture but brought in by the *Macassans* (Thomson 1954). In fact remains of pottery brought by the *Macassans* appear at many sites across northern Australia from Arnhem Land to the Kimberleys (Morwood and Hobbs 1997; Mulvaney and Kamminga 1999:415-7; Bulbeck and Rowley 2001).

[133] Harper 1898:421,427-9,432; Horne and Aiston 1924:36; McCarthy 1962:37; Anon. 1969:20-2; Tolcher 2003:37.

western Victoria, large balls or bricks of clay (10-15 cm in diameter) were placed in the "ovens" (large cooking fires on elevated mounds) as heat retainers. These were baked prior to use, or became baked in use, often turning into a form of coarse pottery, and were re-used until they fragmented.[134] Deliberately created baked clay objects include a finely crafted phallus over 17 cm long and 4-5 cm wide and fishing sinkers, both having been reported from north west New South Wales.[135] In addition large (29 cm by 24 cm), spindle-shaped, sun-dried, "bricks" of iron (and manganese) oxide were produced by the *Waramanga* of the northern part of the Corners Region.[136] Sun-dried vessels and storage containers were also in use. Seed storage containers "wrapped up in grass and coated with mud" and holding up to 50 kg have already been mentioned. Utensils may have been made this way as well. Duncan-Kemp, for example, reports uncovering a "cracked bowl, a sort of rough pitcher" made from sun-dried clay at an abandoned "gunyah" in south west Queensland.[137] In conclusion it is apparent there are strong parallels between the use of clay and the baking of clay for utilitarian, ceremonial and recreational purposes in the areas where it has been argued the Early Australian Neolithic was in effect, and the early Neolithic of south west Asia.

Turning to the issue of domestication, it has already been established that the *Nhanda* were cultivating a domesticate, *Dioscorea hastifolia*, if one accepts the argument that it was originally an introduced species of yam that was probably already a domesticate.[138] While yams often do not actually become "true cultigens",[139] nevertheless, once introduced, it underwent further speciation, as noted in Chapter 4. This occurred either through its isolation and the "founder effect", by becoming a naturalised form of the original species or, as happens when domestication occurs, as a result of the selection regime applied by its *Nhanda* cultivators.

Unfortunately the domestication status of the plants being cultivated in the Corners Region is not known at the present time. Indeed, with *ngardu*, because it is a rather unusual food plant with no common relatives in cultivation, it is difficult to assess what features would characterise it as being domesticated. Another of the major food plants of that region, *munyeroo* [*P. oleracea*], may not be domesticable as this species, despite having been cultivated, as a vegetable mostly, in other parts of the globe for 4000 years, nevertheless remains indistinguishable from wild specimens in those places.[140] Consequently the focus here will be on *P. decompositum*, because it is a millet with cultivated and domesticated relatives elsewhere, and of the same family as most cereal crops. Furthermore, there is at least some relevant evidence available on its exploitation in traditional contexts that has a bearing on the issue of its domesticability. Before reaching a conclusion on this, however, the processes leading to domestication need to be considered.

When plants become "domesticated" as the result of a human induced selection regime, they undergo changes in form and structure to such an extent they often become new species. Genetic change takes place in this process and the subject plants frequently become dependent on humans for the continuance of their life cycle. If humans cease to plant the species it may either revert to its former wild form or die out because it is human-dependent. In some of the cereals, such as einkorn and emmer wheats, and foxtail millet, for example, the principal changes that takes place are the development of "tough" rachis (the "spine" which holds the seeds), near simultaneous ripening and seed release of the crop, and loss of seed dormancy, so that the seed loses its thick coat and begins to germinate almost immediately when sufficient moisture is present.[141] But "domesticated" cereals may not actually exhibit all these changes. Spelt wheat (*Triticum spelta*), for instance, still retains "brittle" rachis, thought to be a genetic consequence of its adaptive ability to grow in the colder climates of Europe, "close-fitting glumes", whereby the seeds are not easily released during threshing, and uneven ripening.[142]

The cycle of planting and harvesting is probably all that is required for the loss of seed dormancy.[143] Experimental evidence indicates, however, that the harvesting method played a critical role in determining whether cereals such as wheat, rice and millet developed one of the other key characteristics of a domesticate, tough rachis. These experiments began with investigations into the most efficient method of collecting seeds from wild cereal stands. Drawing on archaeological findings and observations of traditional practices five methods of harvesting have been identified in relation to cereals - beating, reaping (cutting), uprooting, plucking and hand stripping.[144] The efficiency of each method depends on the density of the stands of the crop being exploited, with uprooting the most efficient where densities are low, and cutting with a sickle or blade

134 MacPherson 1885:53; Beveridge 1889:32-4,78; Bennett 1893:2:3; Mitchell 1949:190; Gresser 1964:131; Crawford 1965:8; Holdaway, Fanning and Witter 2000:48.

135 Campbell 1921:145-6.
Beveridge (1889:78) also reports that remnants from the clay ball heat retainers were used as net weights.

136 Eylmann 1908:325.

137 Duncan-Kemp 1952:142.

138 The most likely progenitor was D. alata.

139 Coursey 1972:225; Coursey 1976a:387,391.

140 http://www.fao.org/docrep/t0646e/t0646e0t.htm:3.
It is worth noting, however, that in Central Australia munyeroo plants "tend to ripen more or less simultaneously ... ... as if the plant wants its seeds to be devoured at the critical time for collection" (Latz 1995:250). Latz (1995:250) points out that munyeroo may not be *P. oleracea* but a different species from the Portulaca genus requiring proper identification.

141 Zohary 1969:48-9,57-8,60; Wilke et al. 1972:205; Hillman and Davies 1990:161; Lu 1998:906.

142 Helbaek 1960:100,106; Jarman 1972:20; Willcox 1999:484.

143 Hillman and Davies 1990:206.

144 Wilke et al. 1972; Hillman and Davies 1990:172; Anderson 1998; Lu 1998:904-5; 1999:115-6.

where they are the densest.[145] "Green" harvesting appears to have been commonly carried out in undomesticated cereals because this ensured the highest level of seed retention, while still retaining seed viability.[146] Two of these harvesting methods, cutting or uprooting of green cereals, have been shown to lead to tough rachis. This appears to be so for the ancestral wheats as well as foxtail millet.[147] With the ancestral wheats Hillman and Davies calculated they could have been domesticated by these means in the space of 20-30 years, though between 30 and 200 years was more likely, depending on the natural mutation rate.[148] Others have argued that isolation of the mutants, by their cultivators moving outside the natural distribution (translocational planting), is also necessary to avoid the mutants being swamped by the wild varieties.[149] This may explain why the domestication of cereals in south west Asia took longer than calculated by Hillman and Davies.

As currently understood, the sequence of events leading to domestication began with bladelets and flint sickles being used in the early Natufian to harvest wild cereals, with denticulate (toothed) sickles and stone blades so employed to cut millets in Neolithic China.[150] Studies of the cutting edges of the south west Asian sickles (edge-wear analysis) show that the plants being harvested were not fully ripe, but were harvested while "wet" or "green". "Dry" harvesting did not take place for another 1-2000 years, until the beginning of the PPNB, around 10,800 cal. BP, by which time these cereals had become domesticated.[151] But it would appear that uprooting may have been practised as well.[152] Threshing to separate the grain from the ears, still attached to the stalks, was probably done in the "cup holes", an innovation characteristic of many south west Asian Natufian and PPNA sites.[153] Alternatively, rice was probably threshed in holes in the ground using a "wooden pestle" along the Yangtze in the Chinese Neolithic.[154] In the final stages of the harvest in the early Neolithic of south west Asia "sheaf-burning", as evidenced by the high frequency of carbonised seeds and clay-lined pits found at sites in the Levantine Corridor, may have been a part of the repertoire of chaff removal and parching grains while still green.[155]

If one compares these practices with those from parts of the Corners Region some interesting correspondences arise. Kate Langloh Parker's description of the *Yuwaaliyaay* harvesting of *P. decompositum*, which they called *yarmmara*, is most instructive when considered in full:

*"When the doonburr, or seed, was thick on the yarmmara, or barley-grass, the tribes gathered this grass in quantities.*

*First, they made a little space clear of everything, round which they made a brush-yard. Each fresh supply of yarmmara, as it was brought in by the harvesters, was put in the yard. When enough was gathered, the brush-yard was thrown on one side, and fire set to the grass, which was in full ear though yet green. While the fire was burning, the blacks kept turning the grass with sticks all the time to knock the seeds out. When this was done, and the fire burnt out, they gathered up the seed into a big opossum-skin rug, and carried it to the camp.*

*There, the next day, they made a round hole like a bucket, and a square hole close to it. These they filled with grass seed. One man trampled on the seed in the square hole to thresh it out with his feet; another man had a boonal, or stick, about a yard long, rounded at one end, and nearly a foot broad; with this he worked the grass in the round hole, and as he worked the husks flew away.*

*It took all one day to do this. The next day they took the large bark wirrees, canoe-shaped vessels, which when big like these are called yubbil. They put some grain in these, and shook it up ... thus all the dust and dirt sifted to end, whence it was blown off. When the grain was sufficiently clean, it was put away in skin bags to be used as required ...*

Several things are notable in this account. Firstly, the grain was "full in ear yet green", green harvesting was occurring. Secondly they were sheaf burning, releasing the grain and parching it at the same time. Thirdly, the holes used in threshing and initial winnowing, also reported on the Georgina River in south west Queensland and Central Australia,[156] may well be the functional equivalents of the "cup holes" found in the early Neolithic of south west Asia, or the holes in the ground used in Early Neolithic China. This evidence, if considered in conjunction with the flint harvesting knives, equivalents to flint sickles, suggests there are some significant parallels between the reaping, threshing, winnowing and parching practices used in at least some parts of the Corners Region, and those of the Neolithic in south west Asia and China. But the critical issue here,

145 Harlan 1967; Wilke et al. 1972; Jones 1979:144; Hillman and Davies 1990:196; Anderson 1998:150; Lu 1998:904-6; 1999:115-6.
146 Jones 1979:146,148; Hillman and Davies 1990:159-171,175,195-6; Willcox 1992:167; Anderson 1998:148-50; Lu 1999:115-6; Willcox 1999:483.
147 Hillman and Davies 1990:174-6; Anderson 1998:148-52; Lu 1998:906; 1999:63.
148 Hillman and Davies 1990:187-193; 1992:151.
149 Flannery 1965:1251; Hillman 1996:194; Anderson 1998:152; Wetterstrom 1998:42; Bellwood 2005:19.
150 Unger-Hamilton 1989:89; Hillman and Davies 1990:208; Unger-Hamilton 1992:233; Anderson 1998:146-50; Lu 1998:906; 1999:61, 63,68.
Lu (1999:115) showed that blades of only 1.5 cm were required for cutting.
151 Unger-Hamilton 1989:101; Hillman and Davies 1990:195,208-9; Anderson 1992:180; 1998:148-50,152; Willcox 1999:493
152 Hillman and Davies 1990:172,175,197-8.
153 Bar-Yosef and Belfer-Cohen 1989:470,475,480,489; Bar-Yosef and Kislev 1989:636; Bar-Yosef et al. 1991:411; Hillman and Davies 1990:191.
154 Lu 1999:121.
155 Bender 1975:139-40; Hillman and Davies 1990:198; Bar-Yosef 1998a: 171.
156 Coghlan 1898a:48; Cleland 1966:146; Tindale 1974:99; 1977:346; O'Connell and Hawkes 1981:101.

in terms of the domestication of *P. decompositum*, is the harvesting method.

Green harvesting was also noted by Mitchell on the central Darling in 1835, the "grass was beautifully green beneath the heaps and full of seed", with further indications that uprooting was the preferred method there, "one particular kind of grass had been pulled up."[157] He again remarked upon uprooting around the Narran River in 1845, where he had seen heaps of dry straw over a distance of nine miles [14 kms].[158] Bunce, who accompanied the ill-fated explorer Leichhardt in an abortive foray into south west Queensland observed in 1847 that *P. decompositum* was "cut, dried [my emphasis], and threshed,"[159] as did Gregory in 1858, when he had seen the harvesting of expanses of 250 hectares in the same area "by means of stone knives, cutting down the stalk half way,"[160] although in this instance he did not indicate the degree of ripeness. Sturt likewise beheld this species "spread out on the sloping bank of the creek to dry, or ripen, in the sun" in western New South Wales in 1845.[161] Furthermore Gilmore, however her evidence may be viewed, claimed that the *Wiradjuri* collected "grass-seeds ... in the ear" before threshing and winnowing them in preparation for consumption or planting.[162] Therefore, given the information available, it can be concluded that green harvesting of "native millet" was taking place in the Corners Region and that these plants were harvested by uprooting or cutting, unlike elsewhere in Australia where grass seeds were obtained by hand stripping or beating.[163] Consequently, if *P. decompositum* was domesticable, then it is likely the practices in the ECAN, in terms of its planting, would have led to loss of seed dormancy, and harvesting, to tougher rachis, both markers usually associated with domestication.

Evidence from the ECAN also presents an alternative to the hypothesis that genetic isolation by movement of the cultivators away from the wild stands is necessary to avoid domesticated mutants in the crop being swamped. The seeds, as Smith, Duncan-Kemp, Mathews and McBryde attest, were being traded to other areas. This is a simpler explanation, which could unintentionally produce the same result, genetic isolation of the species by physically moving the seeds to other localities.[164]

In conclusion it can be seen that the totality of evidence shows that the planting and harvesting practices in the region, along with the trading of seed stocks, should have brought about the domestication of "native millet", if it was domesticable. If the proposed Australian Early Neolithic did exist then one would expect to see just such practices. Gilmore's evidence of seed selection, if accurate, further supports this claim.[165] And finally, botanical evidence that in some specimens there are indications of reduced seed dormancy,[166] may be seen as proof that domestication of *P. decompositum* had indeed been in progress.

Changes in the division of labour based on gender is the third defining issue in terms of the transition to agriculture to be considered here. Theoretically, where this transition takes place, hunter-gatherers societies, in which stereotypically the men hunt medium and large game while women gather plant foods and catch small animals, are transformed, and a new division of labour emerges whereby both men and women systematically engage in the planting, harvesting and processing of plant foods. A related question is which sex actually initiated agriculture, in the sense of which sex was the first to begin planting seeds, setts of tubers or such like. The critical issue in this context is, however, the change in the sexual division of labour.

Where the transition to agriculture has taken place in particular cultures in the past it is widely believed that the roles of the different sexes changed.[167] But this is usually based on an assumption about what the division of labour was prior to this transition. In such cases it is assumed that the sexual division of labour at that time, in the Paleolithic or pre-neolithic, was virtually identical to the one found in "simple" or "generalist" hunter-gatherer societies known from the ethnographic present or recent history.[168] While this is probably a reasonable assumption, there is, in fact, very limited direct evidence available to support it.[169] One ethnoarchaeological study at a Northern Plains Besant phase site in eastern Montana, the Mini-Moon site (dated 1520 BP), designed to test this assumption, produced results consistent with the assumption. Archaeologist Susan Hughes developed an ethnographic analogy by documenting the division of labour among historic Northern Plains Indians and the Alaskan Nunamiut, and the resultant artefact distribution pattern that was created at their camp

[157] Mitchell 1839:1:290-1.
[158] Mitchell 1848:90.
[159] Bunce 1847:169.
[160] Gregory 1887:132.
[161] Sturt 1849:1:285.
[162] Gilmore 1934:221.
Tindale (1974:105) claims the *Gamilaroi* and *Wiradjuri* used the same harvesting and processing methods as the *Yuwaaliyaay*.
[163] Cleland 1966:146; Cane 1984:77; 1989:104-5; Tonkinson 1991:38,45-6; Institute of Aboriginal Development 1992:23; Latz 1995:240.
Another method was to sweep up seeds collected by ants and laid in a ring around the entrance to their holes (Tindale 1974:22,99; 1977:346; Cane 1989:104-5).
[164] There has been, in fact, a rather dubious suggestion made that *P. decompositum*, along with *Portulaca oleracea*, is an introduced species in the "undisturbed dunefields" of central Australia (Buckley 1981:377)

[165] Gilmore was most insistent that deliberate seed selection of grass seeds was taking place prior to their planting (Gilmore 1934:221-2.) It must be kept in mind, however, that her evidence may well be tainted, it is not clear what trait was being selected for, and it pertains most directly to the adjacent grasslands of central New South Wales, and not strictly speaking to the ECAN/Corners Region.
[166] Common Grasses of Australia: Native Millet: http://www.lpe.nt.gov.au/dlpe/advis/grass/grassprop/planthtml/Millet.html
[167] Crabtree 1991:389-90; Bar-Yosef and Meadows 1995:67,93; Peterson 2002:107.
[168] Conkey and Spector 1984:6; Crabtree 1991:384,386,387; Sassaman 1992:258; Balme and Beck 1993:63-9; Claassen 1997:76-9,83; Peterson 2002:22-3,25,127.
See Flannery 1969:79; 1972:25,31; Bouchud 1987 in Crabtree 1991: 386 for typical examples of this assumption.
[169] Ehrenberg 1989:55.

sites. She then examined the artefact distribution pattern at the Mini-Moon site and found it to be homologous.[170]

Another study, focussing on lower limb morphology, particularly the cross-section of the femur, a skeletal element that is sensitive to change in lifestyle activity, compared a number of Upper Paleolithic skeletons with samples from more recent hunter-gatherer, agricultural and industrial populations. In this instance it was found that the femoral bone morphology of the Upper Paleolithic cohort showed distinct differences between the sexes, with males exhibiting characteristics consistent with more running and walking longer distances, common activities when hunting.[171] The researcher, Ruff, concluded that in this respect the Upper Paleolithic group were "essentially indistinguishable" from recent hunter-gatherers, while differing from the agricultural and industrial populations where there is little difference in terms of that particular marker.[172] He also found increasing convergence in this facet of bone morphology between males and females with the advent of agriculture and then industrialism. Men, in effect, appeared to have been running and walking long distances less and less once the transition to agriculture had commenced.[173]

A paleopathological analysis by Peterson, which endeavoured to ascertain the gender assignment of hunter-gatherer activities in the Natufian, found osteological evidence (Muscloskeletal Stress Markers) in male skeletons consistent with stresses on the shoulder from hurling projectiles when hunting medium to large game.[174] Based on this type of evidence Peterson concluded that there was also a common activity shared by men and women in the Natufian, and, although she could not specifically identify what that activity was, speculated that it may have been some form of food processing.[175]

In agricultural societies, even those in which individuals derive only a limited part of their subsistence or income from growing crops or gardens, we find men are usually, but not always, systematically involved in some or all of the component activities of planting, harvesting, threshing or processing of plant foods.[176] If the supposition, that in Paleolithic or pre-neolithic cultures there was a gendered division of labour identical to that found in ethnographically known generalist hunter-gatherers, is correct then men, at some point in the transition to agriculturally based food production, must have begun to engage with plants in a systematic way. They could have shared all the component activities with women, or a new assignment regime could have evolved, whereby one sex would do the clearing and hoeing or ploughing, the other the planting, another the harvesting, and so on. The crucial point, however, is that while women's role may change in regard to the nature and level of their involvement with plants and the process of turning them in to food,[177] men systematically engage with plants and their processing for the first time. Several archaeological studies, such as the one outlined above, have been carried out in an endeavour to identify the nature and timing of such a transformation. One in particular, an osteological study in the south east of the United States, noted women's tasks became more varied, with one particular set of markers showing a marked increase in an activity consistent with the intensive pounding of corn. Conversely men developed greater bone strength, especially in the legs, interpreted as them taking up clearing, planting and harvesting, the type of activities undertaken by their male descendants in their ethnographically observed traditional context.[178]

Similar research regarding the transition to agriculture and the Neolithic in south west Asia, while able to provide general indicators, and some specific suggestions, lacks sufficient resolution to be able to directly link any particular activity to one or other of the sexes. The most pronounced change that occurred during the transition was in the musculature, manifested by its impact on the skeleton, of males.[179] Evidence relating to the division of labour in the Neolithic appears to show that the division was less pronounced than that found in the preceding period, the Natufian, before becoming much more obvious in the Early Bronze Age.[180] As in the Natufian there were osteological indications of a common, unidentified, habitual activity in the Neolithic, but there were other signs that men may have been doing some form of heavy repetitive work consistent with an agriculturally oriented activity, such as using a hoe, an axe or an adze.[181] Furthermore, skeletal changes consistent with those seen in individuals who do reciprocal grinding, such as seed grinding, were seen not only in females but also some males.[182] However, according to Peterson, no "associations" between particular Neolithic artefacts and the gender of the habitual users of those artefacts has been established as yet, at least on paleopathological grounds.[183]

Other archaeological evidence does, nevertheless, provide a basis for making inferences about the division of labour, allowing designation of male activities, in the Natufian and Neolithic of south west Asia, and the Chinese Neolithic. For example, each grave in a mixed gender Natufian cemetery of 50 burials at Nahal Oran had a "pipe mortar", thought to be

170 Hughes 1991.
171 Ruff 1987:397,402-3,411.
172 Ruff 1987:405,411.
173 Ruff 1987:399-400,404,407,411.
174 Peterson 2002:98,101,103-4,128.
175 Peterson 2002:101,104,144.
176 Murdock and Provost 1973:207; Ehrenberg 1989:81

177 While there is speculation as to what impact the transition had on women's role and specific task assignment, the evidence is very limited and conclusions contradictory. See for example Burton and White (1984), Bridges (1989:391), Bar-Yosef and Meadows (1995:93), and Longacre (1995).
178 Bridges 1989.
179 Peterson 2002:128.
180 Peterson 2002:11.
181 Peterson 2002:111.
182 Peterson 2002:111,129.
183 Peterson 2002:33.

used for processing acorns (*Quercus* sp.), as a headstone.[184] It could be inferred from this that men had been using these pipe mortars to pound acorns, commencing in the Natufian. Wood and bone spades, mentioned previously, have been found in male graves from the Chinese Neolithic, again suggesting the possibility that men had been using them in some form of activity associated with plant based food production. The fact that grinding slabs were found in the female graves from the same Neolithic culture, the Peiligang, indicates that gender specific task assignment in the food production regime may have already been in place by that time.[185] Peterson conjectured that in the transition to agriculture, remembering that this is a process taking place over one or two thousand years, men and women may have initially shared many of the activities involved in plant food production before a new task specific division of labour developed.

Before considering Australian evidence in respect to this dynamic, it is necessary to firstly look at the sexual division of labour among hunter-gatherers outside Australia. Already this matter has been raised in terms of the assumption that a stereotypical hunter-gatherer division of labour was in effect in the Paleolithic and other pre-neolithic cultures of prehistory. Certainly there is an overwhelming consensus that "normally" in hunter-gatherer societies women "gather" plant foods, as well as hunt smaller animals, and men, conversely, hunt larger game.[186] Consequently various authorities categorically state such things as: "that women are responsible for plant food gathering in virtually all foraging societies";[187] there is not a single "example of a hunting/gathering society where gathering was a male prerogative";[188] "gathering of wild plants, fruits and nuts is women's work".[189] Yet there are examples cited in the literature of supposed hunter-gatherer groups where men do have an involvement at some point in planting, harvesting, collecting, or processing plants.

Among these contrary examples attention has been drawn to the *Washo* of the Great Basin area in western North America.[190] Here men collaborated with women in collecting pine nuts.[191] *San (!Kung)* men of the Kalahari Desert collect mongongo nuts (*Ricinodendron rautanenii*) with their wives.[192] Eastern *Hadza* men of east Africa spend time in the bush each day collecting plant foods "to satisfy their hunger."[193] The *Mbuti* men of the tropical rainforests of Zaire sometimes collected mushrooms, nuts and berries as they moved through the forest.[194] In Californian groups, such as the *Coahuilla* and *Western Mono* [*Monache*], men, as part of mixed groups of all genders and ages, gathered acorns.[195] The *Coahuilla* men also almost exclusively harvested and processed agave (*Agave desertii*).[196] Men in 46 Californian Indian groups in fact appear to have been involved in harvesting of acorns and pine nuts.[197] And lastly, the *Mikea* men of Madagascar, and other groups in that area, gather roots in the forest as part of a family or mixed sex group.[198] When these examples are examined more closely, however, their validity is questionable.

The *Washo* had permanent settlements, were partially sedentary (i.e. part of the population was fully sedentary) and engaged in large-scale storage of seeds, fruits, nuts and dried fish.[199] The *San* grew crops of maize, sorghum and melons during high rainfall periods, stored, herded, had pottery, used metals and lived in villages.[200] In fact the people of the Kalahari had adopted pastoralism, metallurgy, and "extensive grain agriculture" by 800 AD but de-intensified as the climate worsened.[201] It would appear the Eastern *Hadza* men were only carrying "incidental gathering" (see below) and other *Hadza* groups were agriculturalists,[202] the group studied were in fact "surrounded by cultivators."[203] The *Hadza* group in question also used iron and were at least seasonally sedentary. Likewise the *Mbuti* at times planted crops, engaged with the market economy (trade and provision of labour), lived for periods in villages of multi-roomed mud huts and stored large quantities of dried meat.[204] In the case of the *Western Mono*, they lived in large winter villages, engaged in long-term, large-scale storage, and were "more like village horticulturalists or agriculturalists than stereotypical nomadic hunter-gatherer groups".[205] As for the *Coahuilla*, they planted crops (corn, beans, squashes, melons) on a small scale, stored and lived year-round in large permanent villages.[206] Of the other Californian groups, many lived in permanent villages, often of 500-1200 souls, constructed substantial, rectangular

184 Stekelis and Yizraely 1963:11-12; Bar-Yosef 1983:15,19; Moore 1983:96; Crabtree 1991:389.
185 Lu 1999:61.
186 Berndt 1974:74; Vinsrygg 1987; Ehrenberg 1989:83; Hayden 1992a: 35; Watson and Kennedy 1991:56-7; Jones 1996:245 are just a few examples.
187 Ehrenberg 1989:83.
188 Vinsrygg 1987:28
189 Gough 1975:63.
190 Crabtree 1991:389.
191 Downs 1966.
192 Lee 1968:30; Draper 1976:210.
193 Woodburn 1968:51.
194 Turnbull 1984:52,57,64,76,157,168,306.
195 McGuire and Hildebrandt 1994:45.
196 McGuire and Hildebrandt 1994:46.
197 McGuire and Hildebrandt 1994:45-6.
198 Kelly 1995:26,262,266.
199 Downs 1966:19-21; D'Azevedo 1986:467,472-6,479.
It is possible they also may have planted chokecherries and tobacco.
200 Lee 1968:31,33; Sahlins 1972:20-1; Draper 1975:95-6,100-1,103; Lee 1976:19; Davis and Harpending 1977:275; Denbow and Wilmsen 1986: 1509; Bollong, Sampson and Smith 1997.
Claassen (1997:86) has already pointed out that the *San* did not provide an appropriate analogy for pre-neolithic cultures. The fact that their neighbours engaged in pastoralism and agriculture, and that Lee's original observations, by his own admission, were of a situation that was not "typical" (Lee 1968: 40), must raise doubts about their validity. Moreover they were made at a time of "one of the worst droughts in South African history" (1968:39).
201 Denbow and Wilmsen 1986; Evans 1996:66; Robbins et al. 2005.
202 Woodburn 1968:49-50,52-4.
203 Sahlins 1972:27.
See also Woodburn 1968:54.
204 Turnbull 1984:11,21-5,27,33,51,54,84,179,305.
205 Jackson 1991:303-4,308-9,312,317.
206 Barrows 1900:36,38-9,46,52-3,59,63; Bean and Saubel 1972:10,19-20,33-5,161,167,184,205-7; Heizer 1978:578.

houses, had cemeteries, pottery and engaged in systematic, extensive, large-scale, storage, frequently employing storage structures.[207] Growing tobacco was widespread and some groups were involved in limited planting (broadcast seeding) of crops.[208] Finally, the *Mikea*, it transpires, live in "semi-permanent" hamlets in the wet season, grow corn by "slash-and-burn horticulture" and engage with the market economy (working for wages, selling their roots in the local market).[209]

It can be concluded, from the foregoing, that in generalist hunter-gatherer societies men normally do not engage with plant foods. The contrary examples cited in the literature almost invariably turn out to be referring to groups who are not classic generalist hunter-gatherers. They are perhaps so-called complex hunter-gatherers, or groups in a transitional phase in the development of an agriculturally based economy, possibly people whose lives have been modified by exposure to modern market economies, but they certainly are not generalist hunter-gatherers following a traditional lifestyle. Australia, on the other hand, reputedly being the "continent of hunters and gatherers", may be able to provide more representative evidence of hunter-gatherer socioeconomic systems in regard to the issue of the division of labour, by drawing upon historical observations of groups living in traditional circumstances.

When traditional Australian Aboriginal populations are considered as a whole, statements similar to those characterising hunter-gatherers elsewhere are proffered in relation to the sexual division of labour. The consensus is that traditionally a "clear cut" division of labour existed, whereby plant foods were almost exclusively and systematically gathered and processed by women, while men had almost sole responsibility for the hunting of large game.[210] Included in the assignment to women is seed grinding, an "almost universal role", there being "no observations of men using grindstones for food preparation."[211] In other areas, such as fishing or hunting of smaller animals, more complex or varied customs emerge. Nevertheless, as with most generalisations, there are exceptions. There are numerous instances recorded where men have on occasion gathered plant foods. This is what I call "incidental gathering". Although there are strong distinctions in the division of labour in hunter-gatherer societies, these are not completely rigid, there is always some flexibility to allow for circumstances. Consequently it is possible to find examples where men out hunting, "as they walk along will snatch a few berries from a bush."[212] If a man returned to camp empty handed and there was no food there, and no one, such as his wife, to provide for him, he would dig yams or tubers, even collect seeds, and process and cook them to appease his hunger.[213] Male outcasts in traditional Aboriginal societies also had to provide for themselves, including gathering their own plant foods.[214] But despite these counter examples it is quite evident that overall men in hunter-gatherer societies in Australia did not, in traditional circumstances, systematically engage in gathering plant foods. Yet, as we shall see, there were historically observed situations in Australia where men appeared to be doing just that.

Returning to Parker's account of the harvesting of the *yarmmara* by the *Yuwaaliyaay* it is evident that men were fully involved in the harvesting, threshing and initial winnowing of the grain. Parker refers to "the tribes", "the harvesters", "the blacks", "they", undertaking these activities, not just women. Mentioning that "one man trampled on the seed" and "another man had a *boonal*" removes any doubt that both sexes were involved. Other comments support that view. Following the harvest a "great hunt" took place, a "big feast" prepared and a corroboree was held "night after night",[215] signifying that the harvest had been a significant group activity. Furthermore, the involvement of men appears to have gone beyond just harvesting, threshing and winnowing. *Yuwaaliyaay* men were also engaged in seed grinding, "he begins ... grinding grass-seed".[216] This was in the context of a game but there is additional graphical evidence to support the notion that men ground grass seeds. An illustration drawn "By one of the Euahlayi Tribe"[217] and titled, "A Native Grinding Grass Seed on a Dayoorl-Stone", shows a man grinding grass seeds on a *dayoorl* grind stone.[218] That illustration appears below.

Supporting this proposition are two *Yuwaaliyaay* myths recorded by Parker indicating that the *dayoorl* were men's property and that they used them to grind seeds and make a form of grass seed bread.[219] While it could be argued that

[207] Heizer 1978:60-67,131,151(villages, house construction),32,34, 60-1,131,160,197,209,216,329,357,389,392,455,511,519,522,535(cemeteries),54,56,430,442,452,465,474,542,553,571,578-9,600,602(pottery),130,217,221,242-3,252,338,340,374,388,404-5,409,428-9,431, 443-4,451,493,497,501,516,571,578,597,599,600,636(storage); D'Azevedo 1986:472,479 (villages, construction),269,375,401,421-2(pottery),401(storage).

[208] Heizer 1978:24,130,181,217,222,379,402-3,650,667 (tobacco),217 (broadcast seeding – *Shasta*),600 (horticulture – *Tipai-Ipai*); D'Azevedo 1986:94(garden horticulture – *Death Valley Shoshone*), 371 ("native agriculture", maize, beans, squash – *Southern Paiute* [Calif.]); Shipek 1989 (translocational planting, broadcast seeding, tobacco growing – *Kumeyaay*); Keeley 1995:265 (tobacco); Smith 1995a:17,21 (planting, sowing – Kumeyaay).

[209] Kelly 1995:26.

[210] "A Lady" 1860:228,257; Stirling 1896:55; Basedow 1925:148; Cleland and Tindale 1936:28; Berndt 1974:74; Hiatt 1974:6-7, 10,12; Maddock 1982:169; Berndt and Berndt 1985:119-20; Tonkinson 1991:43 are a representative sample.
Women are not excluded from hunting larger game, at times hunting large game opportunistically.

[211] Hayden 1992a:35; Haaland 1997:378.

[212] Oldfield 1865:271.
See also Worsley 1961:162; Hiatt 1968:209-10; Gould 1969:262; Tonkinson 1991:47.

[213] Warner 1958:140.
See also Spencer and Gillen 1912:2:279; Institute of Aboriginal Development 1992:12.

[214] Bates 1985:242.
See also Berndt and Berndt 1985:119; Tonkinson 1991:45.

[215] Parker 1905:118-9.

[216] Parker 1905:127.

[217] Parker 1905:vii.

[218] Parker 1905:opp117.

[219] Parker 1897:64; Stow 1930:90.

*Figure 17: A Yuwaaliyaay Man Grinding Seeds*
*(Parker 1905:opp117)*

what Parker had recorded was a post-contact adaptation, perhaps a response to loss of hunting grounds, or the artist, being male, wished to portray men as significant actors, it must be remembered that Parker had arrived in the area in 1879, and that her "chief sources" were the "old men."[220] It therefore seems likely that she was recording traditional *Yuwaaliyaay* life from the 1840s, a contention bolstered by references to men seed grinding in traditional myths.

Not surprisingly there are other sources of information, many already referred to in earlier chapters, that document or allude to men's participation in the sowing, harvesting and processing of plants in the Corners Region. Newland, in discussing the *Paakantyi* stated that seed gathering was "principally done by the women and children,"[221] implying that at some point men were involved as well. A number of accounts from the region mention "natives"[222] and "blacks",[223] not women specifically, harvesting or processing seeds and *ngardu*. Broadcast seeding was carried out by men in the descriptions provided by Walter Smith, as it was when the *Wongkanguru* sowed *wirra* and *yaua* seeds. On the fringes of the Corners region, in Central Australia, here Kimber indicated that it was men who did the planting of "native fig", "bush potato", the *moolyardi* spear tree and Sturt's Desert Pea. And a recent oral account indicates that traditionally in the *Jandruwanta* "only the men" ground certain seeds.[224]

When mythological evidence is taken into account there are multiple sources from the Region signifying male involvement not only in the sowing of seeds, as seen earlier in the legends relating to the *muramura Markanjankula*, but many aspects of harvesting and processing. The principal spirit being of much of New South Wales and adjacent areas, *Baiame*, is described pounding nuts and grinding seeds to make "cakes".[225] Other *muramura*s, such as *Ngura-tulu-tulu-ru* and *Ngardu-etya* of the *Jandruwanta* of north east South Australia, are depicted grinding *ngardu* and *munyeroo/paua* on the "magic grindstones on which the men ground their seeds" as well as baking seed cakes.[226] The same *Ngura-tulu-tulu-ru* appears in *Yauraworka* mythology where he "hears the men grinding the seeds the women

[220] Parker 1905:2; Parker and Muir 1982:34.
[221] Newland 1920-1:13.
[222] e.g. Lewis 1975b:20; Brock 1975:111; Gregory 1887:132; Browne in Allen 1972:77.
[223] e.g. Wills 1861:29.

[224] Kerwin and Breen 1981:307.
[225] Mathews 1904:342.
[226] Howitt 1904:794-5; Howitt and Siebert 1904:121-2; Tolcher 1990:5. Bennett (1893:1:24) describes men, probably from the Bogan River area, making seed damper dough, ready for baking.
He also claimed (1893:1:23) that the people from this region did not sow but were "agriculturalists" because of the care they took prior to harvesting in protecting and preserving their grasslands when in seed.

had collected and cleaned."[227] *Karkalina*, a *Wongkanguru muramura* also ground seeds, as did *Kuruljuruna* of the *Diyari*, while *Marluna* of the "*Yeluyanti*" collected and winnowed seeds.[228] One even finds *muramuras* such as the *Ngameni*'s *Pirratintina* and *Darana* of the *Diyari* digging up roots, using a digging stick in *Darana*'s case.[229]

From the foregoing it is clear that men, in parts of the Corners Region at least, were intimately involved in the sowing, harvesting, grinding and cooking of plant foods. These activities certainly do not appear to be "incidental" but reflective of a fundamental shift in the division of labour, men being systematically involved in plant based food production. The reported occurrence of male specialist hut builders among the *Wongkanguru*, the building of shelters normally being "women's work" in mobile groups in Australia, suggests, moreover, that the change in the divisions of labour went beyond engagement with plant foods, and that gender specific assignment had possibly developed.[230]

Realising that the involvement of *Yuwaaliyaay* men in harvesting and threshing of *yarmmara* was unusual, anthropologist Norman Tindale, by way of explanation, claimed that this was taking place because the *Yuwaaliyaay* were a "matrilineally oriented society."[231] This is a curious explanation, inverting the "crude materialist" causality put forward at times to explain descent rules.[232] Given that matrilineality in other Australian Aboriginal groups outside the Corners Region is not correlated with an atypical division of labour, the causal basis for Tindale's position is unclear. Further undermining its validity is the fact that only half the Corners Region had a matrilineal social organisation. The alternative explanation proposed here is that the division of labour was consistent with that found in proto-neolithic and neolithic societies.

When relevant, albeit limited, evidence relating to the *Nhanda* is examined a similar pattern emerges. Surveyor Philip Chauncy, for example, recalled in 1878 that in the earlier days he had "seen both men and women sinking in loose sandy soil for an edible root called *warran*."[233] Explorer Grey, as his hapless party struggled back to Perth, made a journal entry on 11 April 1839 describing how he, unobserved, had seen a man digging yams with a group of women.[234] And again non-gender specific reports of "parties" of "natives" digging yams supports this view.[235] The heavy dependence of the southern *Nhanda* groups on yams and gum, referred to in Chapter 4, whereby Government Resident Burges observed that "they seem very little addicted to hunting and very few of them are even expert at tracking a Kangaroo. This may result from the great variety of edible roots …," and explorer Helpman, that they "seem to depend entirely upon *warran* and gum", is another indication of a local change in the division of labour. Men trading, gifting and paying with yams also points to their intimate involvement with this plant food.[236] Perhaps the clearest indication that yam harvesting by *Nhanda* men of the Victoria District was more than just an incidental activity is provided by Chauncy, who commented:

"*The only bald natives I ever saw are the warran diggers, who are said to wear the hair off the head by pressing it so frequently against the sides of the holes.*"[237]

From this one can readily infer that men in the district were habitually digging for yams. Given such evidence it is reasonable to conclude that there was a division of labour there akin to that found in an agriculturally based neolithic society. As mentioned in Chapter 4, the *Nhanda* were not matrilineal, in fact they had a highly unusual social organisation in Australian terms, one based on consanguinity and reciprocal localities.[238] Consequently, Tindale's matrilineal hypothesis is found wanting in this case as well. The alternative hypothesis, that the *Nhanda* were engaged in agriculture and exhibited the division of labour characteristic of that form of subsistence economy and a neolithic society is, however, borne out.

At the beginning of this chapter the problem of identifying what a neolithic society looked like was alluded to. To address this, a list of characteristic features was developed, and an illustration of an Early Neolithic scene was inserted, to provide some basis for comparison. Nonetheless, this is not necessarily sufficient to provide a complete grasp or conceptualisation of the Neolithic, so that comparable patterns can be recognised as a facet of Aboriginal culture in those parts of Australia where it has been identified. Those with some knowledge of the Neolithic of south west

[227] Howitt 1904:802.
[228] Stirling and Waite 1918-21:118,128,150.
[229] Stirling and Waite 1918-21:138,153.
[230] Horne and Aiston 1924:19-20.
The division of labour in the building of shelters in traditional Aboriginal life has never been studied as far as I am aware, so the generalisation is mine. However there is some support for this in cross-cultural studies (See Murdock and Provost 1973: 213).
In the *Yauraworka* and *Yandruwanta* hut building was a co-operative activity, "both men and women sharing the work" (Tolcher 2003: 14).
There are also indications of hereditary male specialist net makers in the Region (Gason 1895:169; Duncan-Kemp 1961:231; Tolcher 2003:13)
[231] Tindale 1977:346.
[232] Sahlins 1972:54.
[233] Chauncy 1878:2:246.
[234] Grey 1841a:2:54.
[235] Mrs Brown 1851:4.
Although Mrs Brown does not specify the gender of those in the large group she observed there is strong reason to presume, from the brief interactions that took place (such as individuals coming forward to speak to Henry Gregory, leader of Mrs Brown's party), that men were part of this group.
[236] See Drummond 1842:719; Roe 1847:32; Oldfield 1865:236,254.
[237] Chauncy 1878:2:246.
The comment that the men digging the yams were the only bald Aborigines Chauncy had seen, despite experience over 35 years in South Australia, Western Australia, Victoria and New South Wales is an intriguing comment. It can be seen as a small part of the evidence of an unusual genetic influence on the west coast of WA which I have attributed to the presence of Dutch sailors shipwrecked and marooned in the 17th and early 18th centuries (Gerritsen 1994a:70-81,200-2).
[238] See Gerritsen 1994a:144-50 for a fuller discussion.

Asia may well have visions of large town-like settlements such as Catal Hoyuk in Anatolia, or Jericho, in mind, and so may have some difficulty in accepting the dissonant claim of a neolithic in Australia. Of course, such images, of the rectangular cell dwellings and bull cult shrines of Catal Hoyuk, and the tower at Jericho, are of structures from the PPNB or later Neolithic. In reality the early Neolithic was more akin to that seen in Figure 16, except the dwellings were probably plastered with clay. While the advent of agriculture and the Neolithic are linked, it should be realised that even in the PPNB agricultural endeavours were still a secondary part of the economy.[239]

Hunting and gathering continued to be important in the PPNB, as the new forms of subsistence based on agriculture or mixed farming emerged.[240] Large territories were still required to support the resident populations so that people from that period may have roamed considerable distances and spent significant time away, hunting gazelle, oryx and mouflon, fishing with their nets or trying to trap ducks at a local lake or river.[241] Women as well may have gone out every day to harvest wild pistachios, walnuts, acorns and lentils or collect shellfish. Consequently it must be realised that people from places such as Catal Hoyuk and Jericho were probably nomadic to some extent, so that absences of perhaps a month or two at a time, hunting goats in the hills for example, or attending ceremonies at other nearby Neolithic settlements, must be seen to have been the norm at the time.[242]

As for the crops that were grown, initially they were undomesticated or only partially domesticated, so that it was still necessary to harvest them while still green. Domestication was a process that took more than a millennium, as did the full transition to agriculture.[243] But there was also considerable variation in the types of subsistence being pursued.[244] Some Neolithic settlers concentrated on growing crops of wild or only partially domesticated cereals, others grew nothing but fished very successfully.[245] The distribution of wild resources, both animal and plant formed a mosaic across the "Fertile Crescent" region of south west Asia, so that each group developed its own distinctive local economy.[246] But probably one of the most significant changes that took place, however, was relocation. The places where people camped and began to form settlements shifted, from locations such as foothills, open plains and rock shelters, to alluvial sites along rivers, adjacent to springs and *wadi*s, and on the shores of lakes.[247] Here crops presumably grew better and more reliably.

[239] Cauvin 2002:215; Asouti and Fairbairn 2003:1.
[240] Asouti and Fairbairn 2003:2.
[241] Asouti and Fairbairn 2003:4.
[242] Asouti and Fairbairn 2003:4-5.
[243] Willcox 1999:478-80,489-94; Cauvin 2002:51-61,217; Asouti and Fairbairn 2003:1-2.
[244] Brookfield 2002:76; Asouti and Fairbairn 2003:1.
[245] Cauvin 2002:55.
[246] Asouti and Fairbairn 2003:1
[247] Brookfield 2002:68-9; Cauvin 2002:63.

When one takes this type of Neolithic template and views the situation as re-created here of life in the Victoria District or the Corners Region remarkable similarities are apparent. And because the populations from those areas were out hunting emus with nets or spears, fishing with nets and hooks, catching ducks with nets, gathering large quantities of shellfish, or stockpiling a wild resource such as eucalyptus seeds or gum, does not preclude the possibility that they were engaging in agriculture or following a sedentary life. Even with the information that is available at the present time it is still possible to identify that:

1. planting was taking place,
2. that this planting appears to have been relatively extensive in some areas, with the Victoria District being the best documented in this regard,
3. that people were living in villages, and
4. that the peoples involved were seemingly exhibiting a much higher degree of sedentism than normally found in hunter-gatherer societies.

Undoubtedly the degree of sedentism was highly variable across the Corners Region, and certainly not uniform, but it is difficult to achieve a greater level of precision because the evidence available at the moment is quite limited. Clearly there was still some mobility, but the mobility evident in the Neolithic must be kept in mind if one is to compare. Moreover, it must be borne in mind that the very people providing observational evidence, explorers, travellers, "pioneers", missionaries and other intruders may have been the agents, the cause, for mobility, disturbing and disrupting the daily lives of those who they encountered and then recorded.

Perhaps the strongest indicator of the advent of a neolithic type of culture and the advent of agriculture in the Victoria District and Corners Region is the settlement location pattern. It is evident that where apparently permanent or near permanent occupation sites can be identified, through observational or archaeological evidence, alluvial sites predominate. This is true in both the Victoria District and the Corners Region. And so it is with this observation, and the commendation of the preceding body of evidence, that I conclude the case for the conceptual entity posited at the beginning of this chapter, the Australian Early Neolithic. But before closing it may be worth considering as an aside the question of who really did "invent" agriculture?

H. Ling Roth, in 1887, one of the first to seriously consider the question of the origins of agriculture, thought "it was probable" that women "were the first to make steps towards acts which may be regarded as originating agriculture"[248] This view has been echoed many times since.[249] Given women's intimate knowledge and dedicated gathering of plant foods in traditional hunter-gatherer societies, it

[248] Roth 1887:120.
[249] e.g. Stanley 1981:289,292-3; Ehrenberg 1989:77-8,84-90; Bar-Yosef and Meadows 1995:50; Peterson 2002:138.

is not an unreasonable or unexpected proposition. But, as pointed out earlier, while it is often assumed that the division of labour in pre-neolithic societies was along the lines of ethnographically known hunter-gatherer groups, this is not certain because there is such a limited degree of supporting evidence. Certainly there does not appear to be any archaeological evidence to support the notion that women, or men for that matter, "invented" agriculture.[250] Consequently we are only able to speculate on this at the present time, although Australian evidence may provide some insights.

Despite the lack of direct evidence Stanley claimed that women not only devised intentional cultivation and instituted the domestication of plants but also invented digging sticks, carrying bags, reaping knives and sickles, as well as being responsible for a dizzying plethora of other innovations.[251] However, it has also been suggested that the supposition that women invented agriculture may not be correct. Crabtree has argued that with the apparent advent of cooperative procurement in the Natufian, men may have played a role in the evolution of early agriculture.[252] An alternative view proposed by Ehrenberg, was that women did indeed initiate agriculture but that men took over in the later Neolithic, presumably as the economic importance of cultivated plants became more pronounced.[253] So what does the Australian evidence suggest?

If planting in the Corners Region and adjacent areas is considered it is evident, in cases where the sex of the planter is known, that both sexes had "responsibility" for planting. It would appear from Walter Smith's evidence that men were widely engaged in sowing seeds in the western portions of the Corners Region. Similarly, it is men seemingly carrying out the planting ceremonies in Horne and Aiston's account of broadcast seeding of *wirra* and *yaua*. Men appear to be involved in cooperative harvesting, threshing and initial winnowing as well, if notice is taken of Parker's description of the *Yuwaaliyaay*, and other relevant accounts given credence. The agricultural founding myth of the *muramura Markanjankula*, also from the western part of the Corners region, reinforces the perception that men may have had a critical role in the initiation of cultivation in parts of the Corners Region. However, in two other instances of planting in or adjacent to the Corners Region it is women who are carrying out the act. Duncan-Kemp's description of the *katoora* planting ceremony in south west Queensland is one, and Gilmore's suspect reportage of *Wiradjuri* women planting is the other. Interestingly, in Gilmore's account this appears to have been carried out under the direction of a male "chief".

So what can be concluded from this? Firstly, it is quite likely that the assertion that women initiated agriculture, or at least planting, in pristine situations is not necessarily correct in all circumstances. While this is counter-intuitive to some extent, it must be remembered that men in hunter-gatherer societies usually have an extensive knowledge of plants. This is acquired not just through everyday interactions and observations but because men, as impressionable children accompanying their mothers in their daily food search, inevitably absorbed knowledge of many aspects of plant life, and the processes involved in procurement, harvesting, preparation and processing of plant foods. Secondly, it is likely that the species most likely to be planted by men initially included those that had particular value to them. Those species that contain psychoactive substances, or relate to such plants, are a case in point. The *wirra*, for example, valued because the ash from its leaves was used in the consumption of *pitjuri*, was one of the plants sown by the *Wongkanguru* men. Although not necessarily a traditional practice, some of the species Kimber claimed to have witnessed being planted in Central Australia, such *moolyardi* spears trees and Sturt's Desert Pea, required for adornment, may well fall within this category.

With these observations and deductions in mind a hypothetical scenario could be that planting was an undertaking that could have been initiated by either gender. If women initiated planting and it produced a food or plant product valued by men it is more than likely it would have been expropriated by men given the dominance of men in gender relations in hunter-gatherer societies. A parallel for this is fishing in the Sydney region as observed in the late 18th century when the first British colony became established. Women caught fish using fish hooks but handed over most of their catch to their husbands.[254] However, where plant food staples are involved and become a more significant, and in men's eyes a more productive component of subsistence, it would seem logical that men would begin to involve themselves in this instead of hunting. Moreover, women's labour alone may not be sufficient to plant enough plants or to complete the harvesting, threshing, winnowing, parching, storage and finally processing of the subject species. It is at this point that co-operative production may commence, as perhaps exemplified by the *Yuwaaliyaay*. As the scale and importance of plant-based food production increases and this economic mode becomes established, a further change takes place and specific tasks or steps in the process are assigned to either sex, a feature of many societies dependent on subsistence agriculture. Clearing, as alluded to the legends relating to the *muramura Markanjankula*, and planting of staples, such as *ngardu*, grass seeds and *yaua*, as described by Smith, and Horne and Aiston, and even grinding, as indicated in the legends pertaining to *Baiame* and other *muramura*s, are examples that suggest that gender specific task assignment may even have been in effect in some areas.

Finally, it is worth noting that co-operative or collective subsistence endeavours involving the whole group also

250 Pryor 1986:887.
251 Stanley 1981:293.
252 Crabtree 1991:389.
253 Ehrenberg 1989:99-107.
254 Bowdler 1976:256; Walters 1988:110.

represents a fundamental departure from typical practices in generalist hunter-gatherer societies, not only in regard to the sexual division of labour but socioeconomic organisation as well. In these instances the productive unit seemingly extends beyond family and kin, and, at times, embraces the broader social group instead. This, it would appear, occurs not only in the transition to agriculture, as well as in mature agricultural systems, but where other forms of food production are commencing or are in place, as will be contended in the following chapters.

# Chapter 7.
# Big Men and Wannabe Chiefs – Part 1

In prehistory, as well as in reference to some contemporary indigenous populations in some parts of the world, groups have been identified who did not grow crops or keep animals, yet they were sedentary or "semi-sedentary". Such groups are often referred to as Complex Hunter-Gatherers (CHGs), sometimes as Affluent Foragers or some combination of Sedentary Hunter-Gatherer-Fishers. In Australia the *Gunditjmara* of south west Victoria have been specifically identified as CHGs.[1] Apart from having permanent or nearly permanent settlements and exhibiting higher levels of sedentism, CHGs usually had:

- higher populations densities
- more "complex" and "highly developed" technologies
- "intensified" subsistence exploiting a wide-range of species but "concentrating" on a few staples
- increased economic and occupational specialisation
- changes in the division of labour, often beyond age or gender
- storage or food preservation systems[2]

But the notion of CHGs usually goes beyond the characterisation of their subsistence mode, technology, trade, population and settlement pattern and incorporates elements of what is normally termed "sociopolitical complexity". Consequently, descriptors for CHG societies may include expressions such as: "varying degrees of economically based status competition accompanied by socioeconomic differentiation"[3] or, "more formalised leadership."[4] In summary, this facet of CHG societies is seen as embracing features such as:

- more elaborate rituals, beliefs and ceremonies
- more complex religious symbolism
- simple social ranking
- differential distribution and non-egalitarian assignment of wealth, status and authority
- erection/creation of special structures or areas (for "feasting" or the display of status objects)[5]

Although originally thought to be rare and anomalous, and largely restricted to the "Pacific North West" coastal regions of North American, it is now apparent that CHG societies were relatively numerous. In prehistory the Natufians have been nominated as the first CHGs by Henry.[6] The Jomon of Japan, particularly the Eastern Jomon, were another early example. They engaged in coastal marine and pelagic fishing, hunting with bows and arrows, keeping of wild pigs, and heavily utilised storable plant foods such as nuts. The population lived in permanent pit-dwellings in villages of 60-80 people, had storage pits, grinding stones and, as stated before, made some of the earliest pottery known. In eastern Japan their primary specialisation was fishing (using nets, fish hooks, weirs, open ocean watercraft) whereas plant resources (harvesting nuts and possible small scale planting), along with some freshwater net fishing, seems to have been more significant in western Japan.[7] Socially there was an "increasingly normative social order" as the period progressed, with an attendant growth in ritual activity as evidenced by increasing numbers of ritual objects, ceremonial arrangements and buildings.[8] In north western Europe the Ertebølle culture are also proffered as a CHG entity, one that lasted from 8000 to 5500 BP. Here the population lived, at times, in very large permanent settlements by lakes, streams and on the coast, and used stones axes, nets, fish hooks and canoes. Their main sources of food were fish and seals, these providing 70-90% of their diet.[9]

Overall there are dozens of examples now identified as CHG cultures, either in prehistory or in more recent times. These include the *Calusa* of Florida, pre-ceramic coastal communities in Peru from the 3rd millennium BC, Amur and Primodj regions of eastern Russia, south of the Aral Sea, in the Ukraine, sago harvesters of lowland New Guinea, the *Ainu* in Japan, the *Nuaulu* of Ceram in Indonesia, some groups in the Great Basin in the United States, central and

1 Lourandos 1983,1985:404-8; Williams 1987, Lourandos 1988:153-4; Williams 1988; Builth 2000a,b.

2 Price and Brown 1985:5; Keeley 1988:373-4; Hayden 1994:238; Fagan 1995:167-8; Ames and Maschner 1999:25-7.
Some (Price and Brown 1985:5; Keeley 1988:134; Hayden 1994:238) include regional exchange and significant trade in "exotics" as additional criteria.

3 Hayden 1990:33.

4 Palsson 1988:191.

5 Price and Brown 1985:5; Keeley 1988:373-4; Hayden 1994:238; Fagan 1995:167-8; Ames and Maschner 1999:25-7.

6 Henry 1989; See also Shnirelman 1992:187; Bellwood 2005:51-2.

7 Akazawa 1981; Kotani 1981; Shnirelman 1992:187; Akazawa 1986; Imamura 1996:56-120; Tsutsumi 2003:250-261.

8 Imamura 1996:116-120.

9 Rowley-Conwy 1986:25; Zvelebil and Rowley-Conwy 1986:79-88; Shnirelman 1992:187; Price 1995:133.

*Figure 18: Groups and Places Mentioned in Text*

lower Mississippi (such as Poverty Point in Louisiana), Western *Apache*, Tucson Basin, the *Chumash* of the Channel Islands and adjacent Californian coast, and coastal Thailand (such as $2^{nd}$ millennium BC Khok Phanom Di).[10] The north west coasts of North America were particularly rich in societies designated as CHG societies, from northern California up to Alaska. Some examples in this region were the *Eyak*, *Tsimshian*, *Haisla*, *Nuxalk* [*Bella Coola*], *Kwakwaka'wakw* [*Kwakiutl*], *Makah*, *Quileute*, *Chamakun*, *Kwalhioqua*, *Clatskanie*, *Chinook*, *Kalapuyans*, *Alseans*, *Siuslawans*, *Coosans*, *Athapaskans*, *Takelma*, *Tolowa*, *Tututui*, Coastal *Yurok*, *Nuu-chah-nulth* [*Nootkans*], *Salish*, *Haida*, *Gwaii*, *Tlingit*, *Nunamuit*, *Tutchone* and *Lilloeet*. Their distinctive origins have been traced back at least 5000 years.[11]

A feature of the Pacific North West Coast was the existence of economies based on intensive fishing and exploitation

[10] Watanabe 1968:69-71; Gibson 1974; Harris 1977a:220; Bettinger 1978:36-42; Brown 1985; Feldman 1987; Higham 1987:51-8; Yesner 1987:303; Ellen 1988; Marquardt 1988; Widmer 1988; Higham 1989: 70-2; Arnold 1991; Lambert and Walker 1991; Jackson 1991; Arnold 1992; Minnis 1992:130; Shnirelman 1992; Hayden 1994:237; Fritz 1995:3,9-12; Pringle 1997; Higham and Lu 1998:872-3; Higham and Thosarat 1998:44-63; Arnold 2001; Bellwood 2005:34,178; Roscoe 2005.

[11] Piddocke 1968; Suttles 1968; Harris 1977a:219-20; Schalk 1977; Rolland 1985:255-7,261; Cashdan 1989:35-39; Suttles 1990a:ix; Ames 1991; Maschner 1991; Hayden 1992b:526; Shnirelman 1992:187; Ames 1994; Smithyman and Kalman 2004:18,23.

of food from the sea and rivers, or sea mammal hunting. Salmon was a critical resource because of the regular and predictable annual up-river spawning migrations by five species. Fish weirs and fishing platforms in conjunction with nets, basket traps and "rakes" were commonly employed and the salmon, caught by the thousands each day, were dried on racks or "smoked" in purpose built smoke-houses to preserve them. Food, principally salmon, was stored in large quantities, to tide the population over the winter. Other abundant, predictable and/or seasonal resources such as cod, herring, halibut and sturgeon were also exploited. Shellfish were another significant resource and consequently very large middens are evident at prehistoric occupation sites. Where sea mammal hunting took place seals, porpoises, sea lions and humpback whales were caught. While the indigenous economies in some of these regions "specialised in the production, processing and exchange of foodstuffs," predominantly marine and aquatic resources, many and varied wild foods, such as deer, 30 species of waterfowl, 40 species of plant, were also utilised.[12] Fish hooks, copper headed harpoons, ground stone tools, socketed points, bow and arrows, slate and copper knives, even mirrors were in use, often produced by specialist craftspeople. Canoes up to 18 metres long, with two masts and sails and capable of carrying 6-8 tonnes of freight plied the waters in some areas.[13] One unusual feature of the traditional technology of the Pacific North West was their use of metals, principally copper and iron, in weapons, exchange, ceremonial objects and personal decoration, whereas pottery was not used apart from in tobacco pipes.[14]

Many parts of the Pacific North West Coast were marked by unusual settlement and mobility patterns. The region was well populated, one estimate putting the numbers at 188,000 people prior to extensive contact with outsiders.[15] There was considerable variation in the settlement pattern from region to region but large, permanent, villages and towns of up to 2,000 people with "semi-sedentary or sedentary" populations were in evidence throughout when Europeans arrived. Residential structures, rectangular "plank houses" being the norm, housed 40-50 people each, sometimes as many as 100. Villages and towns consisted of from one to a dozen such houses, with the largest having 40. In the southern region conjoined "shed" type houses were common, forming longhouses up to 192 metres long, 18 metres wide and 5.4 metres high. The large residences had, at times, some internal structural features or differentiation and were often fronted by forests of totem poles up to 15 metres high. Numerous villages were fortified with palisades or ditches around them, with those situated on hills being, in effect, hill forts. Cemeteries were common and burial mounds were constructed in the southern parts. An internal settlement hierarchy had also evolved, with status differentiation between dwellings within settlements.[16]

The settlement and mobility, and subsistence, patterns in the Pacific North West were strongly influenced by the very wet stormy winters, convoluted fjord-like coastline, rugged terrain, vast temperate rainforests and seasonal nature of the main food sources, particularly salmon. There were large scale movements of the population to the entrances of the many rivers in the region to harvest the enormous runs of the different species of salmon that occurred in late spring, summer and early autumn.[17] Often the residents removed the planks of their houses and used them to set up temporary villages near the salmon runs and other abundant seasonal resources. Though the degree of sedentism in the Pacific Northwest has not been closely studied, it would appear that a "logistic foraging" strategy was followed, with the different populations probably exhibiting, in accordance with the classification developed earlier in Chapter 5, "sequential", "partial" and "seasonally mobile" and "full" forms of sedentism.[18]

The social and political structure of Pacific North West societies has received considerable attention as a "test case" in debates about the emergence of hierarchies and status differentiation in human societies, the development of sociopolitical complexity. The linkages between households, their salmon resources, storage, sedentism and social stratification have been of considerable interest and the basis for the development of a number of models for the evolution of sociopolitical complexity.[19] Another crucial element was the fact that there appeared to be definite social ranking and political hierarchies in what were ostensibly hunter-gatherer societies. Specifically, there were three classes in evidence in the Pacific North West cultures, "Nobles", clan or lineage chiefs and wealthy high status individuals related to the chiefs, "Commoners" and "Slaves".[20] Despite their attribution as chiefs, their ability to command was very limited and normally restricted to

[12] Plants were also grown in small plots tended by women and slaves. Camas (Camassia sp.), a plant with a bulbous root, was "reseeded" by the Central Coast Salish, while Pacific silverweed (Trifolium sp.) and springbank clover (Potentilla wormskoioldii) was grown in "gardens" on Vancouver Island. Riceroot (Fritillaria spp.) was also grown in the region (Ames and Maschner 1999:120; Hayden 1990:40; Suttles 1990d:459). Potatoes were introduced in the 19th century and rapidly spread through the region, becoming a high status food, probably because of the limited sources of carbohydrates there (Hayden 1990:40,43).

[13] Arima and Dewhurst 1990:394-5; Kennedy and Bouchard 1990:444; de Laguna 1990a:192; 1990b:208; Suttles 1990c:25, 1990d:457-59; Ames and Maschner 1999:113-123.

[14] Codere 1990:369; Cole and Darling 1990:120,122,128-9; de Laguna 1990a:192; Holm 1990a:603-4; Miller and Seaburg 1990:588; Ames and Maschner 1999:102-104,171,175,181,221,Plates 37,44.

[15] Ames 1994:215; Ames and Maschner 1999:53,113-17

[16] Blackman 1981:11; Kennedy and Bouchard 1990b:446; de Laguna 1990b:206-8; Suttles 1990b:4; 1990d:462; Ames 1994:210,212-3, 217-8; Ames 1995:159; Ames and Maschner 1999:26,147-154,160-63, 186-93,210-11; Smithyman and Kalman 2004:9,13.

[17] Suttles 1990c:16,18,25; 1990d:462; Ames and Maschner 1999:44-46, 120-21.

[18] Cashdan 1989:38; Boyd 1990:135; de Laguna 1990b:206; Kennedy and Bouchard 1990:446; Renker and Gunther 1990:422; Silverstein 1990: 538; Suttles 1990b:4; Ames 1994:218; Ames and Maschner 1999:120- 1; Smithyman and Kalman 2004:9
See Ames and Maschner's (1999:120-1) description of Coast *Tsimshian* sedentism and mobility as an example.

[19] Suttles 1990:4; Ames 1994:211-15; Ames and Maschner 1999:28.

[20] Ames and Maschner 1999:178.

the "household".[21] There is debate and uncertainty as to whether there were broader political structures, though some researchers recognise "town" chiefs, "confederacy" chiefs, with informal Great Chiefs known in some areas.[22]

Another notable cultural feature found in much of the Pacific North West in former times was the *potlatch*. This phenomenon has also been the subject of extensive debate because it appeared to have a central role in the sociopolitical structures of the Pacific North West, and at times involved spectacular "conspicuous consumption".[23] There were two main types of *potlatches*, one a gift-giving ceremony marking an auspicious occasion, such as marriage or the completion of a house, the other involved elaborate feasts and ostentatious displays of wealth through the liberal distribution of gifts to the guests. The first type usually acknowledged normal social landmarks, rites of passage and acted as a social lubricant.[24] The second common type of *potlatch* is the one most well known and the one subject to much of the research and speculation. In this form chiefs, or other nobles, issued invitations to other chiefs and influential individuals from other tribes to join in a *potlatch*. On these occasions an elaborate feast, accompanied by dancing and singing, was provided with ceremonies being conducted in which considerable numbers of valuable gifts were distributed and some highly valued possessions, particularly copper shields, known as "coppers", were deliberately damaged or destroyed. The chiefs who gave these *potlatches* engaged in self-aggrandizing speeches, with the whole event intended to assert their high status.[25]

Cultural features such as *potlatching* have been cited as a prime example of "competitive feasting", underpinning a "prestige economy". This forms the basis for one type of a set of theories, "Social Competition" theories, developed to explain the origins of CHG and agricultural economies and societies. But many theorists further argued that CHG economies and societies are inherently unstable or transitory and that they rapidly evolve into agricultural economies and societies,[26] although in the Pacific North West, as mentioned earlier, CHGs appear to have had a long history.

Having detailed the characteristics that form the basis for ascribing CHGs, and considered various prehistoric and recent examples, a detailed examination of traditional life in south west Victoria is now apposite. This will focus principally on the *Gunditjmara* but will touch upon other groups from the region, the *Jarwadjali*/eastern *Bungandit*j [*Bungaditji*], *Djabwurung, Giraiwurung, Djargurdwurung*.[27]

Australia is a continent noted for its ancient rocks and tectonic stability. South west Victoria is an exception to this because it lies within a region which was tectonically active until recently. Volcanic plains began forming in western Victoria (Earlier Newer Volcanics) about 4.5 million years ago with a recent episode (Later Newer Volcanics) in south west Victoria and an outlier in the south east of South Australia (Mt. Gambier, Mt Schank). The most recent eruptive episode commenced around 30,000 BP and did not cease until 4000 BP (Mt Gambier), leaving a series of striking cones jutting up from the plains, Mt. Eccles, Mt Napier, Mt Rouse, Tower Hill, to name a few of the more prominent ones. The lava flows from these extended for 50-60 kilometres in some cases, blocking rivers and leaving tongues of scoria, known locally as the "Stony Rises" criss-crossing the landscape. Due to the earlier period of orogenesis, soils are very rich in south west Victoria, but there are also many swampy areas as a result of the normally plentiful rainfall and disruption to drainage patterns arising from the recent volcanism. Overlaying this is a cline in precipitation, becoming much drier toward the Wimmera in the north western part of the region. Consequently seven vegetation zones occur in south west Victoria with heath, grassland, dry sclerophyll woodland, savannah woodland, woodland and mallee being the most important.[28] Within this environmentally diverse landscape numerous microniches occur as well, largely as a result of the disrupted drainage pattern.[29]

As far as is known Europeans did not have a direct presence in western Victoria until 1800 when they stumbled on Port Phillip Bay, where Melbourne now stands. In 1803 a short-lived convict settlement was established at Port Phillip Bay. Before the settlement was abandoned the following year, convict William Buckley, a former soldier, escaped with three others. Only Buckley survived, being accepted into *Wathaurung* society in the region on the western side of Port Phillip Bay. He lived with them for 32 years before rejoining British society as the "Wild White Man" when Melbourne was founded in 1835.[30] In his account of his years with the *Wathaurung* Buckley mentions visits to western Victoria to trade, roots for eels.[31] Undoubtedly there were also unrecorded contacts between sealers and whalers and coastal Aboriginal populations in western Victoria from the late 1790s but the first dateable documented contacts took place when the Henty family from Tasmania set up an unofficial settlement at Portland Bay in November 1834.[32] Two years later the "Australia Felix" expedition, led by the

[21] Feinman 1995:269.
[22] Earle 1987:284; Ames 1995:169-70; Hayden 1995b:64-5; Ames and Maschner 1999:177-179.
[23] Ames and Maschner 1999:16.
[24] See Blackman (1981:18-9; 1990:252), de Laguna (1990b:220) and Suttles and Jonaitis (1990:85) for examples of the occasions for such potlatches.
[25] Codere 1990:368-9; Sutton and Jonaitis 1990:84-5; Smithyman and Kalman 2004:18-19.
[26] Pryor 1986:889; Henry 1989:5; Hayden 1994:238; Price and Gebauer 1995:7; Belfer-Cohen and Bar-Yosef 2000:23; Bellwood 2005:34,40.
[27] There continues to be ongoing debate and discussion as to the correct group, language and dialectal names in western Victoria. See Blake 2003:1-2.
[28] Coutts et al. 1978:39; Coutts 1981:43,53.
[29] Coutts et al. 1976:9-10,19, Lourandos 1980b:163.
[30] See Morgan 1852.
[31] Morgan 1852:49.
[32] A local Aboriginal tradition relating to first contact with Europeans is contained in Dawson (1881:105) but cannot be securely dated.

intrepid Major Thomas Mitchell, marched through the south west of Victoria, fleetingly recording aspects of traditional life as they progressed. Then the colonial frontier began to move out from Melbourne, reaching the edge of south west Victoria in 1839.[33] The intruding land-hungry "squatters" seized land on which to run their sheep. As they "took up" "runs" the next wave leapfrogged them and the "frontier" crept forward.

Conflict between the Aboriginal population and the British intruders in western Victoria began as early as 1834.[34] Sustained, organised, resistance to the encroachment did not develop, however, until early in 1842 with an attack, reportedly involving 150 warriors, on "Eumeralla", the first station established in the central part of the "Western District". The conflict intensified in 1844-5, becoming known as the "Eumeralla War", and was followed by another two years of guerrilla warfare.[35] Despite the tensions the Chief Protector of Aborigines for the Port Phillip Protectorate, George Augustus Robinson, travelled extensively in the region during this period, visiting much of the local Aboriginal population and recording language information and aspects of their culture in considerable detail. Other contemporary observers and rapporteurs were traveller William Westgarth and squatter Tom Browne.[36]

By 1867 traditional life had largely ceased and most of the remaining Aboriginal people from south west Victoria and some adjacent areas were forced to move to the Lake Condah and Framlingham Missions. However, a local farmer, James Dawson, friend, advocate and Local Guardian of the remnant populations, along with his daughter Isabella, had maintained close ties with a number of surviving elders such as Kaarwin Kuunawarn [Hissing Swan], Wombeetch Puuyuun, Weeratt Kuyuut [Eel Spear], Yarruun Parpur Tarneen [Victorious] and Wombeet Tuulawarn. Unusually for the time, the Dawsons made their enquiries in the language of their informants.[37] Subsequently Dawson published *Australian Aborigines* in 1881, a systematic and detailed amateur ethnography of traditional life in the region. Although it contained "several glaring errors,"[38] it is now viewed as being "reasonably accurate,"[39] and shows a high level of consistency with other contemporary sources.

From the late 1890s through to 1930 a number of publications appeared making reference to "stone circles", being the foundations of dwellings roofed with boughs and bark, and fish races in the vicinity of Lake Condah.[40] Little further was written about traditional life in south west Victoria until two papers, one on a system of fish traps on Darlot Creek, and another on artificial channels at Toolondo, in the north of south west Victoria, were published in 1968 and 1976 respectively. The Toolondo system was reputedly created to control the movements of eels (*Augilla australis occidentalis*) and maximise the numbers caught.[41] These reports were followed by a sustained period of archaeological research investigating the extensive fish and eel trapping systems and "stone circles" in the Lake Condah area, as well as occupation mounds, the remains of dwellings and possible village sites in other parts of south west Victoria.[42]

The first clear sign to outsiders that there was something a little unusual in the economy and settlements in south west Victoria came in 1836 with the aforementioned "Australia Felix" expedition led by Major Mitchell. On 26 July 1836, as Mitchell was traversing around the western end of the Grampians [*Gariwerd*], a mountain range in central western Victoria,[43] he reached a place now known as White Lake.[44] Here he:

"*noticed some huts of very different construction ... being large, circular, and made of straight rods meeting at an upright pole in the centre; the outside had first been covered with bark and grass, and the entirety coated over with clay. The fire appeared to have been made nearly in the centre; and a hole at the top had been left as a chimney.*"[45]

Mitchell then went on to comment that the "place seemed to have been used for years, as a casual habitation."[46] From the material left behind it would appear to have been occupied at the time of Mitchell's visit, although the residents were not present, probably having absented themselves to avoid him and his party.[47] But the significance of this description only becomes apparent when considered alongside an account recorded on the same day by Mitchell's deputy, Granville Stapylton. In his journal entry Stapylton wrote, "Passed today several Guneaks [dwellings] of very Large dimensions one capable of containing at least 40 persons and of very superior construction."[48] Some years later, confirmation of this was provided by people at the Lake Condah Mission, who had come from all over the western

[33] This was actually preceded by an isolated expansion around the Wannon River, initiated by the Hentys in 1837, with 6 runs established by 1840 (Critchett 1990:22).
[34] Critchett 1990:120-123.
[35] Gerritsen 2000:5.
[36] Westgarth 1846, 1848, 1851, 1888; Boldrewood 1884; Clark 2000. Browne was a young "squatter" who wrote under the pseudonym of "Rolf Boldrewood", achieving renown as the author of Robbery Under Arms, an Australian classic.
[37] Dawson 1881:iii-v; Critchett 1981:Introduction to Facsimile Edition in Dawson 1981; Critchett 1990:3,155.
[38] Curr 1886:1:55.
[39] Corris 1968:19.
[40] Worsnop 1897:78,104-6; Kenyon 1912,1930
[41] Massola 1968; Lourandos 1976.
[42] Coutts et al. 1976; Coutts and Witter 1977; Coutts et al. 1977a, 1977b; Lourandos 1977; Coutts et al. 1978; Clemens 1979; Lourandos 1980a,b; Wesson 1981; Williams 1984; Geering 1985; Clark and Geering 1986; Williams 1988; Penney and Rhodes 1990; Clarke 1994; Schell 1998; Builth 2000,2002.
[43] The Grampians are a rugged mountain range, 500-1,000m [1,600-3,100 ft] high, which, along with the Mt William and Serra Ranges and the Pyrenees, are part of a chain running east-west through central western Victoria.
[44] According to Thornly (1878:2:61) White Lake was originally known as *Jerrywarrook*.
[45] Mitchell 1839:2:194.
[46] Mitchell 1839:2:194.
[47] Mitchell 1839:2:194-5.
[48] Andrews 1986:146.

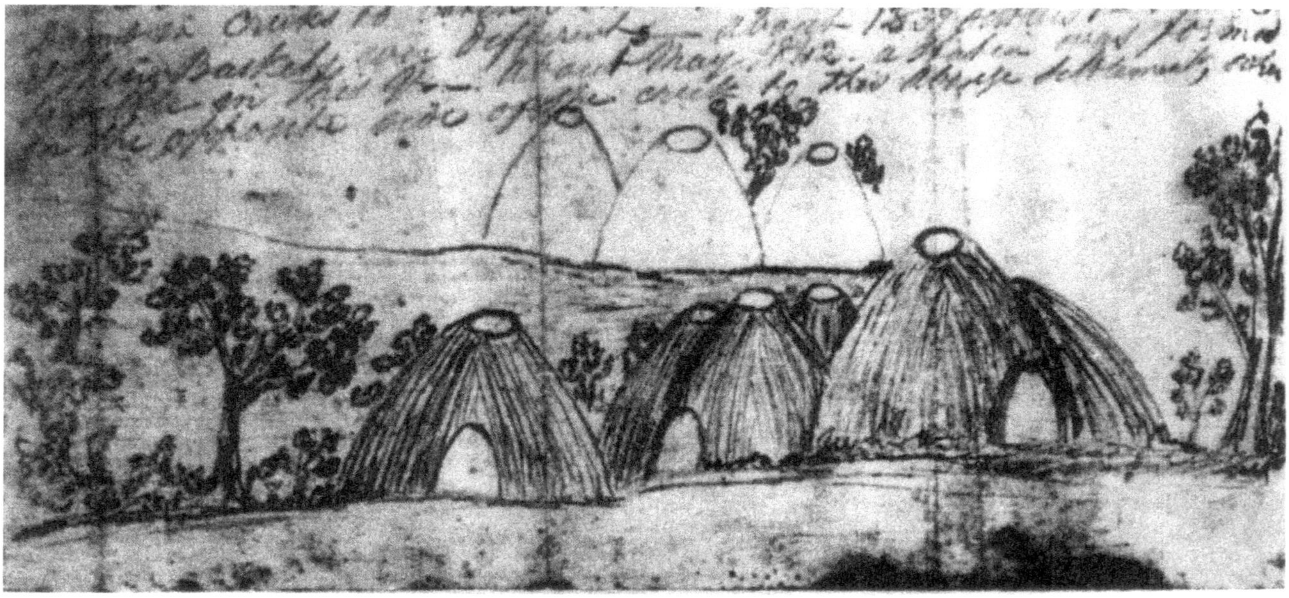

*Figure 19: Sketches of Village near Caramut*
*(Thomas n.d.b., attributed sketch in R. Brough Smyth Papers)*

part of the Western District. They told the manager, the Reverend J. Francis, in 1868 that they had formerly lived in, "communities of 30-40 and even more, occupying one Mia mia [dwelling]."[49] This information, and perhaps that of his own informants, led James Dawson, in discussing the traditional habitations of the region in *Australian Aborigines*, to conclude that, in "what appears to be one dwelling, fifty or more persons can be accommodated."[50] Chief Protector George Augustus Robinson also made a rough sketch of a structure in June 1841 that may have been one of these larger residences.[51]

As the colonists began to force their way into the district they encountered other types of structures and settlements. One of these was a "village" of "beehive" and "sugarloaf" shaped abodes near Caramut, in the central part of the Western District. Based on descriptions provided by an informant, identified as George Arabin,[52] a number of relatively detailed descriptions were recorded by Sub-Protector William Thomas, along with a rough map of the location and sketches of the village itself.[53] In his most extensive description Thomas stated:

"*... by Mustons and the Scrubby Creek to the westward ... first settlers found a regular aboriginal settlement. This settlement was about 50 miles* [80 km] *NE of Port Fairy. There was on the banks of the creek between 20 and 30 huts in the form of a beehive or sugar loaf, some of them capable of holding a dozen people. These huts were about 6 feet* [1.8 m] *high or little more, about 10'* [3.0 m] *in diameter, an opening about 3 feet 6* [105 cm] *inches high for a door which they closed at night if they required with a sheet of bark, an aperture at the top 8 or 9 inches* [20-22 cm] *to let out the smoke which in wet weather they covered with a sod. These buildings were all made of a circular form, closely worked and then covered with mud, they would bear the weight of a man on them without injury. These blacks made various well constructed dams in the creek which by certain heights acted as sluice gates in the flooding season ...*"[54]

Accompanying this description were two drawings of the village.

According to Thomas the people of the village were "perfectly harmless and stationary."[55] Unfortunately this village was burnt down late in 1840 or early 1841 by recently arrived "squatters," – "while the Blacks were from their village, up the creek, seeking their daily fare ....white people set fire to and demolished the aboriginal settlement ...".[56] Based on the number of dwellings, and number of residents each could accommodate, its population probably lay between 120 and 240.[57]

Another Indigenous settlement, what may have been the largest in western Victoria, also came under notice in the early Contact Period there, a settlement notable not only for its size but the fact that the inhabitants may have constructed stone houses. My research indicates that this settlement was located at the end of a scoria tongue, a "stony rise," that

49 J. Francis to J. Dawson 14 April,1868, In Williams 1985:75.
50 Dawson 1881:10.
51 See "Emu Creek Huts" in Clark 1998-2000:2:298,Fig. 6.25.
52 Williams 1984:177-8.
53 Thomas n.d.a.:537 - "Sketch Map of Village at Caramut".

54 Thomas 1858.
55 Thomas 1858.
56 Thomas 1858.
57 Assuming that the population lay between a minimum of 6 residents occupying 20 dwellings (120) and a maximum of 8 residents occupying 30 dwellings (240).
Duff (n.d:3) indicates that the "Muston's Creek tribe" numbered 600, which is consistent with calculations of the size of Scrubby Creek Village.

projected into Gorrie Swamp, formed by disruption of the flow of the Eumeralla River.[58] One of the earliest and most extensive descriptions of this village was published in 1846 by William Westgarth following an excursion there in June 1844. In this account Westgarth stated:

"*There was a 'native township' as it was termed, on the banks of the Eumaralla Lake or swamp where the stony rises in that part of the country commence. The Aborigines generally encamped there during a portion of the year, for the purpose of fishing, with occasional rambling over the neighbouring country. Mount Eeles* [Eccles], *an adjoining volcanic hill, with a large and romantic crater, appears to have been a favourite resort, their repeated visits having worn a distinct track to the summit At the period above alluded to* [June 1844], *these Eumaralla blacks were stated to be about two hundred in number; but two years previously* [1842], *when this locality was first taken up for pasturage the 'township' was said to contain five hundred.*"[59]

Chief Protector Robinson appears to have actually visited this village, accompanied by Sub-Protector Sievwright, shortly before Westgarth, as he recorded in his *Journals* at the time:

"Sat. 19 March 1842 *The native name of Hunter's place or river is Eumer.ral.lar. ... The station very pleasant ... ... - called 'Eumeralla'. The river is like a lake, is two miles long: serpents* [eels], *plenty fish deep water.*

20 March 1842 *decided to stay and go to natives Set off south through wooded land ... - crossed a swamp (dry) and at five miles came to the stony rises .... masses of lavre, steep stone ... Led our horses into the stony rises: ... plenty ash hills* [mounds], *round sharp layrs, plenty huts of dirt* [clay] *and others built of stone. Stone houses, stone weirs. Saw Mt. Eel* [Eccles]. *Mt Napier bore north and Mt Eels WNW.*"[60]

and later reported to La Trobe, the Superintendent of the "Port Phillip settlement":

"*On the 19th I visited Mr Hunter's station* ["Eumeralla"] ... *Mr H was absent ... Whilst in the neighbourhood I deemed it advisable to effect a communication with the natives, and on the 20th crossed a swamp to some strong rises, and succeeded in conferring with the blacks; they had a sort of a village and some of their habitations were of stone. I passed several stone and wooden weirs for taking fish, also places for snaring birds; their dwellings are among rocky fragments and loose crags, thickly wooded and bound by swamps.*"[61]

Commissioner of Crown Lands, Captain Foster Fyans, leading a small force intent on suppressing the *Gunditjmara* resistance emanating from this locality, also appears to have visited this settlement some months after Robinson. According to Fyans:

"*the party proceeded on foot, and waded with considerable difficulty through the marshes nearly one mile and entered a great stone Range to a point called the Township.*"[62]

Not surprisingly, with armed conflict in progress, the inhabitants were not to be seen. When Robinson visited this location Sub-Protector Sievwright counted only 33 people,[63] although, as they approached, Robinson noted that "Natives fled".[64] However, in early June 1839, when passing just to the south west of this site, Fyans had been "surrounded by, I suppose, 150 natives, following us with spears."[65] Similarly Hunter claimed that when his "Eumeralla" Station had come under attack in August 1842, 150 warriors had approached from the direction of the "Township". Applying population ratios of adult males to the total population, including females and children, calculated to be between 3.29 and 4.20,[66] these figures suggest at least 493 people were present in the area. While such calculations have many uncertainties,[67] they do, nevertheless, provide some support for Westgarth's claim that the "township" had had a population of around 500 in 1842.[68]

Based on the preceding evidence and other lines of enquiry I endeavoured in September and December 1999 to identify both village sites, sites which others had had difficulty in locating and investigating previously. I was able to locate what I believe to be the correct site of the "Scrubby Creek Village" near Caramut and what may be the remains of

58 Gerritsen 2000:13-20.
According to Dawson (1881:lxxxi-lxxii) Mt Eccles was called "*puutj beem*" (*budj bim*) while the stony rises around it was known as "*tung att*".

59 Westgarth 1846:8n. See also Westgarth 1848:100-101; 1851:707n.
The actual source of information has been identified as Robert Craufurd, brother of Lord Ardmillan and manager of "Eumeralla East" at the time of Westgarth's visit. See Gerritsen 2000:17.

60 Clark 1998-2000:3:47-8 (19-20 March 1842).
NOTE: Dates have been added to references from Clark 1998-2000 to assist in locating the relevant passages in other published transcriptions of Robinson's Journals.

61 Robinson 1844b:209.

62 Fyans 1844:235.

63 Sievwright 20 March 1842.

64 Clark 1998-2000:3:48. (20 March 1842)
Initially all the residents except some of the women fled when Robinson, Sievwright and three others approached. Some probably returned but Sievwright noted that they had met only "a portion of the tribe". Robinson mentions that after the "Natives fled" that the "Women came and went for the men" and that when leaving 30 of those present had agreed go to the Protectorate Station at Mt Rouse.

65 Fyans 1898:117; Peel 1996:260n.
Fyans was a Police Magistrate at the time, on his way to investigate a massacre of *Gunditjmara* at Portland. He became a Commissioner of Crown Lands the following year.

66 Gerritsen 2000:19,142n.

67 Gerritsen 2000:19,143n.

68 Sub-Protector Thomas (Journal of Proceedings for period 1.12.1843 - 1.3.1844 Note) mentions "Settled Villages", "on the Banks of Thorn's River," of "well built slab huts some capable of holding 15 or 20 persons", one village "supposed to contain 400 Inhabitants." This, however, appears to refer to a reports in Sydney newspapers in late September/early October 1838 (Clark 2001:24). I have deduced from those reports that "Thorn's River" is in fact the Richmond River in northern NSW.

*Figure 20: Complex Stone Arrangement – "Eumeralla Village"*
*(The author, 4 December 1999)*

the fish traps associated with it.[69] Similarly, I visited and surveyed the site I had concluded was the location of the "Eumeralla Village".[70] Regarding the latter, many stone circles, some conjoined, the remnants of what may have been a stone walled dwelling, and a complex stone arrangement, possibly originally forming part of a multi-roomed structure, were observed.[71] A photograph of this last feature can be seen below. It is noteworthy in this context that Robinson had stated that "some of their habitations were of stone".

A number of other accounts of settlements of varying size, and references to "villages", were documented in the early Contact Period in Western Victoria, particularly by Robinson. For example, as he made his way into western Victoria, Robinson came across, on 16 April 1841, a small hamlet of nine dwellings, "large enough to contain seven or eight persons (adults),"[72] at Lake Elingamite in the eastern part of the Western District, giving an estimated population of 60 to 70. On 2 May 1841 he reported he had been told by "Mr Sutton", overseer of "Spring Creek" in the central part of the Western District, that, "there was a large number of huts on the river when they first came."[73] Then, on 9 May 1841, at "Forlonge's Dairy Station", near Hamilton, Robinson commented in his journal that, a "whole village, therefore, has been forced away [by the colonists] from their ancient pool [spring]."[74] On his way from "The Grange", also near Hamilton, to Mt. Napier, Robinson reported that he had seen "at least 20 well built worns or native huts."[75] Robinson then encountered a "village" (*munn'yah* in *Djabwurung*)[76] which consisted, "previous to its occupation by white men," of "13 large huts built in the form of a cupola," near Mt. Napier on 10 May 1841, although only 3 of these huts appear to have been recently occupied.[77] On the same day he noted that "three women went ... to their village,"[78] and then, in the "evening the natives went to their village."[79] On the following day, he referred to another village in this area, "One of the old men went, pm, to his village but returned again in the evening."[80]

[69] Personal Observation, 18-19 September, 4 December 1999.
Scrubby Creek has been identified in the course of that research as Tea Tree Creek, with Scrubby Creek Village on that creek, a little to the north of its confluence with Muston's Creek (Gerritsen 2000:6-12).

[70] A range of explanations for this complex arrangement were considered but none provided a viable alternative. See Gerritsen 2000:52n137.

[71] Personal Observation, 4 December 1999.

[72] Clark 1998-2000:2:136 (16 April 1841)

[73] Clark 1998-2000:2:184 (2 May 1841)
Sutton was referring to April-May 1840 when Spring Creek was first "taken up".

[74] Clark 1998-2000:2:195 (9 May 1841)

[75] Clark 1998-2000:2:201 (11 May 1841)

[76] Dawson 1881:xliv.

[77] Clark 1998-2000:2:196 (10 May 1841).

[78] Clark 1998-2000:2:196 (10 May 1841).

[79] Clark 1990:107 (10 May 1841).

[80] Clark 1998-2000:2:203 (11 May 1841).
Massola (1969:53) also reported many habitation sites in this area.

*Figure 21: Cupola on Mound*
*(Robinson in Clark 1998:2:243,Figure 5.11)*

Near Port Fairy, according to Robinson, there was:

"*a large swamp on the east of the Port Fairy River* [Moyne River] *where the natives get their chief support, roots etc. and near to the small eminence on the edge of this swamp, called by Campbell's men 'Tower Hill', is a native village: an assemblage of huts.*"[81]

Upon his return there the following year Robinson referred to it as a "native township", and appeared to indicate the population may have been up to 150.[82] Sievwright, who accompanied Robinson on this occasion, reported that they, "visited a tribe who constantly reside in the neighbourhood of Port Fairy."[83]

Clearly, given this evidence, there were settlements of significant size in south west Victoria. However, it is not just the size of settlements, but also the settlement pattern, the degree of settlement permanence and the level of sedentism, that has a significant bearing in classifying the *Gunditjmara*, and other groups in south west Victoria. Were they indeed Complex Hunter-Gatherers, or something else?

Consideration of these factors begins by drawing attention to the variety of habitation types that have been mentioned thus far - the "tent" or "teepee" form from White Lake, the "beehive" and "sugarloaf" from Scrubby Creek Village, and the "cupola" type frequently identified in other reports.[84] The "cupola" refers to a dome shaped dwelling, which Robinson also referred to as a "kraal",[85] probably the same as the "beehive", often built on mounds, as illustrated in a drawing done by Robinson.

The form referred to as the "sugarloaf" was also described by the Chief Protector, who observed that shelters, "like a cupola are sometimes double and have two entrances."[86] He recorded an example in his travels, at "The Great Swamp" near Mt. Napier, "a fine large double hut ... with two entrances ... I went in at one door and came out the second."[87] Charles Griffiths, in commenting on the original dwellings around Port Fairy and Portland, wrote in 1845:

"*they construct a kind of hut for the winter season, which is more durable in character. They do this by heaping sods and clay on top of the original mi-mi* [dwelling] ... *when they remain in one place for any length of time, these earths reach considerable size: I have seen one fully fifteen feet* [4.5 m] *long, and high enough for a man to stand upright in.*"[88]

This too could have been a "sugarloaf" type of dwelling,

81 Clark 1998-2000:2:154(28 April 1841) [Tower Hill is an extinct volcano east of Port Fairy]
Its adjacency to Port Fairy suggests it may possibly have been a post-contact development, but archaeological investigations suggest this locality had been utilised since 5000 BP (Gill 1977: 13-14). Moreover, Robinson's comments, that the people from Port Fairy he met at Spring Creek, "had never seen a white man write before" (Clark 1998-2000:2:181), that the people at the village "had not a word of English nor any of the manners and customs", and that "they had not any names from the English, all were pure, original" (Clark 1998-2000:2:154), as well as other elements in the account of his visit to this village, strongly indicate that it was a traditional settlement.
82 Clark 1998-2000:3:54 (23 March 1842).
83 Sievwright 23 March 1842.

84 Examples mentioned in this chapter are only selection of reports. See Gerritsen 2000 for other examples.
85 Clark 1990:108 (13 May 1841)
86 Clark 1998-2000:2:196 (10 May 1841).
87 Clark 1998-2000:2:196 (10 May 1841).
See also Robinson 1844a:240.
88 Griffiths 1845:152.

Griffiths' indication that it was "fifteen feet long" suggests it was not circular as the other types were. Dawson, in discussing what appears to have been a cupola-type "family wuurn [dwelling]" capable of accommodating 12 or more people, added that, "when the family is grown up the wuurn is partitioned off into apartments, each facing the fire in the centre."[89] Although these descriptions are not very detailed it is possible to conclude that larger, multi-roomed structures were another form of habitation in the central parts of the Western District, which will be denoted as the "sugarloaf" type.

Although concentrated in the central part of south west Victoria, larger habitations requiring significant labour, often described as "permanent" or "substantial"[90], were to be found as far north as White Lake, north east to Fiery Creek, east to the "stone *mia-mys*" at Lake Purrumbete,[91] and in adjacent regions in South Australia. Robinson, for example, observed a cupola-type residence in South Australia, north of Mt. Gambier, "Left Sturt's accompanied by Adelaide police ..... Past a Native hut, made like those of Tapoc [Mt. Napier], with logs and turf."[92] Confirming distribution into South Australia, Gideon S. Lang, who took up squatting at Kentbrush on the coast west of Portland in 1842, described not only the habitations in rare and significant detail, but their construction as well:

"*The huts are generally about nine feet* [2.7 m] *in diameter, five feet* [1.5 m] *high, and in the shape resembling half an orange. They are built in the first place of ... dry stiff branches ... the lower row set in the ground, and the rest interlaced above in the manner of a bird's nest. Upon this they place branches of trees, reeds, or long grass; over this they again place grass, turf, and above all sand if they have it, the top being rendered around and smooth like the Esquimaux winter hut. There is one low opening or door at one side of the hut, and in the opening is placed a fire. The largest of these huts I ever saw was on the Koorong* [the Coorong], *an arm of the sea behind the coast sandhills* [Younghusband Peninsula], *between Adelaide and Portland; it was fourteen feet* [4.2 m] *in diameter and quite eight feet* [2.4 m] *in height inside, and rose perpendicularly at the sides, and could accommodate an unusually large number of people.*"[93]

In addition to these types of dwelling there were temporary structures, such as the "half cupola",[94] and "windbreaks", used when only a short stay was anticipated. The "half cupola", probably only made for one or two individuals, were called a "neich" or "niech"[95] [niche] by Robinson, who observed, "a vast number," of "old" ones, along with, "thousands of dead eels," at the eastern end of Lake Bolac in April 1841, at the close of the eeling season.[96] By collating all the ethnographic accounts, such as those preceding, it is evident that there were a number of types of dwellings.[97] These can be classified as:

**Temporary**

- windbreak (boughs and grass, or stone wall)
- half cupola (limbs, bark, grass)
- temporary cupola (limbs, bark, grass) var. stone foundations.

**More Permanent**

- cupola ("substantial", timber and turf, often coated with clay, sand or earth)
- stone walled* (stone walls of at least 1m, roofed with branches and bark)

**Complex**

- sugarloaf (double cupola)
- tent (enlarged cupola with central support)
- stone gallery* (galleried)
- stone* (all stone cupola?)

---

* Provisional Identification

While Robinson's journals have their shortcoming they are almost unique in providing frequent comments on the locations across the region where such habitations were situated. When Robinson was in the Spring Creek/McArthur Creek locality, for example, he wrote:

"*Passed by a rivulet ... where there is a spring of excellent water which runs in the driest season. The natives are deprived of this. At the springs and water courses the natives had their fixed residence or villages or homesteads.*"[98]

Other informative comments were provided by Robinson relating to "fixed residences" or "homesteads" associated with the "mounds" commonly found in this region. Near the Hopkins River in April 1841 he recorded:

"*we saw a large mound of earth at least four feet high and 10 feet long, five wide. My native companion said it was a black man's house, a large one like what* [sic - white] *man's house ... it appeared the whole had been burnt down. A short distance from this, about 200 yards* [180 m], *was the remains of another hut of similar description.*"[99]

89 Dawson 1881:10.
90 e.g. Clark 1998-2000:2:194,196 (9,10 May 1841).
91 Chauncy 1878:234.
92 Clark 1998-2000:4:295 (1 May 1845).
93 Lang 1865:26.
94 These were demihemispheres with a parabolic axis, made of branches, bark and grass. A photograph of one can be found in Williams (1988:Figure 6.9a).
95 e.g. Clark 1998-2000:2:196 (10 May 1841)
96 Clark 1998-2000:2:119,164[Fig. 4.3] (1 April).
97 Gerritsen 2000:23-6.
98 Clark 1998-2000:2:184 (3 May 1841)
99 Clark 1998-2000:2:145 (24 April 1841).

Not all these "houses" had been on mounds. Robinson mentioned another setting where he encountered, near an outstation of "The Grange", "the frame of a small native hut; made very substantial and neat and placed on the slope or declivity of a hill with an oven [mound] at the back."[100] In the end Robinson concluded:

"*Some* [wuurns] *were placed near the river, others on aclivity of the hills and some on the top of an eminence. One on top of an eminence was erected on a mound of earth thus*: [Figure 21]"[101]

In general terms Dawson concurred with Robinson's assessment, associating habitations with sources of water, watercourses and certain swamps:

"*These comfortable and healthy habitations are occupied by the owners of the land in the neighbourhood, and are situated on dry spots on the bank of a lake, stream, or healthy swamp, but never near a malarious swamp, nor under large trees,*"[102]

A more extensive and systematic analysis of settlement locations by Lourandos also showed a strong link with wetlands, with a preference for siting in woodlands.[103] On this basis, and with the historical ethnographic evidence in mind, it can be concluded that permanent habitations and settlements were normally located in close proximity to watercourses, water bodies, water sources and wetlands. Every example cited thus far, where the location is known, conforms to this pattern. Within those parameters there also is some evidence for a preference, in terms of siting, for woodlands and higher ground or vantage points, "eminences" as Robinson called them, possibly for defensive reasons.[104] Interestingly, this distribution pattern is virtually identical to that for mounds. The mounds found in south west Victoria, largely made up of ashes and clay, cluster along major drainage systems close to streams, lakes, lagoons and swampy areas subject to flooding, and are found in places where there is good vantage, often just inside the timber line.[105] While Bird and Frankel claim that mound, "clustering cannot be construed as evidence of large-scale settlements or sedentism,"[106] this does not preclude a correlation between mounds, habitation and settlement locations. A correlation such as this would not be surprising given the archaeological and ethnographic indications that mounds were used as habitation and camping sites, activity areas, for cooking, and for burials.[107] Clearly there was some sort of association between mounds and residential activities.[108] Even if the mounds themselves were only infrequently used as habitation sites, the association still holds.

On the basis of this correlation, between mounds and habitation sites, and drawing on a compilation of the ethnographic reports, it is now feasible to propose a traditional settlement hierarchy for south west Victoria. Utilising a modified form of the "lower limb" of Grossman's rural settlement hierarchy,[109] and applying it to structures previously identified as "More Permanent" or "Complex", the classification, which may provide the "best fit" for the available evidence, is:

**Homesteads** Single structures occupied by twenty persons or less, at least one kilometre from any other structure. (e.g. Cupola on Mound)

**Lodges** Single structures occupied by twenty to forty persons at least one kilometre from any other structure. (e.g. White Lake Structure)

**Dispersed Settlements** Non-contiguous structures, or contiguous clusters of structures with less than forty persons, placed at least 150 metres apart, but closer than one kilometre. (e.g. "20 well built native worns" fringing the Great Swamp")

**Hamlets** Contiguous structures having a total population of forty to one hundred persons. (e.g. Lake Elingamite cluster)

**Villages** Contiguous structures having a total population of more than one hundred persons. (e.g. Scrubby Creek Village)

Direct validation of this settlement hierarchy is not possible without more extensive regional archaeological studies being carried out. However, a survey of a large sample of mounds (207) by Coutts *et al.* (1976) may provide indirect validation of the proposed settlement hierarchy. In this survey data was presented in histogrammatic form, representing various mound parameters. When the number of mounds within a 1 kilometre radius (Figure 22) is considered a tri-modal distribution is apparent.

Given the relationship between habitations and mounds postulated before, this data would suggest that the largest grouping of 1-3 mounds is indicative of the numerically smaller sites, the "Homesteads", "Lodges" and "Dispersed Settlements", while the other two peaks, 5-7 and 9

[100] Clark 1998-2000:2:194 (9 May 1841).
[101] Clark 1998-2000:2:201 (11 May 1841).
[102] Dawson 1881:10.
See also Chauncy 1878:2:232.
[103] Lourandos 1980b:156a,b,162-3; Williams 1988:62.
[104] Robinson 1844b:209; Boldrewood 1884:49; Williams 1985:78; Clark 1989:3; Personal Observation:18-19 September 1999, 4 December 1999.
[105] Chauncy 1878:2:232; Smyth 1878:1:239; MacPherson 1885:50; Coutts et al. 1976:1,9-13; Coutts et al. 1977a:21; Williams 1988:12; Bird and Frankel 1991:7.
[106] Bird and Frankel 1991:8.
The claim may well be flawed, see Gerritsen 2000:55-6n210.
[107] Chauncy 1878:2:232; Smyth 1878:1:239; Dawson 1881:103; MacPherson 1885:55; Coutts et al. 1976; Coutts and Witter 1977:62-6; Williams 1988:16-21,135,206; Bird and Frankel 1991:7; Clark 1998-2000:2,3.
[108] Coutts and Witter 1977:66; Williams 1985:79.
[109] Grossman 1992:1-2.

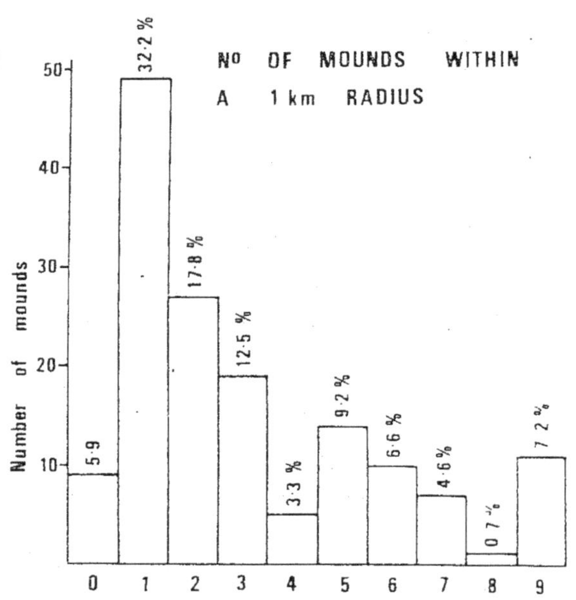


*Figure* **22**: *No. of Mounds Within a 1 Kilometre Radius (Coutts et al. 1976: Figure 3)*

respectively, may represent larger types of settlements, possibly "Hamlets" and "Villages". Although speculative, support for this proposal can be found in the data on mound diameters and volumes which also show a tri-modal distribution.[110]

Finally, before moving to consideration of the traditional economy in south west Victoria, its significance and the attendant theoretical implications, the degree of sedentism should be formally considered. As with the case studies in earlier chapters, this will begin by examining some of the correlates of sedentism, where there is evidence bearing on the question. Unlike the other case study areas, there have been extensive archaeological investigations undertaken in south west Victoria, so that it is not necessary to rely so heavily on historical ethnographic evidence. The archaeological investigations have not been without their difficulties, however, as will become apparent.

One common approach to assessing sedentism is to determine the seasonality of occupation by looking for remains of animals and plants at a particular site that are characteristic of a specific season or seasons. Elizabeth Williams attempted this at a site known as Gorrie Swamp Hut, on the southern edge of the Mt Eccles (*Butj-peem*) lava flows (*tungatt*),[111] four kilometres west south west of the Eumeralla Village site. Ten structures with common walls up to one metre high were excavated. However, high soil acidity (pH 3.5) defeated her efforts to determine the seasonality of occupation at Gorrie Swamp Hut.[112] Similar problems were experienced by Williams at another site, McArthur Creek, where she endeavoured to determine seasonality by employing palynological analysis.[113] High soil acidity is a general problem in south west Victorian sites, causing most organic remains to rapidly decompose and disintegrate.

Higher levels of sedentism can also be inferred where the separation distance between structures is less than the dimensions of the structures themselves. This would seem to have been the case at Gorrie Swamp Hut where the structures, two to three metres across, had common walls. Similarly, some of the structures at the Eumeralla Village had common walls.

Another correlate of sedentism is the presence of partitioned and multi-roomed habitations. Again the galleried arrangement of the structures observed in the vicinity of the Eumeralla Village fit this criterion. When applied to ethnographically observed residences from the region a number of examples can be found of partitioned or multi-roomed habitations. Robinson's double cupola would seem to fit the description of a multi-roomed dwelling, and partitioning was another feature reported by Dawson.

For the remaining indicators of sedentism - communal structures (such as kivas and mounds), extractive facilities (such as fish traps), cemeteries and thick cultural deposits, relevant evidence is also available. Mounds, discussed earlier, are a form of cultural deposit and were present in great numbers in south west Victoria. The *Gunditjmara* and others in this part of Victoria began creating mounds around 2500 BP.[114] They appear to have been complex formations, possibly involving the deliberate construction of the basal layers and then accretion through everyday occupational, residential and subsistence activities, along with the probable long-term accumulation of habitation remains.[115] Many mounds have been destroyed or suffered significant deflation since the 1840s.[116] Observers in the colonial period often remarked on them, and occasionally gave specific dimensions. One near Mount Elephant, for example, was reported to have been 27-36 metres long, 13-18 metres wide and reached a height of 6.1 m when first seen by the squatter who settled there.[117]

One of the more remarkable features of traditional life in south west Victoria was the extent of landscape modification, produced by fairly large labour investment

[110] Coutts et al. 1976:Figure 3.
Williams (1985:79) also shows a tri-modal distribution from a sample of 117 mounds.
[111] Clark and Heydon 2002:26.

[112] Williams 1988:141.
[113] Williams 1988:80,96.
[114] Coutts et al. 1976; Bird and Frankel 1991:9.
[115] Dawson 1881:103; Coutts et al. 1976; Coutts and Witter 1977:62-6; Coutts et al. 1979:Fig.1; Williams 1988; Bird and Frankel 1991:7.
[116] Four mounds measured in the 19th century were found to be on average only 25% of their original height (Williams 1988:15,Table 2.2).
[117] Currie 1858-9:62; Smyth 1878:1:240.
Mathew (1899:91) claims to have seen a mound of similar height in Victoria. See also Smyth 1878:1:239; McPherson 1885:55-7; Williams 1988:13,Table 2.1.

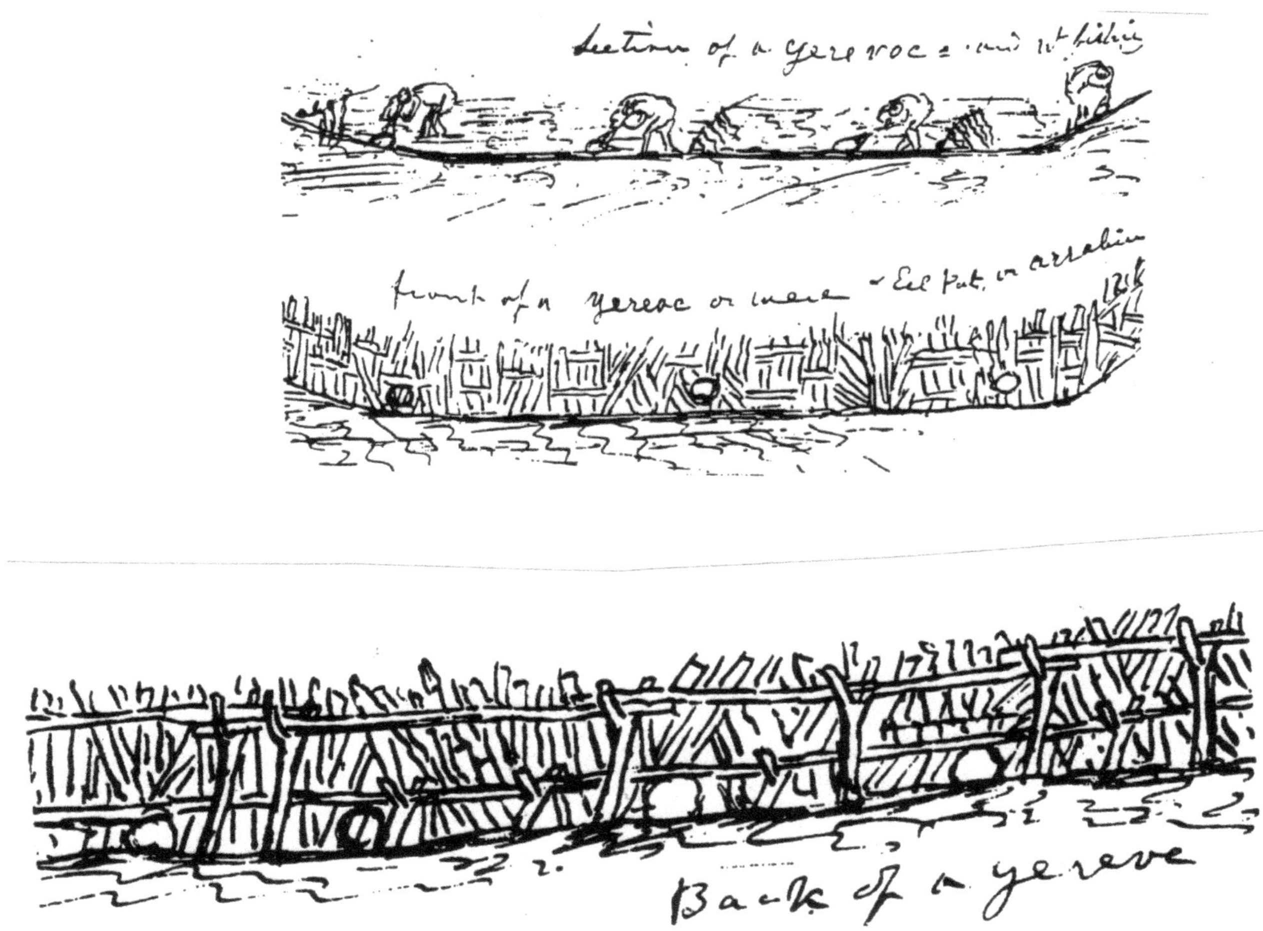

*Figure* **23**: *Robinson's Drawing of Yere.roc Traps*
*(Clark 1998-2000:2:177, Figs 4.37-8,178,Fig.4.40)*

in procurement infrastructure. While details of the region's economy will be discussed a little later, some examples make it clear that the communities there must have been engaging in some very extensive cooperative enterprises to produce those extractive facilities. Fish traps for catching fish and eels, known as *yere.roc*, were especially prevalent. They were variously described by the colonists, as "dams" on the water course adjacent to Scrubby Creek Village, "stone and wooden weirs" near Eumeralla Village,[118] and a "woven fence" type on lower Muston's Creek. Nicholas McCann, who began squatting lower down Muston's Creek early in 1840, noted these at the time, "the great number of bridges along the waterholes and the large number of woven fences across the shallow portions of the Creek."[119] On the Moyne River, Chief Protector Robinson drew and measured a wooden weir type fish trap, 60 metres [200 ft.] wide and 1.5 metres [5 ft.] high, and described and drew in detail another of 90 metres [100 yds.] on a tributary of the Hopkins River.[120]

These examples pale in comparison, however, to the kilometres of complex traps for eels and fish found at Lake Condah and Darlot Creek. These had first come to notice in 1897 but were not systematically investigated until the 1970s.[121] Initially a system of races, walls (of up to 1.3 m), dykes and weirs extending for 3 kilometres as part of the Lake Condah outflow was identified.[122] It has since been realised that this complex was much more extensive and encompassed significant portions of Darlot Creek. It included culturally constructed water bodies to manage the eels in the course of their lifecycle, as well as what are believed to be walled storage pits up to two metres by five metres.[123]

Another major artificial procurement system, for catching eels, involving considerable landscape modification had also been developed at Mt William, at the foot of the Mt William Range in the north of south west Victoria. As Robinson described it, in July 1841:

"*At the confluence of this creek with the marsh observed an*

[118] These stone "dams" or barriers and their operation were described by Dawson (1881:94).
[119] N. McCann quoted in Critchett 1990:63.
[120] Clark 1998-2000:2:145,162-3,174,Fig.4.25,177-8,Figs 4.38-41 (24 April,30 April 1841).
[121] Worsnop 1897:78,104-6; Kenyon 1912,1930.
[122] Coutts et al. 1977a:24; Coutts et al. 1978.
[123] Kenyon 1912:110; Massola 1968; Builth 2000:6-8; 2002:211-274.

*immense piece of ground trenched and banked, resembling the works of civilized man but which on inspection I found to be the work of the Aboriginal natives, purposely constructed for catching eels. ... ... These trenches are hundreds of yards in length. I measured at one place in one continuous trepple line for the distance of 500 yards* [450 m]. *These treble watercourses led to other ramified and extensive trenches of a most tortuous form. An area of at least 15 acres* [6 ha] *was thus tracd over. ... ... These works must have been executed at great cost of labor to these rude people ...*

*The plan or design of these ramifications was extremely perplexing ... ... curious curvilinear winding and angles of every size and shape ... At intervals small apertures were left and where they placed their arabine or eel pots. These gaps were supported by pieces of the bark of trees and sticks. In single measurement there must have been thousands of yards of this trenching and banking ... ... Some of these banks were two feet* [60 cm] *height, the most of them a foot and the hollow a foot deep by 10 or 11 inches* [25-27 cm] *wide. The main branches were wider.*

*Around these entrenchments was a number of large ovens or mounds for baking, there were at least a dozen in the immediate neighbourhood. They were the largest I had seen: the one i measured was 31 yards* [27.9 m] *long, two yards* [1.8 m] *high and 19 yards* [17.1 m] *broad.*[124]

Robinson drew rough sketch plans of the earthworks which show a highly convoluted system of artificial waterways.[125] Dawson confirms details of the purpose and dimension of these channels, called *vam*,[126] and their mode of operation. Such landscape "engineering" on both Darlot Creek and Mt William, it has been argued, was a form of artificial niche expansion which probably increased the numbers of eels.[127] It has also been pointed out, these facilities would have required considerable social cooperation to construct and maintain.[128]

These features, along with "Lodges" may, consequently, fall within the rubric of communal structures. The presence of what have been interpreted as substantial storage structures on Darlot Creek is a further indicator of higher levels of sedentism in the region.[129] There have been no cemeteries *per se* identified in these parts,[130] although this is not surprising given the soil acidity problem. Historical ethnographic evidence, moreover, indicates mortuary customs were quite complex and involved cremation, platform burials and in-ground burials in different parts, with mound interments in the form of considerate burial only occurring in particular circumstances, sometimes after a mound had been abandoned.[131] Nevertheless, burial in mounds, including at least one possible status burial,[132] appears to have had a long history in south west Victoria and were a relatively common occurrence. There is much anecdotal evidence of skeletons being found in mounds in the 19th century, and three of the six mounds investigated by Coutts and Witter contained three graves each.[133]

To assess the degree of sedentism exhibited by traditional owners in this part of Victoria, the indicators outlined previously need to be taken into account. Superficially they are certainly consistent with what would be expected if there were higher levels of sedentism. However, a more rigorous assessment, based on the classification developed earlier, requires additional evidence that directly addresses that issue as far as possible.

One of the usual correlates of higher levels or full sedentism is that residences are permanent, as are the habitation sites where they are located. But it needs to be recognised that there may be misleading exceptions to such associations. In the Pacific North West, for example, the dwellings, or at least elements of them, were seasonally relocated, although the actual framework of their plank houses remained *in situ* and the habitation sites themselves still appear to have been permanent. Alternatively, the degree of permanency may be inadvertently misrepresented by the one-off impressionistic comments made about dwelling and settlement permanency in historical ethnographic accounts. Ideally these types of judgments ought to rely on extended observations. In regard to south west Victoria some of the pertinent documentary evidence does indeed provide at least intermittent, if not extended observations, as well as evidence deriving from Indigenous sources, that has a direct bearing on the question of dwelling and settlement permanency there.

As in other parts of Australia, where explorers and others in the colonial vanguard briefly encountered what appear to have been permanent habitations, in south west Victoria comments initially referred to residences as being, "very substantial" "very superior", "very substantial construction", "well constructed", "well built", "permanent"

124 Clark 1998-2000:2:308 (9 July 1841).
125 Clark 1998-2000:2:345-6,Figs 7.8,7.9
126 Dawson 1881:94; Clark 1998-2000:2:231,324 (19,23 July 1841).
127 Lourandos 1980b:255;1988:153
128 e.g. Builth 2002:277.
129 Builth 2002:208,230-4,245,268.
At Eumeralla Village I observed 2 "wells", about 30 cm across, perhaps 1 m deep, with stone blocks 30-60 cm fallen across or laid over them. There are a number of possible explanations for these features (see Gerritsen 2000:52n137) which may possibly have been pits for the live storage of eels.
130 Though anecdotal evidence suggests there may have been a burial ground in a sandy lunette at Victoria Valley (Massola 1969:56)
131 Dawson 1881:63-5,103.
Robinson (Clark 1998-2000:2:249,257,284,368,379) observed (2:257; 8 June 1841) what seems to have been a platform burial on the Wannon River and reported it as a local custom. He also reported (2:344-5,368; 28 July 1841,6 August) in-ground interment in the vicinity of Mt. Cole in the north western part of south west Victoria (*Djabwurung*). Buckley (Morgan 1852:53) described a platform burial in the eastern part of western Victoria.
132 Burial in Mound FM/1, south west Victoria. The individual was buried with "twenty-five bandicoot mandibles and other items clustered at the right shoulder." (Coutts and Witter 1977:64)
133 Godfrey 1858-9:62; Chauncy 1878:2:234; Smyth 1878:1:239; Dawson 1881:103; Williams 1988:16,19; Schell 1998:113; Coutts and Witter 1977:63-4.

or "fixed residences".[134] Presumably this is the reason terminology such as "villages" was used to describe the larger settlements they encountered. Clearly they intuitively recognised that considerable effort was put in to the erection of these abodes. This is readily apparent in Mitchell's and Stapylton's accounts from White Lake, Robinson's and Dawson's descriptions of the construction of the cupolas, and Lang's record of the building of a large cupola on the Coorong.[135]

The solidity of structures, usually with stone walls, has also been used in some contexts as sufficient grounds for assuming permanency of dwellings and being indicative of sedentism.[136] This would apply to such structures at Eumeralla Village, Gorrie Swamp Hut and possibly some of those on Darlot Creek.[137] However, while certainly suggesting the possibility that the inhabitants were sedentary, the dangers involved in making this assumption were demonstrated in regard to the stone circle structures from the sites at Lake Condah, Condah Swamp [Palmer or Allambie site] and Louth Swamp [Kinghorn site] area, some of which had "walls" up to 75 cm high.[138] These were the sites that first caught the attention of ethnographers from 1897, and archaeologists in the 1970s. Detailed investigations at the Allambie and Kinghorn sites showed that distinguishing between artificial structures and natural formations was problematic, occupation had only been of brief duration, and that European artefacts such as pipes, glass and metal were present in some of the occupation floors.[139] In addition, all radiocarbon dates were found to be "modern".[140] From this it was concluded stone circle habitations were not of pre-contact provenience.[141] It has since been argued that these dwellings were qualitatively different from those found at Eumeralla Village and Gorrie Swamp Hut, and that they were probably only temporary structures built when the *Gunditjmara* and others withdrew to the area as a refuge following the Eumeralla War.[142] Historical evidence also indicates this was the last area to be occupied by the colonists.[143] Furthermore, temporary structures may have been constructed in the vicinity because accommodation at the nearby Lake Condah Mission was grossly insufficient in its initial years (1867-70) and most of the 70 residents were forced to "live in mia-mias".[144] So it is unsafe to conclude at this point that the stone-walled dwellings at Gorrie Swamp Hut and Eumeralla village were necessarily permanent or that the residents were sedentary, other lines of evidence need to be considered.

According to Dawson's informants, permanent habitations were built on mounds and these "formed homes for many generations,"[145] being rebuilt on the same spot in the event of their destruction by fire.[146] Mitchell too had commented that the White Lake structure "seemed to have been used for years, as a casual habitation," though the exact basis for his conclusion is uncertain. Griffiths may have provided supporting evidence by stating that although the "sugarloaf" sod and clay huts were built just "for the winter season," nevertheless, "when they remain in one place for any length of time these earths reach to a considerable size," implying that the structure itself and occupation may have lasted well beyond the winter season.

Archaeological evidence from Gorrie Swamp Hut, the only stone structure outside the Allambie and Kinghorn sites to be dated, provides significant support for the proposition that at least some of the habitations in south west Victoria were permanent. The anthropogenic sediments in the structure investigated by Williams were 25 centimetres deep. A date of 380±150 BP [1570 AD ± 150 years] was obtained from the 10-20 centimetres interval.[147] Calculations based on the anthropogenic sedimentation rate of that same cultural deposit conclude the structure was utilised for a period of between 83 and 459 years, the latter figure being more likely if the dating of Gorrie Swamp Hut is accurate and the site had been in use until abandoned when the colonists occupied the district.[148] Whatever the case, the duration of usage of the site strongly suggests that Gorrie Swamp Hut was a permanent structure. Following analysis of the lithics and debitage from Gorrie Swamp Hut, Williams tentatively concluded that this site was used to "repair tools" and was not continuously occupied, that "the main living areas of the cluster were located in other dwellings."[149]

Overall, the preceding indicators are consistent with higher levels of sedentism in south west Victoria. However, if more precise specification of the nature and degree of that sedentism is to be achieved then other evidentiary perspectives need to be considered within the context of the typology of sedentism developed previously. As before, this will largely depend on the detail found in historical ethnographic sources for the necessary insights to inform judgements and conclusions. For example, the inhabitants of the permanent mound habitations, Dawson reported, "abandon them for a season in search of a variety

[134] Mitchell 1839:2:247; Robinson 1844a:240; 1844b:209; Dawson 1881: 10-11; Darlot 1940:72; Andrews 1986:146; Clark 1998-2000:2:194-6 (9-10 May 1841).
[135] Robinson 1844b:209; Lang 1865:26; Dawson 1881:10; Clark 1998-2000:2:196 (10 May 1841).
[136] Kenyon 1957:52-3; Henry 1989:38-9; Habu 1996:38-9.
[137] Builth 2002:208,231,245.
Circular stone walls of a metre, believed to be habitations, were found in association with the storage structures along Darlot Creek.
[138] Clarke 1994:10.
[139] Coutts et al. 1977a,b; Coutts et al. 1978:1,16; Wesson 1981; Geering 1985; Clark and Geering 1986; Clarke 1994.
[140] Coutts et al. 1977a:38; Coutts et al. 1977b:201; Wesson 1981:49, 76.
[141] Coutts et al. 1977a:33.
[142] Wesson 1981:96-7; Bird and Frankel 1991:8; Gerritsen 2000:4-6.
[143] Gerritsen 2000:4-5.
[144] "Sixth Report of the Central Board of the Aborigines" cited in Massola 1970:97; Gerritsen 2000:4.
The pursuit by the Mission's residents of some traditional subsistence activities for "several days a week" in this area in the following 30 years may also explain the presence of the post-contact structures (See Gerritsen 2000:5,48n54).
[145] Dawson 1881:103.
[146] Dawson 1881:11.
[147] Williams 1988:145.
[148] Gerritsen 2000:33-5.
[149] Williams 1988:151.

of food."[150] Consequently, it was only "in summer, or for shelter while travelling,"[151] that the more temporary forms of accommodation were utilised. Summer was also the period in which the "Great Meetings" were held, such as the one at the marsh of "Mirraewuae", which Dawson calculated was attended by 2,500 people.[152] Late summer and early autumn, when mature eels were migrating back to the sea, seems to have been part of this summer mobility, although this could have been a second episode of regional mobility in the summer/autumn period. Whatever the case, as Dawson relates, for "a month or two the banks of the Salt Creek presented the appearance of a village all the way from Tuureen Tureen, the outlet of the lake [Bolac], to its junction with the Hopkins [River]." Robinson, indeed, had witnessed the aftermath of this phenomenon at Lake Bolac on 1 April 1841 when he encountered "mounds, thousands of dead eels" at "the old camping places."[153] He reported that 800-1,000 people participated in the eeling.[154]

While it would seem the *Gunditjmara* and others from south west Victoria living in more permanent habitations became mobile at times, the nature and extent of that mobility may have varied locally. When Westgarth was discussing the Eumeralla Village he stated:

"*The aborigines generally encamped there during a portion of the year, for the purpose of fishing, with occasional rambling over the neighbouring country. Mount Eeles* [Eccles], *an adjoining volcanic hill, with a large and romantic crater, appears to have been a favourite resort, their repeated visits having worn a distinct track to the summit.*"[155]

Westgarth's comments lack some precision as to what portion of the year the residents were domiciled, and his reference to their apparent visits to the crater of Mt Eccles is curious. Additional evidence, based on the reports of the presence of the inhabitants of Eumeralla Village, does shed further light however. They appear to have been in residence in early January in 1842 when Hunter's "Eumeralla" station was first attacked, later March in 1842 when Robinson visited, in June when Fyans was confronted by the warriors in 1839, also in June of 1844 when Westgarth came, in August of 1842 when further attacks on "Eumeralla" took place and in October 1842 when Fyans attempted to lead a raid on the village. Although these observations of residency are essentially random events, they do indicate a high frequency of the presence of people in the vicinity in the course of an annual cycle.

When the Scrubby Creek Village is considered in terms of how sedentary the inhabitants were, there is greater clarity. According to Thomas the "first settlers found a regular aboriginal settlement", presumably indicating it was a permanent settlement. The residents were "perfectly harmless and stationary", appearing to indicate they were sedentary. There can be little doubt they were sedentary for much of the year. Their fish and eel traps operated during the "flooding season," that is, in winter, and they still appear to have been in residence in summer when their village was destroyed because it was at this time that the "grass got bare or scarce."[156] Nevertheless, on the day their village was burnt down they were, "up the creek, seeking their daily fare." And although Thomas explicitly states that they were "stationary" they may well have been seasonally nomadic, in late summer to early autumn during the eeling season. No direct evidence exists for this contention but the proximity of Scrubby Creek Village to Lake Bolac and Salt Creek, a matter of 15-20 kilometres away, makes this a distinct possibility.

In summary I would propose that in terms of the traditional lifeways, and in accordance with the classification developed previously, several forms of sedentism were in apparent in south west Victoria. Various lines of evidence clearly point to the presence of well-constructed and, at times quite substantial, permanent habitations. Where multiple dwellings occurred these formed permanent settlements, such as the Port Fairy Village, here the occupants, Sievwright reported, "constantly reside in the neighbourhood of Port Fairy". Some groups were at least seasonally sedentary, being in residence throughout winter, probably longer, as movement away from these habitations appears to have only taken place outside that period. Other groups, such as those at Eumeralla Village, were at the very least exhibiting multi-season sedentism. Lastly, populations such as those at Scrubby Creek [now known as Tea Tree Creek], perhaps also Eumeralla Village, were probably seasonally nomadic, becoming only truly mobile for whole or part of a season. In these cases they probably became mobile in late summer and early autumn, to attend, for instance, the Great Meeting and the eel migration at Lake Bolac. It is possible, however, that the Scrubby Creek and Port Fairy villagers were fully sedentary. It is also reasonable to infer in the cases of Eumeralla and Scrubby Creek Villages that "logistic foraging" was the principal mode of mobility being followed. Thomas and Westgarth may well have been attempting to convey this with expressions such as "up the creek, seeking their daily fare" and "occasional rambling over the neighbouring country." To gain an understanding how this functioned, how higher levels of sedentism were supported, and to explain some of the ethnographic observations and archaeological findings presented so far, it

[150] Dawson 1881:10.
[151] Dawson 1881:10-11.
[152] Dawson 1881:3.
Robinson (Clark 1998-2000:2:363) reported just such a "grand meeting", attended by "chiefs and delegates of tribes" in the northern part of the Western District on 3 August 1841, which possibly involved 300-600 individuals ("100 fires").
Williams equates Dawson's famous Mirraewuae Swamp, with Black Swamp, "not many miles west of Caramut" (1881:3). I have reservations, however, about the identification of Black Swamp as Mirraewuae and am inclined to believe, as did Massola (1969: 53), that it was Buckley Swamp, the "Great Swamp" (Clark 1998- 2000:2:196) near Mt Napier.
[153] Clark 1998-2000:2:119 (1 April 1841).
[154] Clark 1990:105 (27 April 1841)
[155] Westgarth 1846:8n.

[156] Thomas 1858.

is necessary at this point to examine the regional economy in more detail.

Mention has already be made of some of the procurement facilities constructed by the Indigenous groups of south west Victoria. There was the Lake Condah and Darlot Creek systems, wood and stone weirs near Eumeralla Village, "dams" (probably stone weirs) and "brush fence" fish traps on Muston's Creek, the Mt. William eel channel system, along with the smaller example near it. It would appear systems such as those at Mt William were common along the margins of many of the rivers flowing from the Grampians and other ranges.[157] To the west, 20 kilometres from White Lake, at Toolondo, a 3.6 metre wide channel, approximately 1.2 kilometres long, had been dug joining two swamps in which elaborate systems of weirs, traps and trenches were located.[158] Further west, just over the border, at Naracoorte in South Australia Robinson saw "some of the fishing bank spoken of, same as under Mt William."[159] Other recorded fish traps were located on the Merri and Hopkins Rivers,[160] and on the margin of Lake Bolac.[161] Robinson described yet another fish trap on a lagoon eight kilometres up the Moyne River, but also noted and had drawn an additional type of procurement facility at that location:

"*The lagoon is called Tow.wer.deet.mole. There are fish in this lagoon, eels, and round the banks I counted six places the natives had made for fishing thus:* [drawing], *round the edge of the lagoon supported by fork sticks. I also saw a were* [weir] *across the river for catching eels.*"[162]

What Robinson was describing and drew were fishing platforms, very similar to those constructed on some of the rivers in the Pacific North West. This is probably what McCann meant when referring to the "great number of bridges along the waterholes" on Muston's Creek. These were used to fish for fish and eels,[163] both Dawson and Buckley provide a slightly different or more elaborate description of this activity. As Dawson noted:

"*by tying a bunch of worms, with cord made of the inner bark of the prickly acacia, to the end of a long supple wand like a fishing-rod ... when swallowed by the fish* [eels according to Buckley][164], *it is pulled up quickly before the fish can disgorge it.*"[165]

In procuring fish and eels there were times when the Aboriginal people of the Western Districts caught fish or eels *en masse*. The eel migration from Lake Bolac in early autumn is an example. At certain times the Lake Condah fish traps too produced large numbers of fish and eels. Sections were constructed at different levels to ensure that if a fish or eel escaped one trap it would most likely be caught in the next. But, only when there was sufficient rain for the lake to overflow could this system operate with full effect, as the fish or eels attempted to make their way downstream.[166] Thomas described the "dams" on Scrubby Creek in similar terms, which at "certain heights acted as sluice gates in the flooding season". Seasonal exploitation of migrating fish also extended to the mouths of the rivers where, Dawson reported, "long baskets", presumably the *arabine* [*ngarraban*] type trap, were used to catch fish called *tuupuurn*.[167]

While various types of facilities were constructed to encourage the breeding of eels in particular, as well as ensure high procurement rates of eels and fish when they migrated, other strategies relied on obtaining smaller but more constant yields. Fishing from platforms when the fish and eels were in non-migratory periods was one of those strategies, as was the capture of eels by detecting their air holes in waterless swamps during dry periods.[168] But, where employed, fish traps and eel channels appeared to form the foundation of the local economy. The "very fine and large" 60 metre fish trap, for instance, that Robinson measured at Tarrone, on the Moyne River, was owned by an unnamed man who took him to "several spots where he had resided and had worns or huts."[169] This man was a member of several families that resided in the area and, it appeared, operated the trap.[170] They may have been like the Scrubby Creek Villagers who went up the creek "seeking their daily fare" but nevertheless remained in the area. Locally there was a "mosaic of different environments, including woodland, swamps and small patches of both forest and open country,"[171] and as a consequence they were well positioned to exploit various ecotones. But they were, in effect, "tethered" to the facilities they had constructed. These presumably provided a significant proportion of their food supply but they still needed to find sustenance from other sources such as, according to the minimal information supplied by Thomas, "grubs [witchetty grubs] and small animals."[172]

South western Victoria was a relatively rich environment. At the time of the first European incursions the plains abounded in kangaroos, emus, cockatoos, parrots and brush turkeys. Apart from fish and eels the swamps and water bodies provided ducks and other aquatic birds,

---

157 Hall 1898:219.
158 Massola 1963:255; 1969:101-2; Lourandos 1976.
159 Clark 1998-2000:4:300 (8 May 1845).
160 Gill 1970:30; Clark 1998-2000:2:149,162 (25,30 April 1841).
161 Dawson 1881:94; Coutts et al. 1977a:23.
162 Clark 1998-2000:2:161,177,Fig. 4.35 (29 April 1841). See also Dawson 1881:95.
163 Dawson 1881:95; Clark 1998-2000:3:63.
164 Morgan 1852:40,48.
It is quite likely eels were caught from these platforms, with both Robinson (Clark 1998-2000:3:63) and Dawson (1881:95) saying they were used at night, which Buckley (Morgan 1852:40) recounted was the preferred time for catching eels by the method he described.
165 Dawson 1881:94.
Robinson (Clark 1998-2000:3:63) also mentions this as a method for catching "fish".

166 Coutts et al. 1978:24-5,28.
167 Dawson 1881:19.
168 Dawson 1881:95.
169 Clark 1998-2000:2:162 (30 April 1841).
170 Clark 1998-2000:2:162-3 (30 April 1841).
171 Williams 1988:23.
172 Thomas 1858.

along with their eggs, in addition to mussels and tortoises. Vegetal resources were also prolific, chiefly *murnong*, the daisy yam (*Microseris lanceolata* - plains), rushes (various names - *Typha* spp. - swamps) and common ferns (*mukine* - *Pteridium esculentum* - stony rises). In terms of exploitation of these resources the Eumeralla Village was well situated. Located on a "stony point", adjacent to Gorrie Swamp, it overlooked plains to the east, north east and south. The Eumeralla River at this juncture was, as Robinson had observed, "like a lake."

On the plains Mitchell noted in 1836 that, "kangaroos were more numerous in this part of the country than any we had traversed."[173] Robinson also remarked upon the "abundance of kangaroo"[174] in the vicinity of Eumeralla Village, but Fyans in 1839 was even more effusive - "the most magnificent arable and pasture countries in the world ... kangaroos and emu running before us, crossing in every direction; quails, parrots, cockatoos, and various kinds of the feathered tribe here."[175] Obviously various means of capture of these creatures were set up, although we do not know the exact methods. In his report to La Trobe on his visit to Eumeralla Village Robinson mentioned "places for snaring birds," corroborated by Sub-Protector Sievwright's observation on the same day, that they "passed many native snares on a very large scale for kangaroos, birds, etc.,"[176] as they crossed the plain back to "Eumeralla". An ingenious method for catching brush turkeys was also employed in this area, involving the hunter camouflaged by a "bush shield" using a pole about 3-4 metres long with a noose at the end and a lure of a butterfly or feathers. Mesmerised by the lure the birds were easily caught.[177]

The root of *murnong* was a "favourite" food in southwest Victoria, particularly in spring and early summer.[178] It was not confined to that region, reportedly growing by the "millions" in northern central Victoria,[179] as well as being prevalent west and north of Port Phillip, parts of New South Wales and South Australia.[180] Mitchell had first come across this plant on the Bogan River in the Corners Region where children were taught to dig for the root with a special shovel.[181] Resembling a dandelion, he observed that "in many places the ground was yellow with flowers" as he crossed the northern plains of the Western District in September 1836.[182] Similarly Robinson watched women harvesting *murnong* in this northern part in July 1841:

> "*Today the native women were spread over the plain as far as I could see them, collecting pannin, murnong, a priveledge* [sic] *they would not be permitted* [by the "squatters"] *except under my protection. I inspected their bags and baskets on their return and each had a load as much as she could carry.*"[183]

*Murnong* was known by a variety of local names in the region around the Eumeralla and Scrubby Creek Villages, *muurang*, *myrnong* and *yerat* for example, with the root itself called *keerang*, *pannin*, *pun.nim* and *pun'-yin*.[184] It would seem that *murnong*, which grew "chiefly on the plains" was probably utilised by the residents of the Eumeralla and Scrubby Creek Villages. Though not specifically referring to the Eumeralla Village, Westgarth wrote that "Various roots are resorted to, particularly that called *murnong*, a small root ... ... of which they are very fond."[185] Robinson, on the day he visited the Eumeralla Village, had noted that at "the native camp they had oven [oven mound] baking roots."[186] A digging stick, often called a *murnong* stick, was used to extract *murnong*. Dawson reported that "every family is provided" with one,[187] although that is not unusual, virtually every Indigenous adult female carried her digging stick. Such substantial quantities of *murnong* were collected, Dawson related, that when "several families live near each other and cook their roots together, sometimes the baskets [purpose made rush baskets] form a pile three feet [90 cm] high."[188]

*Murnong* roots contain inulin, a polysaccharide carbohydrate also found in the Jerusalem artichoke (*Helianthus tuberosus*), and have almost the same

[173] Mitchell 1839:2:252.
[174] Clark 1998-2000:3:51 (22 March 1842).
A similar statement was made by Robinson on 20 March 1842 (3:48). Robinson (Clark 1990:106) mentions "droves of kangaroo" in the Spring Creek area.
[175] Brown 1986:230.
A number of early colonial intruders made comparable comments. See for example Boldrewood (1884:36,39,49) and Hall (1898:216-7). It also seems the bulk of kangaroos and emus disappeared from the plains within a year or two of colonial occupation of these areas, an observation made by a number of early settlers such as Hall (1898:221) and Munro (as reported by Robinson in Clark 1998- 2000:1:135) .
[176] Sievwright 20 March 1842. See also Clark 1998-2000:3:48 (20 March 1842)
The trapping method may have been that described by Robinson on 9 May 1841 (1998-2000:2:194-5) with accompanying drawings (2:242, Figs 5.7,5.8).
[177] Clark 1998-2000:2:147-8 (25 April 1841),175 (Fig. 4.26)
See also Griffiths 1845:154; Lloyd 1862:409; Smyth 1878:1:192; Hall 1898:219; Robinson noted that brush turkeys were occasionally seen, but he encountered around 100 in one flock around 29-30 June 1841 (Clark 1998-2000:2:197,201; Clark 1990:111).
[178] Winter 1898:278.
[179] Clark 1998-2000:1:135 (22 January 1840).
It was in that part of Victoria that squatter (and later compiler of The Australian Race) Edward Curr (1886:1:240) said that the wheels of his dray "used to turn them up [*murnong*] by the bushel", Eyre (1984:127) observed early in 1838 "the plains were full of native yams" and Sub-Protector Parker reported to Robinson (Clark 1998-2000:1:134) "large heaps of *murnong*" in a camp he had visited in the area (21 January 1840).
[180] Gott 1983:8,10,Figs 9,11.
[181] Mitchell 1839:1:336.
[182] Mitchell 1839:2:272.
[183] Clark 1998-2000:2:326 (24 July 1841).
On another occasion (5 August 1841) Robinson drew women digging for *murnong* (2:388,Fig.8.6). Two colonists, Wedge (1835) and Godfrey (1843), also made drawings of women digging for *murnong*. This is highly unusual as they are probably the only Australian examples of drawings made in the colonial era of women engaging in plant gathering.
[184] Smyth 1878:2:183,210; Dawson 1881:xxvii; Clark 1998-2000:2:324, 326 (23-24 July 1841).
See Gott (1983:14-16) for a list of all variant names in different languages and for different parts or states of the plant.
[185] Westgarth 1851:709.
[186] Clark 1998-2000:3:48 (20 March 1842).
[187] Dawson 1881:12,20.
[188] Dawson 1881:20.

nutritional value. When cooked it produces fructose.[189] It appeared to form a staple from the latter part of winter through to early summer,[190] although it may have been harvested at all times of the year except late autumn and early winter when the tuber was shrivelled and bitter.[191] When in season it must have provided an ample supply of food. Robinson, it will be recalled, had observed that the each of the women returning from harvesting *murnong* had "as much as she could carry". Near Echuca in northern Victoria, Curr claimed that *murnong* was "so abundant and so easily procured, that one might have collected in an hour, with a pointed stick, as many as would serve a family for the day."[192] Referring to *murnong*, Buckley reportedly told settler James Malcolm that "a man may live on the root for weeks together".[193] Where *murnong* was abundant it could, based on this evidence, yield a substantial part of the food supply for lengthy periods.

*Murnong* can grow in many soils types but seems to thrive in open country with well-watered, loose, soil.[194] It is thought the act of digging for *murnong* promoted its propagation, through loosening, aerating and enriching the soil with humus.[195] One of the early colonists in Victoria in fact commented that the "soil dug like ashes" in an area where *murnong* had been growing.[196] As mentioned in Chapter 3, this has been called "natural cultivation". Elsewhere it has been termed "accidental gardening".[197] Because it can vegetatively reproduce, if a portion of the tuber were left *in situ* the plant would regenerate.[198] According to the classification of human-plant interaction developed in Chapter 3, this is "Selective Propagation". If a portion of a tuber were deliberately dropped in a hole where the plant had been dug up, "Replacement Planting", it would also lead to a new plant growing.[199] There is, however, no definite evidence that either took place, although Gott suggests that the Indigenous people of Victoria had a role in the distribution of *murnong* as a result of it being carried back to camp or traded.[200] This would be "Incidental Planting", or if being deliberately spread, "Translocational Planting". Whatever the case, Gott has claimed that the harvesting practices "bore a sufficient resemblance to agriculture/horticulture to be regarded as a sort of natural gardening."[201]

It has been suggested that it is possible, though not inevitable, that in time the groups of the Western District of Victoria would have gone on to develop agriculture, with all the preconditions for that form of food production being present.[202] *Murnong* may have been the plant that provided the basis for that, although it should be kept in mind that planting and agriculture may arise from a myriad of circumstances. An interesting example comes from near the Eumeralla Village. While returning to "Eumeralla" on the day he visited the Eumeralla Village, Robinson noticed "plenty parrots and a fire. The ground is clear all round and seeds put."[203] In this instance, based on the context, Robinson is more likely to be describing a method for catching parrots than the planting of seeds. Nonetheless, activities like that may lead to seeds sprouting and an eventual realisation it may provide a more reliable source of food than catching parrots.

Generally all sedentary societies rely on storage, but that does not mean societies which engage in storage are necessarily sedentary. Sedentism and storage, where they are evolving, nevertheless appear to be strongly correlated, they co-evolve and are in effect dependent variables.[204] As mentioned earlier, features interpreted as storage structures, including substantial constructions with thick walls, and pits, have been identified in the Darlot Creek area. These appear to be associated with habitation sites on the edge of wetlands.[205] A prerequisite for storage, as pointed out in earlier chapters, is the development of a means of preserving food that is produced. It has only been realised recently that one of the major food sources of south west Victoria, eels, were being smoked. Archaeologist Heather Builth has determined that eels were smoked at the base of "cooking trees", which have distinctive charred areas on the lower trunk and lipid residues are a result.[206] These "cooking trees" are also present at the Eumeralla Village site.[207]

Eels were not, it would seem, the only food preserved and stored in south west Victoria, however. Dawson pointed to wattle (*Acacia* sp.) gum being "produced", consumed and stored:

*"The gum of the acacia, or common wattle, is largely consumed as food, as well as for cement; and each man has an exclusive right to a certain number of trees for the use of himself and his family. As soon as the summer heat is*

189 Gott 1983:10.
190 Mitchell 1839:2:270,272; Smyth 1878:1:209; Armytage 1898:141; Clark 1998-2000:1:134 (21 January 1840); 2:201,324,326,366 (11 May, 23-24 July, 5 August 1841)
191 Dawson 1881:20; Gott 1983:4-5,9; Clark 1998-2000:2:139 (18 April 1841).
192 Curr 1886:1:240.
193 Malcolm 1845:13.
194 Gott 1983:5.
195 Frankel 1982:44; Gott 1982:65; Gott 1987:45.
196 Edward Page quoted in Frankel 1982:44.
197 Isaac Batey quoted in Frankel 1982:44.
198 Gott 1983:11.
199 Gott 1983:11; Kirkpatrick et al. 1995:2.
200 Gott 1983:12; 1987:45.
201 Gott 1992:43.
202 Williams 1988:222; Hayden 1990:42.
203 Clark 1998-2000:3:48 (20 March 1842).
204 Keeley 1988; Rowley-Conwy 1999:136-7; Varien 1999:33; Eerkens 2003.
205 Builth 2002:231.
206 Builth 2002:152-210.
207 Personal Observation 4 December 1999.
I noticed examples of these trees, burnt low down on one side, at the Eumeralla Village site at the time of my visit and photographed one. I recognised that they were unusual but was unable to ascertain their significance until I became aware of Builth's research.

*over, notches are cut in the bark to allow the gum to exude. It is gathered in large lumps and stored for use.*"[208]

Another possible stored food in south west Victoria was a red berry known variously as *muntharri*, *munteri*, *mantari*, *ngurp* or *nurt*, which came from a creeper (*Kunzea pomifera*) growing on the sand dunes of coastal South Australia and western Victoria.[209] According to Dawson it grew "abundantly" on the dunes at the mouth of the Glenelg River where "it is very much sought after, and, when ripe, is gathered in great quantities by the natives, who come from long distances to feast on it."[210] Where it was harvested in adjacent parts of South Australia (*Bungaditji*, *Milipi*, *Narriyerri*) it was reportedly pounded into cakes, stored and traded,[211] though Dawson did not specifically report that activity in south west Victoria.

As mentioned previously, it has been proposed that "logistic foraging" becomes a significant procurement mode where there is emerging sedentism, with storage taking on a significant role.[212] This may mean there is a central settlement, more than likely of a permanent nature, and procurement of food comes from exploitation of sources adjacent to it, within reach of it, or through the utilisation of temporary outstations. Crucially, food is brought back to the central settlement, rather than people moving to the food sources as is the usual practice in hunter-gatherer societies, and a proportion is converted to a storable form. A detailed hypothetical reconstruction of food procurement activities and the utilisation of facilities by the inhabitants of the Eumeralla Village can provide some understanding of how "logistic foraging" operated in south west Victoria.

Three types of exploitable food sources were available to the Eumeralla Villagers, constant, seasonally predictable and variable food sources. Major, constant sources were ferns, immediately available from the stony rise on which the village was situated, and rushes, close at hand in Gorrie Swamp. Fish and eels could also been caught on a daily basis from Gorrie Swamp or the Eumeralla River, which, it will be recalled Robinson described as being "like a lake ... two miles long: serpents [eels], plenty fish deep water." Significant seasonal food sources would have included eels in late summer and early autumn, *murnong* from late winter until early summer, and perhaps *nurt* in summer, requiring expeditions of 20 kilometres to the coast. Both eels and gum, and possibly *nurt*, were preserved and stored, however, and could be relied upon for many months. While eels may have been obtained when the migration from Lake Bolac took place, the traps on the Eumeralla River made it likely many were caught in the vicinity of the settlement.[213]

There would appear to have been a variety of significant food sources within reach of the Eumeralla Village. Parrots were one, with kangaroos and emus also available when in the vicinity. Many "native snares on a very large scale", as Sievwright observed, were set up on the plains adjacent to the village for catching kangaroos, emus, parrots and other birds and animals. The structures at Gorrie Swamp Hut, on the southern edge of the stony rises around Mt Eccles, overlooked the plains to the south. In this context they could be interpreted as a "station" for hunting kangaroos and emus on the plains, explaining the permanent but non-sedentary nature of the site as well as "tool repair" being a predominant feature of the site's lithics. A short journey was all that was required to reach it.

Another variable food source was fish, and the Lake Condah traps may have had a significant role here. The traps only worked when there were significant falls of rain and the lake overflowed. While most rain fell in winter, such falls could happen at any time in any season. Paradoxically no dwellings or settlements of any sort were reported there in any of the accounts from the earliest colonial period.[214] Close scrutiny of Mitchell's and Stapylton's journals, for example, showed that although they spent a whole week at Condah Swamp in September 1836 there is not the slightest indication of any habitations there at that time, even though Mitchell crossed and recrossed the swamp on two occasions.[215] So who built and operated the Lake Condah traps?

Returning to Westgarth's original account of his visit to the Eumeralla Village in June 1844, one sentence is of particular interest:

"*Mount Eeles* [Eccles], *an adjoining volcanic hill, with a large and romantic crater, appears to have been a favourite resort, their repeated visits having worn a distinct track to the summit.*"

There was no obvious reason for the residents of Eumeralla Village to be regularly visiting the rocky crater of the extinct Mount Eccles. It did, however, lay on a direct route to the Lake Condah fish traps. The implication in this instance is that they crossed over Mt. Eccles to find their way to the fish

---

[208] Dawson 1881:21.
This may have been a common practice in northern and western Victoria and south east South Australia. Referring to northern Victoria (Lake Boga) Edward Curr (1965:171) mentioned that there "were bags full" of this gum, "manna", "in almost every camp", and (Berndt, Berndt and Stanton 1993:109) indicate the *Narrinyeri* stored wattle gum in baskets which they consumed over winter.

[209] Dawson 1881:22; Zola and Gott 1992:21; Bonney 1995:14; Clarke 1998:17.

[210] Dawson 1881:22.

[211] Zola and Gott 1992:21; Bonney 1995:14; Clarke 1998:17.

[212] Binford 1980:15.

[213] There may also have been an eel bank system close to Eumeralla Village as Williams (1988:166) found 50 metres of ditching and banking in the area.

[214] Supporting the proposition that the remains of "stone circle" dwellings investigated in the 1970s and 1980s were not constructed until later.

[215] Mitchell 1839:2:245-252; Andrews 1986:177-84 (1/9-8/9/1836).
Mitchell (1839:2:247) nevertheless noted two "very substantial huts" around Byaduk North, about 10 kilometres from Lake Condah, on excursions he made to Mt Napier during that period, showing he was not necessarily failing to record sightings of habitations.
Robinson also passed through this area on 27 May 1841 and did not report seeing any settlement of any sort.

traps when necessary, a distance of about 10 kilometres. This does not mean they necessarily had exclusive rights or were the people who constructed the traps, it may well have been a cooperative undertaking involving two or more groups. Similarly they may have operated the fish and eel traps on Darlot Creek, 18 kilometres distant, in conjunction with other groups. There are indications that there was significant inter-group cooperation in south west Victoria, one of the most striking examples being the "Grand Battue", 24 to 32 kilometres across, that drove all the kangaroos and emus to a point close to Scrubby Creek Village.[216] According to Dawson "several tribes" were involved, and calculations suggest that well over 2,000 people participated.[217]

Thus one can see that Eumeralla Village could be maintained as a permanent and substantial settlement through logistic foraging, utilising in the first instance an array of food sources close at hand through direct exploitation, the construction of extractive facilities and the fabrication of specialised procurement technologies. Further infrastructure was constructed in the form of the Lake Condah - Darlot Creek systems to exploit seasonal runs of fish and eels at some distance from the Village. There, the *Gunditjmara* from the Eumeralla Village, perhaps others from elsewhere, may have erected temporary shelters, like the ones seen at Lake Bolac, while they caught and processed eels in their traps. Finally, opportunistic procurement, again possibly involving the creation of some infrastructure and excursions by hunting parties, as exemplified by the structures as Gorrie Swamp Hut, rounds out the hypothetical reconstruction of the sustenance regime of the Eumeralla Villagers. But, the question that still needs to be answered is, were they CHGs?

At the beginning of this chapter a number of identifying features of CHGs were listed. The first was "higher population densities". It is difficult to make comparisons and determine whether population densities in south west Victoria were higher relative to the rest of Indigenous Australia. As in a number of areas of Australia, smallpox epidemics prior to contact with the British colonisers seem to have had a devastating effect, so that the extent of the impact on population and population densities is very uncertain.[218] Nevertheless, comments by the colonial vanguard convey the impression of higher than "normal" population densities in the Western District. Dawson's assertion that the Great Meetings were attended by over 2,500 people is supported by calculations of the numbers involved in the "Grand Battue". Fyans estimated that there were 3,000 Aboriginal people in southwest Victoria in 1845, at a time when conflict with invading colonists, disease and starvation had already taken a considerable toll.[219] Curr thought that the Aboriginal population of Victoria was "greater per square mile than was to be found in any other part of the continent."[220] MacPherson, who had studied oven mounds and their distribution late in the nineteenth century, concluded that the population was "very much greater" in the Western District than elsewhere in southern Victoria.[221] Critchett arrived at a pre-contact population of 8,000, certainly a greater population density than found in the arid zone in Australia.[222] Thus, although the evidence is not conclusive, there are strong indications that there were higher pre-contact population densities in south west Victoria relative to many other parts of Australia.[223]

"Complex" or "highly developed" technologies were more clearly in evidence in south west Victoria, if the fish trap and eel bank systems, and specialised storage, qualify. "Intensified" subsistence was also a feature, eels, fish and *murnong* being particular examples. The extent of "investment" in procurement and storage infrastructure for fish and eels is also indicative of "increased economic specialisation". While many activities, such as day-to-day hunting and digging for *murnong*, appear to have been conducted in accordance with the division of labour characteristic of generalist hunter-gatherers,[224] others tasks reflected "changes in the division of labour beyond age and gender." The collection of gum is a case in point, this was one form of procurement that was "plant" based (actually the gum came from a small tree) in which men were involved.[225] The harvesting of *nurt* also seems to have been a collective enterprise, involving the whole group, across age and gender.[226] Similarly, the "Grand Battue" involved the whole group,[227] and it would appear that "several families" operated the fish trap on the Moyne River. The evidence,

[216] Dawson 1881:79.
Robinson (Clark 2000:126) reported that another "great hunt" had taken place in June 1841, about 30 kilometres west of Hamilton, in the vicinity of the Wannon River, providing some corroboration of Dawson.
[217] Dawson said that adult males stood "about 200 yards apart", [representing 536 individuals] while the old men, women and children waited at the closing point to help despatch the game [based on population structure calculations representing a total population of 2278].
[218] Butlin 1983:131-142; Critchett 1990:76-85.

[219] Critchett 1990:69.
[220] Curr 1886:3:469.
[221] MacPherson 1885:57.
[222] Critchett 1990:75,85.
[223] Other contemporary figures were:
- Buckley (Morgan 1852:41) indicating that in the region where he had resided there had been a confrontation with 300 warriors, representing a population of between 987-1260.
- Buckley reporting that he had attended meetings involving 700 people (Fyans 1845:44).
- Robinson (1844b:210; Clark 1990:105) had met with 3-400 people about 18 kilometres east of Port Fairy in 1841.
- Robinson indicating that 800-1,000 assembled for ceremonies in the eeling season at Lake Bolac (Clark 1990:105)
- Robinson (Clark 1990:105) reporting 800 people congregated at Port Fairy Swamp for ceremonies during the traditional whaling season.
- Robinson's (1845:48) estimate that the population of the Port Phillip Protectorate in 1845, although only the "remnants of tribes", was "five thousand (at least)".
- Duff's (n.d.:3) claim that the "Mustons Creek tribe" alone reputedly numbered 600.

[224] Dawson 1881:20-22,89-96.
[225] Another method of harvesting gum in which men were involved employed a special long slender twig with a hook on the end, called *inyen*, which was thrust through the gum and pulled away from the tree (Clark 1990:114; 1998-2000:2:319,349,Figs 7.14,7.15 - 18 July 1841)
[226] Based on the comment by Dawson, that "the natives", not just women, of south west Victoria were involved in harvesting *nurt*, and that both *Narrinyeri* men and women were involved in harvesting *nurt* (*manteri*)(Berndt, Berndt and Stanton 1993:569).
[227] Dawson 1881:79.

limited as it is, indicates the building and operation of fish traps indeed involved all family members.[228] Lastly, "food preservation and storage systems" were a feature of the economy in south west Victoria as well, involving the preservation and storage of eels,[229] gum and quite possibly *nurt*.

The second dimension to societies labelled as CHGs revolves around indicators of more complex social relations, hierarchy and ranking, accompanied by elaboration of associated beliefs systems, reflecting and reinforcing those elements. Simple social ranking, the appearance of differentials in the ownership of goods and access to resources, the accumulation of wealth, usually concentrated in the hands of a few, are emergent properties characterising such developments. In practical terms that may mean, using the Pacific North West as the exemplar, the presence of classes such as Nobles, Commoners and Slaves. Ownership there ranged from the large dwellings, weirs, drying racks and slaves to ceremonial bowls and "coppers",[230] while privileged access was manifested in control of stretches of river or good fishing locations in many instances.[231] Wealth, usually objects having value beyond their utility, was accumulated by individuals, but a cohort of "chiefs" mobilised their clans and allies to amass food and objects which were then ostentatiously displayed, distributed or destroyed during *potlatches*.[232] The Pacific North West groups had an elaborate ceremonial life, typified by the extensive *Kwakiutl* "Winter Ceremonies", and a rich decorative artistic tradition redolent in symbolism.[233] Carving of designs in wood, such as house interiors and boxes, sacred painted front boards of houses, and intricate patterns woven into cloth were a distinctive regional tradition.[234] Most striking were the enormous totem poles, erected in the front of houses, often the culmination of a mortuary *potlatch*, or *potlatch* marking the erection of the dwelling.[235]

One of the critical features of what are deemed to be complex societies is the presence of hierarchies, particularly formal hierarchies, and social ranking, both being institutionalised forms of inequality in power, authority, roles, rights and privileges. This not a relative change, it represents a fundamental shift in the organisation of society. There are numerous types of hierarchical positions that have been identified, and classifications of those types, usually in accordance with the degree of formalisation, scale of integration, centralisation of powers and process of ascension. As the size of the polity increases, so, it would seem, does complexity, with nominated formal positions usually beginning with Clan Headman, Paramount Headman, Chief, Paramount Chief and so forth. Complicating this neat classification are "Big Men", found in many traditional Melanesian societies, as well as elsewhere.[236] Big Men have an informal but central role in their local groups, and so do not wield any formal authority or power, their role being non-hereditary and reliant on their personal qualities and influence.[237] The critical indicators of centralisation, formalisation and institutionalisation of decision making where more complex sociopolitical structures exist can therefore be summarised as:

- inheritance of the office, either by direct or lineage descent
- central, formal, role in arbitration of disputes
- central, formal, role in judicial proceedings
- important role in redistribution
- central, formal, role in mobilising, coordinating and/or directing community or collective productive activities
- power to initiate and direct armed conflict with other groups

Depending on local circumstances and the degree of centralisation of roles and functions, such leaders may incorporate and consolidate in the one office other roles, such as ceremonial, religious and totemic leadership roles, as well as controlling external trade and exchange. Accompanying the formal role, the hierarchical leaders may be accorded special status, rights and privileges, such as special burial arrangements reflecting that status, right to high value items, to the best of the spoils of war, to receive tribute, the right to display, wear or carry badges of office, and retainers to attend to their needs. Specialisation of other roles and functions by individuals within the polity, with craft, sorcery, medical and other specialists being in evidence, is also associated with more centralised, formalised social and political structures. Ownership of increasingly discrete areas of land, such as clan estates, and specific resources, such as fishing sites, food sources or infrastructure, is another feature correlated with increased complexity.

By way of contrast, compare this with the nature of traditional inequality, power, decision-making, governance, integration and hereditary roles generally found in hunter-gatherers, as typified by the generality of Aboriginal groups as they were when contact with outsiders began. There was, and still is, a strong egalitarian ethos in Indigenous societies in Australia.[238] In my view this is a values system, a socio-religious ideology if you will, rationalising the sharing and reduction of risk in uncertain environments, the successful

[228] Dawson 1881:94.

[229] It would appear that the smoking of eels was limited to some degree as the reports of thousands of dead eels would indicate they were not able to preserve all the eels (similar problems occurred in the Pacific North West) or could only do so in suitable localities.

[230] Earle 1987:284.

[231] *Tlingit* (de Laguna 1990b:213), *Tsimshian* (Halpin and Seguin 1990: 270-271, 274), North Coast *Salish* (Kennedy and Bouchard 1990:447), *Chinookans* (Silverstein 1990:536).

[232] Suttles and Jonaitis 1990:84-6; Codere 1990:368-9 (*Kwakiutl*); de Laguna 1990b:220 (*Tlingit*).

[233] Suttles and Jonaitis 1990:84-5; Holm 1990a:602-29; 1990b:378-86.

[234] Holm 1990a.

[235] Blackman 1990:242,252.

[236] Earle 1987:282-3.

[237] Earle 1987:282; Feinman 1995:265-6.

[238] Peterson 1993

hunter, for example, sharing with the unsuccessful because the next day the role may be reversed. In traditional circumstances it probably served to reinforce the social solidarity of the group, thus promoting cohesion and mutual aid, as well as common protection in the event of inter-group conflict. Nevertheless, despite the egalitarian ethos, inequality was an invariant feature of traditional Indigenous culture. The inequality was based on age and gender in terms of social components, and specialist individual social roles in terms of totem group leadership, ceremonial functions, ritual performances, "medical" and magical knowledge, and inter-group diplomacy. Personal qualities could accord individuals higher status - a successful hunter, a strong warrior, a wise woman, a fine dancer, a good storyteller, a skilled teacher, a creative composer of corroborees, a trusted messenger. They could also play a significant part in ascension to specialist social roles – a quick learner, natural leadership skills, experience, a good memory, facility with languages - some of the attributes which could lead the individual to assume or be assigned those roles by their peers. Normally these roles were not hereditary, although they may pass from father to son, mother to daughter, or others in the same lineage as a family tradition.[239]

Social rank in most traditional Indigenous groups in Australia, as stated before, was based on age, gender and specialist social roles. In simple, everyday, normative terms this meant such things as, for women, deferring to the authority of their husband, or for older children and youths, to their parents or "elders". It meant the age-initiated men had privileged access to particular ceremonies and ritual knowledge denied uninitiated youths. Elders often held privileges denied younger members such the right to highly sought after foods, delicacies. As the size and complexity of a polity increases other classes, based on roles engendered in governance (e.g. Chief/commoner) or differential social or economic status (slave/commoner, serf/freeman) may become apparent, depending on the rigidity of the social order. But it is in the development of hierarchies and more centralised decision making that marks the clearest shift to a more complex society.

Governance in traditional Indigenous societies of the generalist kind was diffused and informal, with primary and classificatory kinship providing the underlying cohesion. Ceremonial leaders had authority in relation to specific or sets of ceremonies, sacred sites and objects,[240] while totem group leaders controlled the affairs of their totem group. Individuals may have had special ritualistic roles, such as the *kooradji* or *bulya*, the "sorcerer", who performed an "inquest" to determine who may have been responsible for an unexpected death. Where breaches of customary law, public order and inter-group relations was concerned, in most instances responsibility fell either to the individual and their family or the elders. So, if a man stole another man's wife, it was up to him and his relatives, and any allies he could recruit, to respond. If someone was murdered by someone within the same group it was normally up to members of the victim's extended family to exact revenge. If the murderer came from another group then, depending on the context, it may become a matter for the whole group, and the elders or senior family members would try to decide upon and direct the response, such as a revenge raid, often in the form of a surprise attack, on the other group. If an individual within the group behaved in a highly offensive manner, contrary to customary law, such as committing incest or entered into a "wrong" marriage (one with an individual from a section they were forbidden to marry into, also considered to be a form of incest), then the elders would act. Through discussions, guilt would be determined, a punishment decided and individuals delegated to carry out the sentence. Decisions affecting the whole group, such as whether to attack another group they considered their enemies, were also usually the province of the elders. Because the elders occupied a relatively privileged position and had responsibility for broader group issues and decisions, governance in traditional Australian Indigenous society has at times been referred to as an informal gerontocracy. As can be seen, the reality was not quite that simple.[241]

In light of the preceding characterisation of the conventional sociopolitical structure prevailing in traditional Indigenous societies across Australia, what evidence can we now deduce that there was more formal, institutionalised, centralised complex social structuring in south west Victoria? In the early contact period there were claims that chiefs were in evidence in south west Victoria. Robinson, in compiling lists of members of particular groups, mentioned "chiefs" on numerous occasions.[242] Some he described as "big chief", "principle chief", "chief big one", others he described as "Big man."[243] He even mentions "queens".[244] One could not put much store in these appellations however, Robinson's encounters were brief, he did not explain his use of the terms, and it is unlikely he had much knowledge of chiefs elsewhere. He may have been describing chiefs, clan leaders, heads of extended families, or "medicine men," it is almost impossible to tell.[245]

Dawson, in contrast to Robinson, had more intimate knowledge of the inner workings of traditional society in south west Victoria. He claimed that every "tribe" had a "chief", "whose authority is supreme," and that they could "command the services of everyone belonging to the tribe." No-one "dare contradict or disobey him." No-one

[239] Berndt and Berndt 1988.

[240] To the extent that they could impose death sentences, sanctioned by the elders, on individuals who wittingly or unwittingly committed serious breaches of sacred/secret "business" (See Hiatt 1996:94-5).

[241] Meggitt 1966; Berndt and Berndt 1988; Hiatt 1996:82-100.

[242] e.g. Clark 2000:97-8,109-111,115-118,133,137,150,157-8.

[243] Clark 2000:97-8,109,115-18,158.

[244] Clark 2000:183,185.

[245] At Cape Howe (Clark 2000:204) he met *Cum.mer.rer*, whom he listed as a "Chief". This was possibly a *Gommera*, a type of ceremonial leader/sorcerer (Howitt 1904:314), indicating the role Robinson may have been describing.

could address the chief "without being first spoken to" and "should they fancy any article of dress, opossum rug, or weapon it must be given without a murmur." It would seem there were clan estates and, when the owner died and there were no direct descendants, the chief divided land among "contiguous families". Chiefs also had the power of "commanding the attendance of the tribes" in regard to "matters of discipline." Succession was "by inheritance", although Dawson indicates that the chief, usually the eldest son of the former chief, was elected by the other chiefs. Dawson also details a series of customs entailing great deference being given to the chief.[246]

It is unclear whether chiefs had the power of arbitration by right, Dawson describing a situation where such matters were referred to a highly respected individual, who just happened to be a "chief", this individual being well rewarded for his role.[247] Nevertheless, there did appear to be formal judicial proceedings in which the chiefs played a role, and there was reputedly a public executioner who acted in accordance with directives from the chiefs.[248] Finally, there is considerable evidence of collective productive activities in south west Victoria, and here Dawson provided some indication that chiefs had a role in directing and organising them. The decision to hold a "Grand Battue" was "agreed by the chiefs", who then sent messengers out. Similarly, when eeling took place around Lake Bolac and Salt Creek, each tribe was "allotted" a portion of the river.[249] So, overall, if what Dawson recorded was correct, then it would be reasonable to conclude that there was indeed greater sociopolitical complexity in the region than that found in typical traditional Indigenous communities in Australia.

As satisfying, consistent and veracious as Dawson's compilation, mostly drawn directly from Indigenous informants, appears to be, there are some significant shortcomings undermining this conclusion. His use of the term "chiefs" is highly questionable, as this term is usually considered to apply to those who have authority over other leaders of smaller polities. There is no evidence of that in his writings. Furthermore, Dawson maintained that each chief and "chiefess" had a retinue of retainers, six bachelors in the case of the chiefs,[250] a most unusual circumstance even in terms of chiefdoms elsewhere. And finally, he refers to Buckley as a chief, when there is little to indicate in Buckley's own story that he was any more than a highly respected individual.[251] As others have noted, something may well have lost in translation, or there may have been in this instance a serious misunderstanding between Dawson and his informants.[252] Dawson's contemporary, Edward Curr, completely dismissed his accounts of chiefs, pointing out that chiefs had never been reported among any other group in Australia, apart from the *Narrinyeri*, that the number of retainers was unsupportable and that the notion of chiefs was contradicted by Buckley's account.[253]

While Dawson was probably incorrect in labelling the group leaders from south west Victoria as chiefs, and the retainers were more than likely an embellishment, strong support for his fundamental claims was provided by his contemporary, the Reverend Stahle, incumbent manager of the Lake Condah Mission. He reported that the *Gunditjmara* had hereditary headmen leading each "tribe" and went on to state that the headman had the power of "proclaiming war", that he "settled all quarrels and disputes", that he received and divided the "spoils" of war, retaining the best for himself, and that "men of the tribe were under an obligation to provide him with food, and to make all kinds of presents to him." Women captured in war had to be taken before the "council of elders and the head man" before the woman's captor could keep her.[254] Sub-Protector Sievwright, also pointed to a possible redistributive role for the headmen, remarking on a marriage ceremony in his Journal on 13 April 1842, that the "hut of the Chief of her [the bride's] tribe which was well stored with provisions which were distributed among the party."[255]

Based on this evidence, while being cognisant of its limitations in depth and detail, and the misapplied terminology of the amateur ethnographic observers, it is possible affirm the conclusion that there probably was greater formal sociopolitical complexity in south west Victoria, with the qualification that the group leaders were not chiefs but lineage-based hereditary clan headmen. For greater certainty one could look beyond south west Victoria to other places in Australia where there was evidence pointing to similar degrees of sociopolitical complexity in traditional circumstances. Noted 19th century ethnographer and compiler Alfred Howitt, for example, documented instances where there were seemingly hereditary headmen among the southern *Wiradjuri* in southern New South Wales (the people Gilmore associated with) and the *Naran-ga* of the Yorke Peninsula in South Australia.[256] The difficulty, however, in relying on observers and recorders such as Dawson, Stahle and Howitt, is that they were outsiders who had limited contact with groups whose sociopolitical structure had broken down or was breaking down. To expect that they could grasp the complexities and nuances of relationships and decision-making in the groups they encountered is not entirely realistic. So, before finally

[246] Dawson 1881:5-7,72,78,81.
This included "chiefs" having a distinctive form of dress at corroborees. Robinson's report (Clark 1998-2000:2:363) of the "grand meeting" he witnessed in the northern part of the Western District on 3 August 1841 indicates it possibly involved 300-600 individuals ("100 fires"), and was attended by "chiefs and delegates of tribes" who were shown "great defference [sic]"
[247] Dawson 1881:75.
[248] Dawson 1881:75-6.
Dawson (1881:74) also claimed there were "educational" specialists.
[249] Dawson 1881:94.
[250] Dawson 1881:5.
[251] Morgan 1852.

[252] See Corris 1968:16-9,35-7.
[253] Curr 1886:1:57-60.
[254] Stahle 1880:276-7.
[255] Sievwright 13 April 1842.
[256] Howitt 1904:303,313.
Sub-Protector William Thomas (1898:94-6,99) also mentions "Chiefs" of the "Barrabool" from the Geelong area.

answering the question as to whether the *Gunditjmara* and their neighbours, or other Australian groups for that matter, were CHGs, another Australian example will be considered. This is another case in which it would appear there was historically recorded evidence of some degree of formal sociopolitical complexity, but, unusually, one in which that evidence was later corroborated by Aboriginal informants in the course of a professional ethnographic investigation. This is the story of the *Narrinyeri*.

# Chapter 8. Big Men and Wannabe Chiefs – Part 2

The *Narrinyeri*[1] were an aggregation of language-name groups, or "tribes",[2] some would say a confederacy,[3] who occupied, and still have a presence in, the lower Murray River, the lakes and wetland region prior to its discharge into the sea, including Lakes Alexandrina and Albert and the 120 kilometre long Coorong. The *Narrinyeri* groups, the *Marunggulindjeri, Naberuwolin, Potawolin, Walerumaldi, Yaraldi, Tangani, Malganduwa, Wonyakaldi, Wakend, Ramindjeri* and Lower *Kaurna*, occupying an area of 5000 square kilometres, comprised 78 clans.[4] Theirs was a very rich and diverse environment, embracing the river, the wetlands, the lakes, the sea, mallee woodlands, coastal scrublands and the unique Coorong formation, a very long, narrow body of water separated from the sea by coastal dunes.

For the *Narrinyeri* contact with outsiders probably commenced with sealers and whalers in the early 19th century, sporadic encounters with explorers in the 1820s and a more forceful intrusion beginning in 1836 with the founding of Adelaide.[5] Detailed observation, interaction and recording of their life and culture by Europeans came initially through Protector Moorhouse, missionaries such as Tiechelmann, Meyer and Taplin, and traveller George Angas.[6] Taplin, who established the Raukkan [Point Macleay] Mission in 1859, was particularly prolific.[7] Relying on aged informants interviewed in the 1940s, such as Albert Karloan, Mark Wilson [Thralrum] and Pinkie Mack, anthropologists Ronald and Catherine Berndt eventually produced a very extensive memory ethnography of the *Narrinyeri*, in particular of the *Yaraldi* clans.[8] The Berndts' investigations in the main verified some of the unusual claims made about *Narrinyeri* sociopolitical structure by Rev. Taplin.

It has been estimated that before smallpox epidemics decimated the *Narrinyeri* population their numbers in 1820 stood at 6000 to 8000, and that they had the highest population density in Australia at that time.[9] They consumed as vast array of foods, 50 types of bird, 11 types of fresh water fish, 13 types of fish from the sea, 3 species of marine mammals (including beached whales), a variety of animals from bandicoots to kangaroos, a range of seeds, roots, and tubers, and all sorts of reptiles, eggs, crustaceans and shellfish.[10] In seasonal terms it would appear that there "no period of hunger,"[11] and they had an "inexhaustible source of food."[12] "We were never hungry," Karloan stated.[13] While at times described as "sedentary",[14] or "semi-sedentary",[15] and living in "permanent"[16] or "semi-permanent",[17] residences, traditionally they only appear to have been seasonally sedentary, although highly circumscribed, with distinct clan estates whose boundaries were marked by piles of stones.[18] In winter they lived in "cupola" type huts, somewhat similar to those in south west Victoria, about 4.2 metres in diameter and housing around 10 people,[19] but Lang's description of the habitation he saw somewhere on the Coorong suggests some clans had larger and perhaps more permanent dwellings. Also indicative of more sedentary or circumscribed populations, a number of cemeteries in the form of burial mounds and grounds were reported in the early colonial period.[20]

The *Narrinyeri* employed an array of "ingenious ... ... apparatus" to procure their food.[21] Their reliance on foods obtained from the lakes and wetlands meant that large canoes,[22] very large reed rafts (capable of carrying 8-10

1 Berndt, Berndt and Stanton (1993:19) suggest that *Kukabrak* may be a more appropriate term.
2 Tonkinson 1993:xvii, Berndt, Berndt and Stanton 1993:19.
3 Jenkin 1985:13,15 but see Tonkinson 1993:xxvii.
4 Berndt, Berndt and Stanton 1993:303-331.
5 Jenkin 1985:25-73; Berndt, Berndt and Stanton 1993:1.
6 Tiechelmann 1841; Meyer 1846; Angas 1847; Taplin 1878,1879a,b; Berndt, Berndt and Stanton 1993:292-3.
Material from Moorhouse is scattered through archives.
7 Taplin 1878:66-7; Berndt, Berndt and Stanton 1993:294-5.
8 Berndt, Berndt and Stanton 1993
9 Berndt, Berndt and Stanton 1993:17-8; Tonkinson 1993:xxiv.
10 Angas 1847:1:89-92,132,134; Taplin 1879a:87; Berndt, Berndt and Stanton 1993:78,553-76.
11 Berndt, Berndt and Stanton 1993:79-80.
12 Moorhouse (December 1840) quoted in Gill 1907-8:223.
13 Berndt, Berndt and Stanton 1993:74.
14 Tindale 1974:133 re *Ramindjeri*.
15 Berndt, Berndt and Stanton 1993:17.
16 Berndt, Berndt and Stanton 1993:17,81.
17 Jenkin 1985:13-4.
18 Angas 1847:1:64,88,93; Taplin 1878:1n; Tindale 1974:25,29,62; Berndt, Berndt and Stanton 1993:17,283,553-76.
19 Angas 1847:1:64; Taplin 1879a:103; 1879b:105; Jenkin 1985:14, opp.80.
20 Worsnop 1897:63; Gill 1907-8:226; Pulleine 1921:279; Simons 2003: 69.
Burial was not the norm traditionally, at least for men. It is quite possible some of these burial grounds were not cemeteries or burial grounds but mass graves consisting of victims of smallpox and other epidemics or inter-group conflict (see Berndt, Berndt and Stanton 1993:18, 278-9).
21 Moorhouse (October 1839) quoted in Gill 1907-8:228.
22 Berndt, Berndt and Stanton 1993:14,83,85-6,Plate 15.

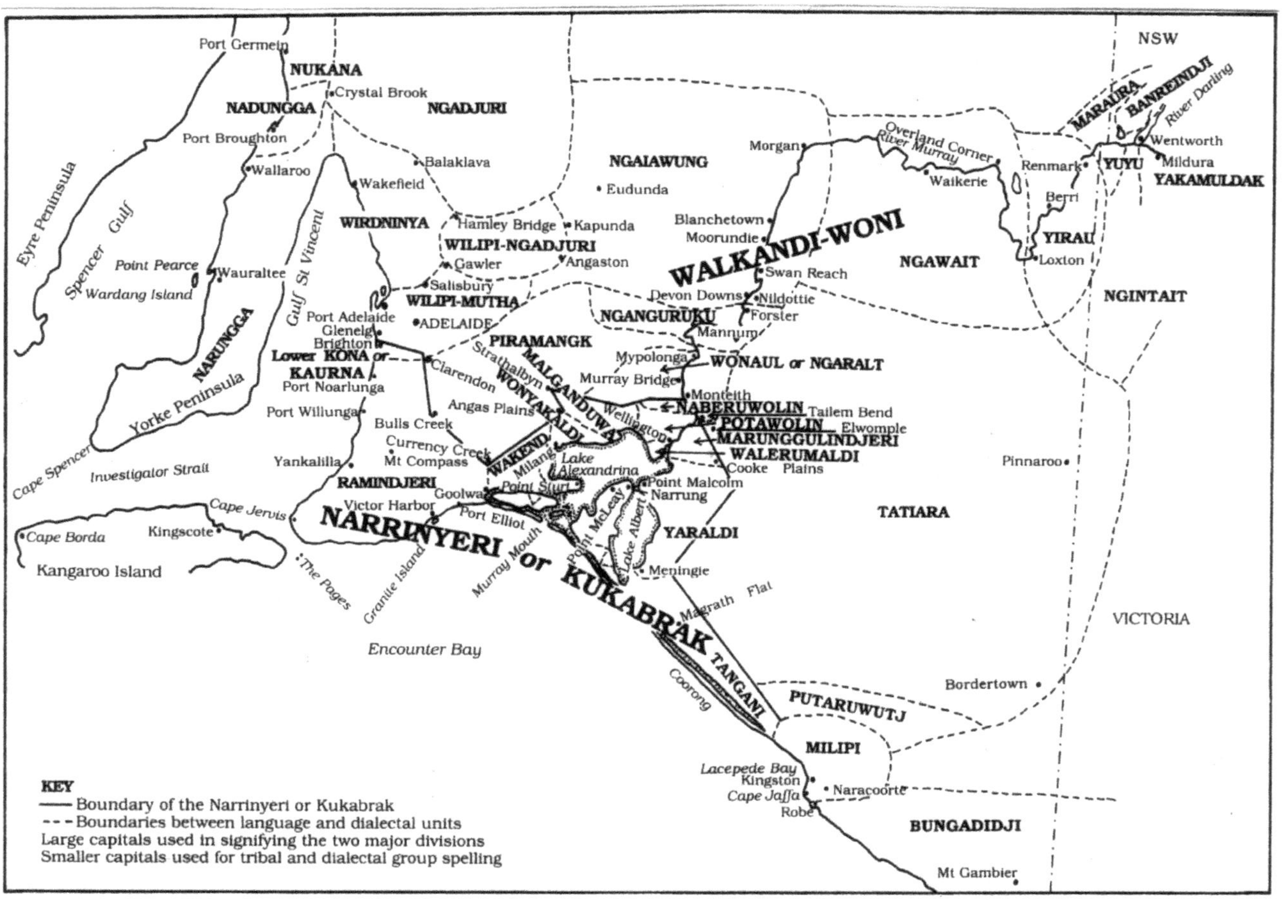


*Figure 24: The Narrinyeri and Other Language Groups of South East South Australia (Berndt, Berndt and Stanton 1993:304, Map 1)*

people),[23] nets,[24] fish traps[25] and fish hooks,[26] as well as many other complex devices, were extensively employed. The *Narrinyeri* were unusual in the extent to which the preserved and stored food. Fish were smoked and kept for 7-8 months, wattle gum was collected and used over winter, as were wattle seeds. *Manteri* [*nurt*], as mentioned previously, were pounded into cakes and kept for months, kangaroo meat was preserved for a week or more, and various roots and tubers such as *manakura* and *tuwaiki* were baked and kept for months at a time.[27] *Tuwaiki*, kept in skin bags after baking and consumed as needed, was also an important source of sustenance when going on journeys, as well as being used in trade, such as with the *Milipi* people, a small group to the east, centred on Naracoorte.[28] Interestingly, *tuwaiki* appears to be what was known elsewhere as *murnong*.[29] Finally, oils were extracted from whales, Murray cod, stingrays, pelicans, emus and Cape Barren geese. Some of these oils, such as cod oil "could be kept for months".[30] All these items, when stored, were kept in baskets, skin bags or containers as required.

The traditional division of labour in gender terms for the *Narrinyeri* was another social feature in which they differed from other, more mobile, Aboriginal populations engaging in typical hunter-gatherer subsistence activities. As with other Australian examples where populations were more sedentary and there was intensified exploitation of food sources, the division of labour was more complex. While women were still the predominant gatherers of plants, men seemingly had exclusive responsibility for the procurement of three plant foods. Both sexes (along with children) engaged in the collection of another five vegetal sources. Women caught two types of freshwater fish and another species in conjunction with men. But men alone were responsible for catching all salt-water fish.[31]

[23] Stephens 1839:75; Angas 1847:1:90; Berndt 1941:27-8; Edwards 1973:9,26.
These were used on Lake Alexandrina and ventured "several miles offshore" (Angas 1847:1:90).

[24] Berndt 1941:24; Berndt, Berndt and Stanton 1993:95-100,562-6, Plates 11,16.

[25] Tindale 1974:18,23,61-2,111;Berndt, Berndt and Stanton 1993:96-7, Fig.15F.

[26] Taplin 1878:139; Gerritsen 2001a:21-2.

[27] Harvey 1943:109; Tindale 1981:1879; Berndt, Berndt and Stanton 1993:79,99,109-111,348,569; Bonney 1995:14.
The baking of the roots was by the *maramin* method (Berndt, Berndt and Stanton 1993:103,110).

[28] Berndt, Berndt and Stanton 1993:19-20,118.
The *Milipi* were also known as *Meintangk* (Tindale 1974).

[29] Identification as *murnong* based on Berndt, Berndt and Stanton 1993:102,Fig. 20D,118,568.

[30] Berndt, Berndt and Stanton 1993:115.

[31] Berndt, Berndt and Stanton 1993:78,553-76.

Nevertheless, the making of large nets was a cooperative activity.[32] As I have suggested elsewhere, as populations start to develop food production economies, collective productive activities are observed, and the stereotypical generalist hunter-gatherer division of labour is supplanted, as a new regime is "negotiated" based on the internal logic of the groups in question.

Trade, and the distinctive manner in which it was conducted by the *Narrinyeri*, has also come under notice in Australian ethnography. Special relationships were established, called *ngengampi*, in a highly ritualised, embedded exchange system. *Ngengampi* partnerships were established with individuals from other dialectal or language groups, often considerable distances away. Intermediaries were used to communicate, and taboos, avoidance rules and other social restrictions applied.[33] Nevertheless, the *ngengampi* system formed a "supra-economic network", a "market" based "trading economy".[34] Through the *ngengampi* system, and more conventional exchange, the *Narrinyeri* traded skin cloaks and rugs, mats, baskets, fibre, smoked fish, oils and vegetables such as *tuwaiki* and *manakura*, receiving in return *pitjuri*, ochre, stone axes, shields, spears and large bark canoes.[35] To the west, direct trade was with the *Kaurna*, to the east the *Milipi* and *Bungaditji*, and northward up the Murray River, with the *Walkandi-woni* and those beyond, as far upstream as the confluence with the Darling River (*Maraura*), 350 kilometres away.[36]

The capacity of the *Narrinyeri* to trade in food and other labour intensive items, such as skin cloaks and rugs, mats, and baskets, would appear to indicate that that the *Narrinyeri* were producing surpluses. Trade in food, particularly if that trade is unidirectional, is normally associated with production of surpluses.[37] The existence of specialist cloak and rug makers, among the *Narrinyeri*, men who required support in the time consuming task of making those items,[38] also points to surpluses. Those surpluses were, by the act of making cloaks and rugs, translated into items of wealth, so that each cloak and rug became "a commodity of considerable value".[39] Some families "accumulated large quantities" of these and, as a result, acquired "a certain degree of prestige".[40]

Many researchers have commented upon the apparent linkage between the production of surpluses and formal sociopolitical structures.[41] This also appears to be the case with the *Narrinyeri*, and probably in south west Victoria as well. As in south west Victoria, traditionally the *Narrinyeri* had clan headman. However, they also had a paramount headman called *Rupulli* ("landowner") or *Mungkumbuli* ("leader, clever fellow").[42] The position was hereditary, usually the eldest son, but if unsuitable, another son from the same lineage was chosen, the final choice being ratified by the *yanarumi* and confirmed by public affirmation.[43] The *yanarumi* was a formal tribal council, called *tendi* by Taplin,[44] and was more than just a meeting of elders, as might have been the case in other Aboriginal groups in traditional circumstances. There was a process of nomination of representatives from the clans, and these representatives were sometimes young men as well as women. There were formal proceedings and structured seating arrangements. Although the *Mungkumbuli* had advisors, the final decision rested with him.[45] The *yanarumi* normally dealt with matters that included:

- Internal Affairs (dispute resolution, coordination of food gathering, allocation of widows, inquests, inter-clan relations)
- Decisions on Going to "War"
- Trading Expeditions
- Receiving "Strangers"
- Inter-tribal Ceremonies[46]

On other occasions the *yanarumi* was constituted as a court in which formal proceedings were followed, there effectively being advocates and prosecutors, the accused could be commanded to attend, with the *Mungkumbuli* acting in these situations as judge. Again structured seating arrangements were in place, though different from when the *yanarumi* acted in its capacity as tribal council. The *yanarumi* decided on guilt or innocence and passed sentences, which included the death penalty.[47]

At the *yanarumi* the clan headmen wore special possum skin cloaks. However, the *Mungkumbuli* had a distinctive cloak made of the best furs which had a fringed hem of mallee bark tassels. This was a mark of his office. A drawing of such a cloak was made by Karloan.

It would seem from the preceding evidence that prior to contact with Europeans the *Narrinyeri* did have a formal hierarchy, institutional forms and simple social ranking. However, their settlement characteristics and degree of sedentism and/or circumscription differs from that seen in

[32] Berndt, Berndt and Stanton 1993:75.
[33] Berndt, Berndt and Stanton 1993:116,118-21.
[34] Berndt, Berndt and Stanton 1993:120,129,283
[35] Taplin 1878:32-3; Tindale 1981:1879; Berndt, Berndt and Stanton 1993:19,113,116-8,120-21,572.
[36] Tindale 1974:81; Berndt, Berndt and Stanton 1993:19,116,118-20, 284,290.
[37] Shnirelman 1992:188.
[38] Berndt, Berndt and Stanton 1993:113-4.
There were other specialists – composers, medicine men, advocates – as well, who received payment for their services (Berndt, Berndt and Stanton 1993:68,130).
[39] Berndt, Berndt and Stanton 1993:113.
[40] Berndt, Berndt and Stanton 1993:113.
[41] e.g. Rolland 1985:255; Earle 1987:294; Kelly 1992:58; Hayden 1995b:20-1; Cobb 1996:252; Yoffee 1998:172.
[42] Taplin 1978:32; 1879b:36; Berndt, Berndt and Stanton 1993:59.
[43] Taplin 1878:32; Howitt 1904:314; Berndt, Berndt and Stanton 1993: 59-61,63-4.
[44] Taplin 1878:34; 1879b:36.
The reason these terms differ is probably because Taplin and Berndt's informants were from different dialectal groups.
[45] Taplin 1878:34; Berndt, Berndt and Stanton 1993:58-9,64-67,71-2.
[46] Howitt 1904:314; Berndt, Berndt and Stanton 1993:58-9,68-9,71.
[47] Taplin 1879b:36; Berndt, Berndt and Stanton 1993:64-69,72.

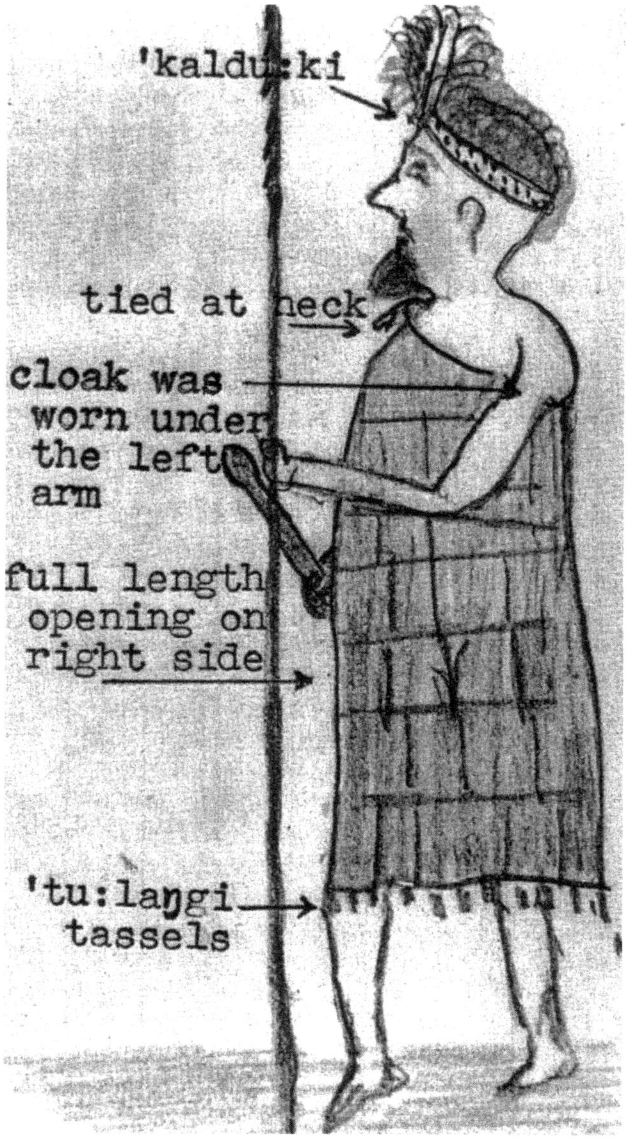


*Figure 25: A Mungkumbuli Wearing His Cloak – Pencil Drawing by Karloan*
*(Berndt, Berndt and Stanton 1993:62,Fig.4)*

the other cases documented in earlier chapters. They do not seem to have had permanent settlements or settlements of any size. Nevertheless, their winter dwellings do appear to have been permanent,[48] and some dwellings on the Coorong, as Lang stated, "could accommodate an unusually large number of people." While economically they seem to have an "intensified" subsistence, they do not appear to have been "concentrating" on a few staples. So were the *Narrinyeri* CHGs or not? To answer that, and the question as to whether the people of south west Victoria were CHGs, it is necessary to reconsider the whole concept of CHGs.

Reference has already been made to the people of Pacific North West of North America, portrayed in much of the anthropological literature as the quintessential, archetypal, CHGs. But, all is not what it seems. For example, the existence of hereditary slavery in the Pacific North West was a highly unusual, perhaps unique, attribute among supposed CHGs.[49] At time of contact, in the latter part of the 18th century,[50] only the northern groups conducted the famous *potlatches*, and on a much more limited scale than in the "golden age" of the Pacific North West.[51] Furthermore, only the *Tlingit* had totem poles when recorded contact with outsiders first took place. And the explosion of elaborate and rich carvings largely came about as a result of contact with outsiders, through the introduction of iron chisels and the wealth generated by participation in the fur trade.[52] Iron seems to have already been in use prior to contact,[53] and the *Tlingit* were already forging iron in a limited way by 1792, not long after first contact with outsiders. The *Kwakiutl* had guns as well by that date, and from the 1790s Chinese coins, molasses, alcohol, tobacco, muskets, sheet copper, along with many other items, were being obtained by trade.[54] This type of material culture is not what one usually associates with traditional hunter-gatherers, of any sort.

There are also more general problems with the concept of Complex Hunter-Gatherers. The relative and highly subjective nature of many key features, such as "more" elaborate rituals, beliefs and ceremonies, "more" complex religious symbolism, and so forth, is one such difficulty. What is meant by "more" in these circumstances, what is the yardstick? Were rituals or symbolism "more" complex or elaborate than in the past? How does one establish that, particularly if there are no records or descriptions of past rituals or symbolic systems? In most instances involving prehistoric cultures, archaeological research is not "fine-grained" enough to shed much light on such things, particularly in regard to ephemeral features and activities such as rituals, beliefs and mythologies. Much symbolic representation is also difficult to detect archaeologically, as the mediums used may have been transitory or highly perishable, such as bark, sand, wood or fibrous materials. Relative statements about such things may easily be compromised by preservational biases, particularly when it is realised that these transitory or highly perishable mediums are probably the most commonly utilised by Indigenous Australians, presently and in the past. Perhaps the characterisation of CHGs rituals or symbolism as "more" elaborate or complex could be viewed in relative terms to those found in generalist hunter-gatherers? But even this is highly problematic as our knowledge of traditional Indigenous cultures in Australia suggests that they universally had a very rich, elaborate and complex ceremonial life. As for special structures or areas, these are not necessarily created by CHGs. It is doubtful that this was the even the case in the Pacific North West where *potlatches* normally occurred in and around their large residences, with

[48] Berndt, Berndt and Stanton 1993:17.

[49] Suttles 1990a:4.
[50] First recorded contacts appear to have begun in 1774 (Arima and Dewhurst 1990:400; Blackman 1990:255; Cole and Darling 1990:119).
[51] Cole and Darling 1990:132.
[52] Cole and Darling 1990:132.
[53] Cole and Darling 1990:122.
[54] Cole and Darling 1990:120-3.

purpose built structures or demarked ceremonial areas a rarity.[55] Conversely in many traditional Aboriginal groups in Australia fixed ceremonial areas were common, often delineated by stone or earthen circles and lines, such as the *bora* grounds used in manhood initiations ceremonies.

One could argue that it is not beliefs systems and socio-religious ideologies that become significantly more elaborate and complex in CHG cultures but their concrete and material manifestations, which are more evident as a result of higher populations, greater material richness, permanence of residential location and surpluses being directed to those aspects of the culture. There are a wider range of more durable materials and objects produced or obtained which are decorated with the full range of religious symbols developed by the people from that culture. More precise and polished symbolic representations are produced by craftspeople with more time to invest in such things and using more technologically "advanced" tools, such as iron chisels. There is, in such societies, greater investment in physical infrastructure of a more durable nature as well, and this includes "religious" or ceremonial structures. This is made possible by the production of surpluses which can be used in materials acquisition and to support those making such structures, as a rational outcome of a greater degree of settlement fixity and higher levels of sedentism, the presence of hierarchies and specialist ceremonial roles.

Finally, the point has frequently been made by others that the characteristics of CHGs are very similar to groups engaged in rudimentary agriculture and identified as being analogous to neolithic societies.[56] For example, if one removes the two criteria specifically relating to food production based on agriculture or animal husbandry (food production and settlement locations on alluvial lowlands) from the list of criteria of neolithic societies developed in Chapter 6 then what remains is:

- The appearance of specialised procurement and/or processing facilities, apparatus, tools or implements.
- The use of ground stone technology, particularly edge-grinding, in the manufacture of stone tools and implements.[57]
- Evidence of conservation of output (preservation and storage as opposed to caching and stockpiling) with a view to providing an ongoing food supply.
- Permanent or near-permanent (i.e. capable of habitation for extended periods of time) dome-shaped or rectangular dwellings with the capacity to house up to 5 people, if not more.
- Permanent settlements, at least on an annual basis, with populations of 50 to 80 initially, growing to 2-300 later.
- Multi-seasonal sedentism at least.
- The presence of cemeteries in the region.
- Continued reliance on hunting, fishing, capture of waterfowl and the procurement of wild food plants.
- The manufacture of nets and weaving of baskets, mats etc.
- High value or exotic goods being traded over distances of at least 500 kilometres.

In applying these criteria to south west Victoria there is a close match, with two small qualifications. Firstly, there is uncertainty as to distances goods were being traded, although the area was certainly linked to the continent-wide trade network, and axes and sandstone implements from western Victoria were reaching in excess of 400 kilometres, as far as the Murray River to the north and *Narrinyeri* country to the west.[58] Secondly, it is not clear that there were cemeteries, as cremation was commonly practiced, although multiple burials in mounds have been reported. Regarding the *Narrinyeri*, again there is a match but with more significant qualifications. There may not have been cemeteries, as reported "cemeteries" could well have been mass burials, platform burial having been the normal form of interment for men.[59] The second qualification is the lack of permanent settlements of significant size in traditional circumstances, with the *Narrinyeri* being only seasonally sedentary, albeit within highly circumscribed clan territories.[60]

The import of this, even with the qualifications, is firstly that there is little difference between so-called CHGs and early agricultural societies, not only in terms of the characteristics listed above, but in the degree of sociopolitical complexity. Similarly, the commonalities among CHGs outlined in the beginning of Chapter 7, in terms of population density, settlement pattern, specialisation, change in the division of labour and storage, are also found in settled agricultural and pastoral groups.[61] Nevertheless, the descriptors for CHG societies include sociopolitical complexity as a defining feature, whereas they do not for neolithic societies. Why is this? I would suggest three reasons:

1. the apparent inability to easily "name" the economic base of CHG societies as a distinct entity, as can be done with agriculture and pastoralism.
2. the belief, engendered in the theoretical orientation

[55] See Suttles 1990a.

[56] e.g. Price and Brown 1985:16; Hayden 1995a:298; Belfer-Cohen and Bar-Yosef 2000:30.

[57] In view of the succeeding discussion it ought to be pointed out that edge-ground axes and tools were in use in south west Victoria and by the *Narrinyeri* (Dawson 1881:24; Davidson and McCarthy 1957:426; Mulvaney 1964:427).

[58] McBryde 1984a,b.
Greenstone axes from the quarries at Mt. William, north of Melbourne (not to be confused with the Mt. William in western Victoria) were finding their way over 700 km to the lower Darling River.

[59] Berndt, Berndt and Stanton 1993:269-79.
"Platform burials" involved leaving the body on an elevated wooden or bark platform held up by four poles.

[60] Perhaps confirming Syrett's (1993) contention that sedentism and sociopolitical complexity are not strongly correlated.

[61] A point made by others such as Price and Brown (1985:16), Hayden (1995:298), and Belfer-Cohen and Bar-Yosef (2000:30).

mentioned in Chapter 7, whereby it has often been argued that CHG economies and societies are inherently unstable or transitory and that they rapidly, inevitably according to some, evolve into agricultural economies and societies,[62] This is in spite of the fact that many CHG societies were remarkably persistent, and some, such as the Eastern Jomon or Ertebølle, were seemingly "resisting" agriculture.[63]

3. the significant degree of hunting and gathering observed taking place outside of the predominant, usually specialised, food procurement activities of the CHGs, while failing to appreciate or acknowledge the persistence of a significant degree of hunting and gathering in early agricultural/early neolithic societies. This arises in part because examples of the first agricultural societies are only known from archaeological studies focussing on their agricultural component, whereas many so-called CHGs have only been documented recently, often by social and cultural anthropologists focussing on their sociopolitical and cultural features. Hence they are put in some form of "hunter-gatherer" box. It is similar to the point made in Chapters 5 and 6, that early agricultural and neolithic societies in Australia have perhaps not been recognised because the people involved still engaged in significant degrees of hunting and gathering.

Paradoxically, this difficulty, I believe, reflects a broader paradigmatic issue, which is impeding the proper understanding of the origins of agriculture. This issue, and a possible resolution of it, will be considered further in the next chapter.

Although the conception of CHGs as a cultural complex incorporating manifestations of sociopolitical complexity is questionable, nevertheless populations identified as CHGs, as well as early agricultural and pastoral societies, do exhibit recognisable formal hierarchies, institutional forms and simple social ranking.[64] They are a result of local trajectories and raise interesting questions as to how and why this develops, as well as why differences emerge? For example, we see in south west Victoria that there was a much greater degree of sedentism, permanent settlements and settlements of some size when compared to the *Narrinyeri*. Yet there appears to have been a comparatively higher degree of sociopolitical complexity in the *Narrinyeri*. Before deliberating on the specific reasons for these differences however, other issues need to be addressed, so consideration will briefly be given to:

1. other observations, correlations and commonalities noted in regard to other complex societies.
2. determining, if possible, whether the Australian groups identified in earlier chapters as being neolithic-type societies also exhibited signs of sociopolitical complexity, as a means of providing further corroboration of those claims.
3. use of the Australian evidence to test some of the ideas and hypotheses about the development of sociopolitical complexity.
4. proposing additional observations about sociopolitical complexity, particularly where they may arise from consideration of the Australian evidence.

Various schema have been proposed to characterise the different types of sociopolitical organisation that have arisen in pre-state societies, bearing in mind some theorists stress the degree of variation in complex societies, or point out that neat categorisation may be misleading and that complexity ought to be treated as a continuum.[65] Feinman, nevertheless, has suggested two types of sociopolitical structures, "Network" (Pacific North West) and "Corporatist" (Polynesian polities) models,[66] while Renfrew has conceived, in respect to European Neolithic chiefdoms, "Individualizing" and "Group-Oriented" types.[67] Where classifications of types of leaders are concerned, a simple, relatively conventional one was employed earlier - the Headman, Paramount Headman, Chief and Paramount Chief. But others, such as Hayden, in keeping with his theoretical orientation, calls such leaders "Aggrandizers", providing a somewhat different classification which includes "Great Men" as well as "Big Men".[68]

The role played by formal leaders in these pre-state societies is also subject to many propositions, and attendant debates. Chiefs, or other formal leaders, are often identified as having a redistributive role in regard to wealth and surpluses.[69] There is some recognition that they may at times control "external" trade and exchange as well as internal credit.[70] Control, or at least mobilisation, of labour and resources, and hence production, has been identified by some as attributes of formal leaders, as well as informal leaders such as Big Men.[71]

Over time a number of observations about the nature of more complex, formal societies have been made by various theorists, attempting to identify what may be the common factors or features, and correlates of those societies. Formal, heritable, ownership of specific resources, resource

[62] Of course some, such as Arnold (1996:83) and Rowley-Conwy (2001: 58-62) have quite rightly criticised these views as being "deterministic", an example of the "progressionist trap".

[63] Akazawa 1986:73; Zvelebil and Lillie 2000:83-87.

[64] Kuijt (2000:314) asserts there is in fact no clear evidence of hierarchies in the Neolithic of south west Asia.

[65] Fagan and Maschner 1991:974; Lourandos 1997:9.

[66] Feinman 1995.

[67] Renfrew 1973.

[68] Hayden 1995b:18.

[69] Sahlins 1972:101-148; Earle 1987:292,296; Hayden 1995b:19; Yoffee 1998:172

[70] Gibson 1974:105; Earle 1987:293; Mauss 1990; Fagan and Maschner 1991:974; Hayden 1995b:19-21.

[71] Sahlins 1972:139; Binford 1983:217; Rolland 1985:257; Earle 1987: 294-5; Feinman 1995:265; Arnold 1996:81; Hayden 1995b:19-20,23;

"zones" and discrete areas of land, vested in leaders or increasingly smaller and specific groups of people, is one such commonality that has been associated with complex societies, distinguishing them from non-complex entities.[72] Craft, economic and functional specialisation are also seen as distinguishing features.[73] The importance of surpluses in underpinning sociopolitical complexity is another factor with almost universal recognition.[74] Attention has also been drawn to research indicating the emergence of political stratification when discrete populations reach a critical threshold, in excess of 1500.[75]

Theoretical discussions and debates about the origins of sociopolitical complexity and its manifestations have a long history, beginning principally with Marx, Engels, Morgan, Durkheim and others in the 19th century. Many modern theories on the emergence of sociopolitical complexity mirror those proffered to explain the origins of agriculture - population pressure, climate or environmental change, and "social competition".[76] One crucial difference in the case of CHGs are the "Garden of Eden" theories that have arisen, suggesting CHGs are "Affluent Foragers", based on the idea that they are people in particularly rich environments. This, it is argued, allowed them to support levels of sedentism and sociopolitical complexity comparable with that found in agricultural societies simply by undertaking the sort of hunting and gathering activities normally associated with generalised hunter-gatherers.[77] The "Garden of Eden" hypothesis has generally been dismissed, however, as it would appear there is no pre-state society that was immune from periods of food shortage, even if there is an abundance or super-abundance at certain times of the year. Furthermore, others have pointed out that all groups identified as CHGs relied on storage to some extent, unnecessary if they really were "affluent".[78]

Apart from the general theories, various causal factors or drivers of sociopolitical complexity have been suggested, including increasing circumscription, greater resource availability, individual ambition, centralisation of power, warfare and ideological transformations.[79] As with most of the common factors that have been identified, there appears to be a series of dependent variables operating, so that it is not possible in some instances to distinguish cause from consequence. Others are particularist explanations generated by specific case studies and cannot be applied generally, or are simply untestable. One other "theory" or explanation is "scalar stress". This posits that "social ranking" and hierarchies are a result of, or response to the growth in population and increasing circumscription of the polity.[80] However, while hierarchies appear to develop in the conjunction with an increase in social scale, this may not necessarily be a "stress" response but a rational outcome.[81]

If one takes a broader view, where there does appear to have been an increase in the scale of societies, be they agricultural, pastoral or "CHGs", it has been the result of more food being available, that is, "output", and with it population, has increased. Associated with this increase in scale is the appearance of systematic storage and the generation of surpluses, the development of procurement infrastructure, initiation of cooperative procurement activities, changes in the division of labour involving age and gender, an increase in the extent of trade, limited resource ownership, and craft, occupational and role specialisation. Rudimentary financial functions including credit, savings, and redistribution also feature in such circumstances.[82]

Rather than explain this phenomenon on the basis of an overarching theory or specific causal factor it may be more productive to consider it as a process, in this instance as one analogous to the process of organisational growth. Numerous researchers have recognised and articulated this in their own way.[83] All small organisations, be they businesses, community based groups or even governments, undergo the same process as they grow. To begin with, the handful of individuals involved have multiple roles (the Secretary is also the Treasurer), parts of the organisation may have multiple functions (the sales department also handles ordering), decision making and governance may be collective with most participants also decision makers, and most of those involved in direct communication with each other. A growing organisation becomes unworkable, however, unless structural changes take place. Because of the increasing size of the organisation, effective, direct, communication between the actors is no longer possible and new mechanisms for decision-making, coordination and communication are developed.[84] Some form of hierarchy of decision-making, separation, differentiation

[72] Sahlins 1972:101-148; Renfrew 1973:543; Earle 1987:284,294-5.
[73] Renfrew 1973:543; Schalk 1977:237; Rolland 1985:255,264; Earle 1987:297; Ames 1991:942; Arnold 1996:91; Lewis 1996:360-1.
[74] Rolland 1985:255; Hayden 1995b:19-20; Cobb 1996:252; Yoffee 1998: 172.
[75] Earle 1987:288; Beaton 1991:949-50; Feinman 1995:260; Keeley 1995:268-9; Varien 1999:144-5.
[76] Keeley 1988:395; Ames 1991:936; Arnold 1991:959; Maschner 1991: 931; Maschner and Fagan 1991:922; Arnold and Tissot 1993; Ames 1994; Feinman 1995:257; Keeley 1995:269; Allen 1996:217.
[77] Testart 1982:524; Yesner 1987; Maschner 1991:931; Maschner and Fagan 1991: 922.
[78] Binford 1983:202; Altman 1987:72; Yesner 1987; Ames 1994:213-4; Burch and Ellanna 1994:148.
[79] Gibson 1974:105; Price and Brown 1985:8; Arnold 1996:96; Cauvin 2002.
[80] Johnson 1982; Earle 1987:289; Higham 1989:86.
[81] Johnson (1982:402-6) draws attention to "spontaneous" informal, temporary, hierarchisation by the *!Kung* when large numbers congregated.
[82] Analyses of economic behaviour in hunter-gatherer societies have often relied on the Immediate/Delayed Return conceptualisation developed by Woodburn (1982). This has been strongly criticised by some (e.g. Altman 1987:99; Allen 1997:23). The most common criticism has been based on the observation that there do not appear to be any Immediate Return societies. I would suggest that it also has little explanatory value as it effectively lumps simple economic concepts, such as saving, investment and credit, together, and in the process obscures the understanding of critical behaviours.
[83] Earle 1987:289,292; Widmer 1988:31; Lewis 1996:360-2.
[84] Johnson 1982:392-6,407-14.
The number of linkages increases geometrically in accordance with formula ½(n2-n).

and specialisation of functions, specialisation of roles and delegation of individual rights and corporate responsibility become the hallmarks of these larger units. The degree of hierarchisation is dependent on the size of the body, and the degree of centralisation of decision making and control the leaders/managers wish to exert, or conversely, their willingness to delegate responsibilities.[85]It is this perspective that will inform further consideration of the evolution of socio-political complexity.

The correlation of more complex economies with sociopolitical complexity is not surprising if it is realised they are dependent variables. To use a simple example, cooperative procurement was not the norm in traditional generalised hunter-gatherers in Australia, each family was expected to be largely self-sufficient. Men went out hunting, sometimes singly, perhaps with male relatives or a couple of friends. Women often gathered collectively, in groups, but this was more for social reasons, such as child-minding, and they were not usually acting in a coordinated fashion. In more complex groups, such as in south west Victoria, when cooperative procurement activities were undertaken, such as a battue, everyone acted in concert. Similarly, construction of infrastructure such as fish races or eel banks required the coordination of the labour to carry out. Sometimes this can be done "organically" but as scale and complexity increases someone has to direct or coordinate those activities. In so doing that person is accorded a role not normally ascribed in generalised hunter-gatherers. These coordinated activities result, however, in a quantum leap in yield, a significant increase in extractive efficiency. This then contributes to an increase in the food supply and surpluses which in turn leads to population growth, with further possibility or necessity of hierarchisation.

There are a number of implications of these types of dynamics. Firstly, social innovation is a significant factor, as important as technological innovation, in increasing output and productivity. Secondly, it is not necessarily easy to recognise the organisation dynamics outlined above because they may be discreet, transitional or obscured by the embeddedness "problem".[86] Until greater differentiation and separation in roles and functions occurs, multiple roles and functions may be invested in individuals or specific groups, particularly when initial formal centralisation of authority occurs. Further obscuring this is the investment of many functions, roles and activities with religious or spiritual significance within the framework of the socio-religious ideology of traditional societies. Thirdly, even though sociopolitical and socioeconomic complexity are effectively dependent variables, they are not necessarily strongly co-variant, thus allowing considerable variation in local trajectories to take place. Consequently, this would allow for the differences that have been adduced between south west Victoria and the *Narrinyeri*. Finally, it is possible to see in early neolithic-type societies the beginnings of modern sociopolitical structures, though in the long-term that trajectory is only followed to its conclusion when there is attendant economic development and a commensurate increase in social scale. Arguably different types or styles of government, "autocratic" as opposed to "democratic" or "inclusive", may even be evident from an early point, echoed in Renfrew's characterisation of chiefdoms as Individualizing or Group-Oriented.[87]

Another relevant aspect of sociopolitical complexity that has attracted attention has been Big Men. This appears to be another example of the trajectory being followed which varies according to local circumstances.[88] The social structure in these instances is not egalitarian, but neither is it stratified as it is in chiefdoms, a situation Hayden has termed "transegalitarian".[89] Big Men are seemingly charismatic leaders who have no claim necessarily to formal authority or heredity right. In situations where there are surpluses being generated and sedentary settlement patterns, conditions normally associated with formal hierarchies, the Big Men become highly influential, using patronage, alliances, marriages, friendships and so forth.[90] They develop a "following", accumulating wealth for themselves and their followers, usually in competition with other Big Men.[91] Feinman identifies two types, ones who occupy a central position in exchange networks and others who mobilise the factors in production - labour, natural resources and capital in the form of procurement infrastructure.[92] Both would seem to be generating or accumulating surpluses which are translated into status and wealth. In modern economic terms Big Men could be viewed as entrepreneurs and middlemen and perform those functions in their societies.

There has been considerable enquiry and debate as to how chiefdoms may evolve from Big Men and Great Men. Clearly major social changes take place in moving away from a kin-based social structure to one embracing some other organising principle that binds the collectivity. Great Men appear to be more overtly political, with some ascribed status and leadership in warfare a common attribute. A trajectory from Great Men to Big Men to chief is usually posited in Melanesia,[93] though conceivably the chiefs could be an amalgamation of those roles or at least some of their functions. As there were no chiefs in Australia, and the question as to whether there were Great Men has probably never been explicitly addressed, it is difficult draw on Australian evidence to add to this discourse on the origins of chiefs. But it is worth noting that in both south west Victoria and the *Narrinyeri* formal hierarchies appear to have been in evidence and yet there are no unequivocal

[85] The choice between centralisation and delegation is a constant tension in all complex organisations as they struggle to balance the conflicting goals of control, flexibility and responsiveness.

[86] For example, mixes of hierarchy and egalitarianism are evident in small scale agricultural societies (Kuijt 2000:313).

[87] A view echoed by Feinman (1995:266).

[88] Earle 1987:282.

[89] Hayden 1995b:18.

[90] Earle 1987:282; Godelier and Strathern 1991.

[91] Feinman 1995:265-6; Hayden 1995b:18,23.

[92] Feinman 1995:265.

[93] Godelier 1991.

indications there were Big or Great Men. Of course there are considerable limitations in the information relating to both those case studies, but it suggests there may be other pathways to chiefdoms in which Big Men and Great Men do not figure.[94] Lastly, higher levels of circumscription, and by extension, sedentism, has been suggested as a factor "forcing" hierarchisation as it does not allow groups to "fission" when there are internal tensions, conflicts or divergences.[95] Both appear to have been the case in south west Victoria and with the *Narrinyeri.*

Having reviewed a broad range of ideas, concepts and theories regarding sociopolitical complexity, it would be apposite at this point to briefly return to the *Nhanda* and the Corners Region, to see what congruence there is with what would be expected in more complex entities. Where applicable, observations from elsewhere in Australia will also come under consideration.

Evidence of sociopolitical complexity among the *Nhanda* at the commencement of the colonial usurpation is limited, simply because the available information from that time period is so limited. However, in one commentary Gregory did refer to a "king" and "queen":

"*The women are far better treated ... ... I have, for instance, known the queen of the district to give her consort the king a sound thrashing with a yam-stick and then escape being speared for such an assertion of co-ordinate authority.*"[96]

As with such references to supposed Indigenous royalty elsewhere in Australia, the looseness of the terminology makes it impossible to ascertain if Gregory was referring to any sort of formal hierarchical position or just using hyperbole. Unfortunately it is the only extant allusion referring to any supposed form of governance in the region. However, there are other observations where it is possible to infer there may have been some sort of overarching structure. Government Resident Burges, for example, wrote:

"*Although the Aborigines here seem thus divided into tribes, they seem to act from a community of feeling and interests like one family. They do not appear to be continually at war with each other like the Southern natives ... ... they seem more united than any other natives I have seen in the colony.*"[97]

Burges's comments were made in the context of the coordinated resistance to the British colonists' invasion of the Victoria District, whereby attacks were launched in the south, and when the police and soldiers rushed to the area, another attack was launched in the north. Such unity at that time may not have been the product of some form of political integration however, but a manifestation of the *Nhanda*'s resolve to avoid the fate of their southern confreres by mounting a more sustained, planned and determined resistance to the colonists encroachment. Nevertheless, there may be grounds for a suspecting there could have been broader sociopolitical integration in the *Nhanda.* Another unusual feature of *Nhanda* society which may have reflected this in some measure was the practice of basing marriages on "reciprocal localities". Whereas marriage rules in most of traditional Indigenous Australia were based on sections or moieties, with the *Nhanda* and their neighbours it depended on the localities in which the prospective partners lived and was explicitly limited by degree of consanguinity.[98] This may have been because of exogenous influences,[99] a product of greater circumscription and sedentism, or a manifestation of more formal estates. Regarding the latter possibility, Gregory recounted:

"*They had their landed estates, the boundaries of which were better defined than many of the squatters leases now are. The daughters inherited landed estates as well as the sons. I have heard an aboriginal lady say, 'What are you? Why, before I married you your land only comprised the miserable range of hills which are not fit to feed an emu. It was my dower which brought you this fine valley, with its yams and springs of water' etc.*" [100]

There is also evidence of collective procurement activities and construction of procurement infrastructure by the *Nhanda*, tasks seemingly requiring coordination of labour and an indicator of complexity. The deep wells adjacent to the yam fields are one such sign, small-animal battues another.[101]

Looking at the Corners Region again there are references to supposed "chiefs", even "kings" by Mitchell, Duncan-Kemp, Bennett and Fraser.[102] They describe individuals who undoubtedly had leadership roles, but claims that certain individuals were "chiefs" do not hold up. Mitchell could not have ascertained from his brief encounters if there was a formal hereditary role such as a "chief", let alone the extent of the hierarchy.[103] Duncan-Kemp similarly appears to be really describing totem group leaders, possibly clan leaders.[104] Bennett claimed:

"*When I was out on the Bogan in 1845 I saw several tribes who showed much respect or deference to their Chiefs or Kings and during a residence of many years in those*

[94] Yoffee (1998) similarly concluded, based on his study on the development of sociopolitical complexity in the south west and south east of the United States, that there was considerable variation in the pathways leading to greater sociopolitical complexity.
[95] Johnson 1982:408.
It has also been claimed (Johnson 1982:417) that the rate of "system evolution" relates to the degree of hierarchy.
[96] Gregory 1886:23.
[97] Burges 1851b:2.
[98] Gerritsen 1994a:136-8,144-8.
[99] It is suggested in Gerritsen (1994:144-50) that marriages based on explicit consanguinity may have arisen as a result of the influences of marooned Dutch sailors.
[100] Gregory 1886:23.
[101] Oldfield 1865:273.
[102] Fraser 1892:38-9,41-2.
[103] Mitchell 1839:1:195,265-6.
[104] Duncan-Kemp 1952:153,221; Watson 1998:46.

*districts the Chiefs always appeared to have absolute sway.*"[105]

While he later indicated "the Chief ... ... is hereditary,"[106] there is no clear indication as to whether he was describing a clan headman or a paramount headman. Other credible sources point to hereditary headmen in the Corners Region, such as Horne and Aiston, who maintained the *Diyari* had headmen whose *poonga* occupied a special position in their settlements.[107] Howitt, while agreeing the *Diyari* had hereditary headmen, went further in pronouncing that the *Diyari* also had paramount headmen, the *Pina Pinaru*, who could summons a "Great Council" which he then chaired. The *Diyari* paramount headman at that time, Jalina Piramurana, had the power to resolve disputes, approve or dissolve marriages, and decide when ceremonies and initiations would take place.[108] What appears to have been some sort of tribute was paid to Jalina, including *pitjuri*, ochre, and skins, by neighbouring "tribes" and groups from as far away as 500 kilometres.[109] Only qualified individuals could attend the Great Council, and there may have been a pecking order, possibly indicating some social stratification as well as hierarchy. The suspicion there was some social stratification within *Diyari* and associated groups is reinforced by the reputed occurrence of insignia of rank for *Diyari* totem group leaders and other "distinguished individuals",[110] as well as possible status burials as detailed in Chapter 5.

Considered more generally, some of the other indicators of sociopolitical complexity are also found within the Corners Region. Examples of substantial collective investment in infrastructure and equipment include the large dam on the Bulloo floodplain, the *mokhani* dams from south west Queensland, fish traps systems such as found at Brewarrina, and 270 metre long nets. A number of instances of collective procurement activities were noted in Chapter 5, such as the *Yuwaaliyaay* harvesting, threshing, winnowing and storing *yarmmara*, and hunting emus using their large nets. Similarly, Mr Poole of Sturt's 1845 expedition in the north west of New South Wales chanced upon "a tribe of about 60 in number engaged in collecting grass seeds," while Gregory mentions "natives" harvesting 400 hectares of *Panicum* grass species and Fraser refers to a battue with a diameter of 16 - 24 kilometres.[111] Occupational specialisation, another indicator of sociopolitical complexity, was noted in passing in Chapter 5 as well, with specialist tool and hereditary net makers, and hereditary hut builders being cases in point.[112]

Beyond the *Nhanda*, the Corners Region, south west Victoria and the *Narrinyeri* elements of sociopolitical complexity do appear elsewhere in Australia. Claims of headman and paramount headman append to the *Wailpi* of the northern Flinders Ranges in South Australia, *Jukambe* of southern Moreton Bay, Queensland, *Kaiadilt* of Bentinck Island in the Gulf of Carpentaria, and *Maraura* at the confluence of the Murray and Darling Rivers.[113] The Murray River, particularly the mid to lower Murray, is an area that has exhibited many signs of socioeconomic and sociopolitical complexity in archaeological and reconstructive research. Hereditary ownership of resources and infrastructure (fish traps) was widely reported in the early colonial period,[114] along with signs of sociopolitical complexity. Stapylton, for example, claimed he discerned a "chief" at Lake Benanee, an anabranch lake, who was distinguished "by a particular kind of cloak he wore".[115] This is consistent with what has been interpreted as a status burials at the Roonka III cemetery on the lower Murray, where one of the individuals interred also had a fringed cloak.[116] Most remarkable of all is the burial ground at Lake Victoria, a large anabranch lake on the lower part of the central Murray. Here there are an estimated 10,000 burials, making it one of the largest burial grounds in the world.[117] But that is another matter, and consideration for now will finally be given to theories and explanations of the origins of agriculture. By necessity such theories and explanations should be able to account for larger scale spatial and deeper temporal dimensions relating to the origins of agriculture in different parts of the globe at different times, the developments in Australia outlined in the preceding chapters, and the local trajectories alluded to above. This is the test to be applied to current theories, and any new theory to be offered in the succeeding chapters.

[105] Bennett 1893:1:15.
[106] Bennett 1893:1:15-6.
As was the *kooradji* [medicine man].
[107] Horne and Aiston 1924:17. See also Reuther 1981:VI,44/501.
[108] Howitt 1891:64-68,72,80.
Gason (1895:173) appears to acknowledge the Diyari had paramount headmen, but that the position was elective not hereditary.
[109] Howitt 1891:66-7.
[110] Howitt 1891:68.

[111] Fraser 1892:41.
[112] Apart from examples in Chapter 5, see also Horne and Aiston 1924: 106-7.
[113] Curr 1965:366[*Maraura*], Howitt 1891:67 ["*Murdula/Mardala*"= *Wailpi*], Uniacke 1825:62; Howitt 1904:318-9 ["*Chepara*"= *Jukumbe*]; Widmer 1988:30[*Kaiadilt*]. See also Hawker 1901:75.
[114] Curr 1965:79-80,243.
[115] Andrews 1986:81 (27 May 1836).
[116] Pretty 1977:305,322,325.
The dating, 3930+120 BP, shows that this may be a custom with a long history.
[117] Pardoe 1995:706; Finkel 1997.

# Chapter 9.
# Food to Trade

It has been argued in earlier chapters that in at least two regions of Australia at the time of British colonisation the Indigenous populations were engaged in agriculture, which is usually considered to be a form of food production. In Chapters 7 and 8, so-called "Complex Hunter-Gatherers" came under consideration, with particular attention being paid to the Pacific North West and south west Victoria. The question arises as to whether they were food producers? Conceptually, food "production" is usually seen as being distinct from food "collection", the latter being the characteristic mode by which more mobile, generalist hunter-gatherers, including many traditional Australian Aboriginal groups, gained their livelihood.[1] In this vein Dennelll has suggested food producers are those who invest "labour and materials in raising the productivity of their environment."[2] Typically, agriculture and pastoralism are viewed as the archetypal and earliest forms of food production. Because of their role in the foundation and formation of the first settled communities in human history they have assumed an iconic status. While the status of agriculture and pastoralism as forms of food production is unquestionable, are there other types of food production? What exactly is food production? Questions such as these need to be addressed if the origins of agriculture are to be properly understood.

Agriculture, particularly in its modern form, may involve a system of monocultural, mixed or sequential cropping, often as part of a mixed economy, an integrated system, whereby a domesticated crop is grown in conjunction with the husbandry of a particular type of domesticated animal or animals. In modern Australia the classic mixed rural economy is based on wheat and sheep, with farmers growing crops of wheat while at the same time keeping large flocks of sheep. This is consistent with notions of what is food production, as pastoralism, as stated above, is also generally considered to be a form of food production. Thus cattle ranching and dairying, involving sedentary pastoralists keeping, or husbanding, herds of tame, captive, usually domesticated, cattle for their meat and the natural products they produce, is seen as a significant form of food production in Australia, as elsewhere. However, there are many ways in which we can manipulate, control and/or harvest living things – plants, animals and other natural, organically based, life forms of many different types - ultimately to provide food or other useful products for ourselves. For example, some Australians, along with people in many other parts of the world, live by means such as fishing or fish farming, or the harvesting of shellfish and crustaceans. In most of these cases special devices or facilities are constructed, oyster frames for instance, so as to optimise the growth in the population of the species to be harvested or to enhance the rate at which they are procured. There are people in some countries who also keep herds of domesticated cattle, but they are not ranchers, they are, or were, nomadic pastoralists. The *Masai* are a well-known example. Then there are the *Sami* of northern Europe, some of whom follow herds of tame, "semi-domesticated", reindeer. These examples raise a series of questions. Are the *Sami* food producers and is nomadic pastoralism a form of food production? Are modern fisherfolk, living in villages, towns, cities, but spending months at sea on large trawlers catching "wild" fish stocks, food producers? What about villagers in India who cast nets from the shore. What about horticulturalists, such as the swidden cultivators of sweet potatoes in New Guinea? They are often seen in the anthropological literature as being engaged in something other than agriculture because of the limited area they utilise and the different production methods employed,[3] are they food producers? What if landholders engage in market gardening? What if the horticulturalist is just growing flowers? Is the sheep farmer a food producer if they are raising sheep, as is often the case in Australia, just for their fine wool? These are not just questions about the finer points of defining agriculture or semantic debates about what is meant by food production, there are fundamental issues involved, issues that have a significant bearing on the origins of agriculture.

Many of the examples just cited come within the ambit of what are termed primary industries. Some of these are specialisations intended to yield food for human consumption. Others are intended to provide useful products. To distinguish between these, and to differentiate food production from food collection, the commoner definitions of food production identify the act of "growing" food or "breeding" animals for human consumption as the critical distinction.[4] Reflecting this view, Leach states that "where people begin to cultivate numbers of wild plants they cross the threshold from wild plant-food *procurement*

1 Ingold 1996:12-13.
2 Dennell 1985:149.
3 Bronson 1977:25-6; MacNeish 1992:10; Leach 1997.
4 e.g. Childe 1934:44.

to wild plant-food *production*" [Her emphasis].[5] There are, however, some difficulties with this proposition. Numerous examples were provided in Chapter 3 of Indigenous people in various parts of Australia who planted "numbers" of seeds, fruits and yams. Yet in most of these cases the groups in question would not normally be seen as food producers, principally because the planting was of limited extent and they seemingly did little else that affected the life-cycle of the plants. Furthermore, this planting did not necessarily provide a "significant" proportion of the food supply. Alternatively, in the Sahel and sub-Saharan Africa, village-based, sedentary, tribespeople harvest various types of seeds found in extensive stands of naturally occurring wild grasses, known as the *kreb* complex. The harvested seed grains are then "stored in warehouses by the ton," and at times sold commercially.[6] If the cultivation criterion is applied, these people would not qualify as food producers, even though they harvest the wild grass seeds on such a "massive" scale that it is "basic to the diet of the local people."[7] In other words it is their main food supply. To resolve these conundrums and arrive at an understanding of what constitutes food production a closer examination of what food producers do may be in order, beginning with agriculture.

A key feature of agriculture is the role the cultivator has in determining where particular food plants will grow and when germination will take place, facilitated by evolved characteristics of easy, more reliable and relatively synchronous germination such as found in domesticated wheat. Sometimes the timing of germination can be more specifically controlled, such as when crops are irrigated. In other cases, such as with broad-acre wheat farming, it may depend on the timing and sufficiency of precipitation, farmers often planting in expectation of this. Nevertheless, by attempting to grow a crop, whether wild or domesticated, at a time and place of their choosing, cultivators provide more optimal conditions than if the subject species were left to the vagaries of nature. This optimisation may initially involve elimination of competing species by clearing, even if only by burning off, denser planting, planting in better soils, planting where there is more reliable, naturally occurring, moisture or planting, as noted earlier, in expectation of sufficient moisture, such as when rain is expected. In this context Walter Smith's comment in Chapter 5, that broadcast seeding was often carried out "just before a shower of rain," is consistent with common agricultural practices. To improve or ensure a crop's survival it may even be flooded, irrigated or, as may have been the case with the yam crops grown by the *Nhanda*, directly watered. Consequently, with cereals for example, there is a greater yield for a given amount of seed and distributional area than would occur naturally, and in effect productivity is increased. Finally, direct human intervention, involving plants being grown in discrete plots and fields rather than being patchily distributed across the landscape, makes harvesting much more efficient, both in terms of yield and the labour required. This is because of the greater concentration of the plants in one location, and higher density of plants in that location, than would occur naturally. The domestication of plants, although probably initially unintentional, may have contributed to this increase in extractive efficiency, the near synchronous ripening and tough rachis of wheat being a case on point.

With animal husbandry, the "breeding" that is normally practised is, in effect, the application of some degree of control over the animal's reproduction while in captivity.[8] Even so, in many cases, particularly in less sophisticated economies, the animals involved are simply allowed to reproduce naturally, there is no control of reproduction. One only has to think of the free-ranging chickens, pigs, goats and so forth running around farmyards and villages in many parts of the world to appreciate this point.[9] Where control of reproduction is exercised it may involve the males being culled or castrated to control the size of the herd, or flock. In other instances the sheep farmer, for example, may only allow the rams and ewes to mate at particular times to control the timing of lambing, thus reducing losses.

"Breeding", is also used in another sense, that of "selective breeding". However, the first animals kept as a food source, sheep and goats, were probably initially undomesticated,[10] and their reproduction was subject to little or no control. Their domestication, which it is claimed took 30 generations for these species,[11] probably arose initially as an unintended consequence of artificial selection by culling, the "harvesting" method, not necessarily selective breeding.[12] Selective breeding, it has been argued, was not practiced until around 4000 BP at the earliest.[13] Where selective breeding is carried out it may involve breeding for desirable characteristics, the introduction of "improved" stock into the husbander's breeding population or the retention of the best animals to improve or maintain the desirable characteristics of a herd or flock. In all cases the intention to "improve," as judged by humans, the husbanded animals is focussed on such things as the utility, productivity and manageability of the animals being kept.

The animals closely associated with humans were originally

---

5 Leach 1997:144.

6 Harlan 1992:22-3.

7 Harlan 1992:23.
Other examples of large scale and/or commercial wild seed plan harvesting can be found in North America (wild rice – *Zizania aquatica*), Europe, India (wild rice) and northern Africa (the kreb complex includes *Panicum* spp., *Eragrostis*, *Digitaria*, *Aristida pungens* etc.)

8 See Clutton-Brock 1989:7.

9 See for example Downs 1964:12,30,35.
There are many other examples of "domesticated" animals in which there is minimal or no control of reproduction, such as the mithan (*Bos frontalis*) of India and Burma (Harris 1977a:225), the Andean llama in Incan times (Harris 1977a:232; Wing 1977:851) and cows of Greece in the Classic period (Russell 1986:33).

10 Bokonyi 1976:20; Legge 1996:241-3,258-9; Uerpmann 1996:233; Clutton-Brock 1999:74-5,78-9.

11 Bokonyi 1976:21.

12 Bokonyi 1976:20-2; Legge 1996:239; Uerpmann 1996:235.

13 Bokonyi 1976:22; Clutton-Brock 1999:40-1.

wild creatures. As a result of human actions some of these animals became kept and tamed, some species even bonding to humans, some treating them as conspecifics, some tethering themselves to the human "home range", most becoming dependent on humans, many ultimately becoming "domesticated".[14] Apart from modern practices, such as artificial insemination and embryo transfer, the mechanism of reproduction is nevertheless the same whether the animals are wild or domesticated. The point in keeping, taming, making dependent, domesticating, even selectively breeding animals is that humans assume a level of control of a population capable of breeding. This is a more efficient, more productive way of ensuring the supply of those animals than if they were wild and roaming free. Under human care and control their survivability increases because they normally have a more reliable food and water supply, they are frequently protected from non-human predation, from extremes of weather and climate, from the vicissitudes of animal life, and their ailments and illnesses are prevented and treated. With humans supplying, providing, making available, their food and water these populations normally well exceed those of their wild, free-roaming brethren. Moreover, they can be "harvested" almost at will. Again animal husbandry, like cultivation of plants, brings about an increase in productivity as well as an increase in extractive efficiency.

The increase in productivity and extractive efficiency arising from cultivation and husbandry reflects the scale identified in Chapter 3 of increasing control and intervention in the life cycle of the subject species, be it plant or animal. According to Ingold this difference in scale gives rise to the distinction between "gathering" and "production," as determined by the "relative scope of human involvement in establishing the conditions of growth."[15] However, as seen in the analysis of cultivation and husbandry, the goal of food producers is not to control plants or animals, that is merely the means to an end. By engaging in a range of activities and tasks that raise productivity and extractive efficiency, food producers increase output, as well as the reliability and predictability of their food supply. But any deliberate intervention may increase productivity and extractive efficiency, even those carried out by generalist hunter-gatherers. For example, "firestick farming", as discussed in Chapter 3, seems to provide an increase in natural productivity and an easier way of catching more prey. A male hunter making a spear, or a female gatherer a digging stick, are also engaging in tasks that presumably result in greater returns than if they did not have those tools and weapons. So at what point in the scale of intervention is it possible to distinguish hunter-gatherers from food producers. The hunting and gathering activities and implements just mentioned reflect a generalised approach to the procurement of food. Food producers, however, concentrate their efforts on one or two species, or small set of similar species, and in so doing they intensify exploitation of those species. As part of that "concentration", food producers acquire and apply in-depth knowledge relating to the species they focus on, including its preferences, behaviours and responses, as well as developing specialised tools, materials, implements, equipment and practices for that purpose. They may also construct special purpose structures, such as storage pits, and other infrastructure, such as corrals, dams, wells and ditches.

Overall, agriculturalists and pastoralists rely on specialised food procurement methods that increase productivity and extractive efficiency, and by extension, output. As food producers they are, in effect, engaging in a form of primary economic specialisation. They are reliant on those specialisations to produce most of their food, although some may rely on more than one specialisation to meet their needs. Over time these food producers tend to become even more specialised. In modern agriculture, for example, the degree of specialisation and level of investment is significantly greater than in the past, and extends to such things as fences, dams, windmills, bores, silos, irrigation systems, tractors, ploughs, seed drills, mechanical harvesters and computers, along with the use of fertilisers, herbicides and pesticides. The most recent extension of this is the introduction of genetically modified crops. All are aimed at increasing productivity and extractive efficiency and ultimately greater, more reliable and predictable output.[16]

From the preceding discussion I would propose that food production is a form of primary economic specialisation, resulting in higher levels of productivity and/or greater extractive efficiency, and achieving levels of output sufficient to provide a significant proportion of the food supply (or income in modern market economies) for the people undertaking those activities. There may be, of course, economic specialisations based on plants or animals which do not produce food but other products useful to humans, such as wool from sheep or cotton from plants. But this conceptualisation of food production, where the proposed criteria are met, means that nomadic pastoralism and fishing can be considered to be forms of food production.[17] Oyster farming, vegetable gardening, horticulture, even large mammal hunting, such as whaling, where they meet the criteria, would also qualify, as would the seed harvesting of sub-Saharan and Sahelian Africa.

A significant feature of food production economies is the

[14] Of course many animal species never become "domesticated", in the sense often used in definitions of "domestication", in that they do not become totally dependent on humans for survival (see Clutton-Brock 1989:7).

[15] Ingold 1996:21.

[16] Food production, as conceived here, could also be formally couched in terms of a basic production function, Y = f(L,N,K,E), where L is labour, N is the natural resource, K is capital and E is enterprise, or, as I prefer, the organising principle.

[17] It could be argued that at least in some instances the fish being caught are "domesticated". Different scales of harvesting which are dependent on the method of catching fish seemingly results in a change in the size of the fish being caught. Through these forms of artificial selection fish may become either smaller or bigger according to the scale of harvesting (Conover and Munch 2002). This is another example of "domestication" as a result of the harvesting method.

fabrication or acquisition of specialist tools, materials, implements and equipment. While possession of such specialist items may not be the exclusive preserve of food producers,[18] and some food producers may lack them, generally this is the case. Neolithic food producers, for example, possessed or created sickles, cupholes, specialised grindstones and bone spades.[19] Fish hooks, nets and watercraft were developed in prehistory by those engaging in fishing as a form of food production. In addition to this, some food producers invest time, materials and effort in building infrastructure. Such things as storage structures, corrals and pens are constructed by those exploiting plants and animals, while fisherfolk may build fishing platforms, weirs, and traps. In the long term more elaborate and more specialised tools, materials, implements and equipment are developed, often in the form of machinery, even computers and satellites, as seen on modern farms and ranches (called "stations" in Australia) or as part of aquaculture. And, as would be expected, investment in infrastructure is considerably greater, more extensive and complex in the modern manifestations of food production.

One of the outcomes of the development of food production economies are large increases in population as these new modes of production greatly increase the amount of food available. A new production equilibrium is achieved. This increase in output, where it is conserved, preserved and stored makes higher levels of sedentism possible. Conversely, increasing sedentism makes strategies to increase productivity and extractive efficiency, and hence output, more feasible. So, Neolithic farmers, for example, grew crops, stored the grain and didn't need to incessantly roam to procure food.[20] That allowed them to spend time looking after their crops, for making tools, and building dwellings and storage facilities, as well as keeping animals. In these terms sedentism, specialisation and output are, broadly speaking, dependent variables.[21] These were not necessarily strongly dependent variables, at least when sedentism and specialisation were evolving, with feedback mechanism linking them being relatively weak, largely because of the complexity of "scheduling" for hunter-gatherers in uncertain environments. Consequently, I would argue, the development of specialisation and sedentism, which took place over thousands of years, involved many very small, incremental, changes before becoming recognisable social features.

At this point it should also be noted that the earliest sedentary, food producing, populations may have relied on more than one specialisation to meet their needs. Beside cropping and animal husbandry, south west Asian Neolithic communities engaged in fishing and even specialist gazelle hunting. Similarly, fishing was an important component of the Chinese Early Neolithic economy.[22] The mix of specialisations would appear to have varied depending on the specific circumstance, such as local resources, climatic suitability and the local trajectory. One advantage of this is risk-reduction, should adverse conditions be encountered in terms of one specialisation the situation may be saved by one of the others. This strategy is still evident in modern times in the guise of mixed farming.

In summary, I would suggest that food producers are those who:

- Rely on specialised procurement, concentrating on one or a few food species or types of species
- Invest in infrastructure and other capital items, usually to increase productivity, extractive efficiency or for the storage of food (to preserve output)
- Engage in collective or collaborative food procurement activities, either in building infrastructure or in direct procurement
- Rely on systematic storage or other forms of preservation of output

One of the implications of the preceding discussion relates to CHGs. If one applies the definition of food production developed in the preceding pages, then many CHGs should be considered to be food producers, or evolving food producers. This is the view of some in regard to the Pacific North West, and Jomon Japan.[23] But there are many other examples of CHGs specialising in fishing. Fishing was a significant activity in the Natufian, although ultimately most people from south west Asia came to rely on agriculture and pastoralism. Nevertheless, fishing became a separate specialisation followed by some in the region. One need only think of Peter in the Bible, the "fisher of men", who was a fisherman on the Sea of Galilee, employing nets to catch his fish from a boat with sails,[24] an industry that persists to this day. This highlights the absurdity of claims that CHGs societies inevitably transform into agricultural societies, different food production pathways are possible, fishing being one of them.

The trajectory followed by the people of the Ertebølle and Eastern Jomon cultures, if considered from this perspective become explicable, as instances of the development of separate specialisations. In both cases theirs were

[18] For example, a slender piece of wood with a hook at the end (called *werkurumi* by the *Narrinyeri*, *tirn* by the *Djab wurrung*) was a common device that seems to have been used in the lower Darling and Murray River region in southern Australia and south west Victoria to extract witchetty grubs (*wiwi* and *peliti* were two types sought by the *Narrinyeri*) from trees (Berndt, Berndt and Stanton 1993:575). While reflective of increasing specialisation many groups using it could not be considered to have been food producers.

[19] In the Corners Region we see the equivalents of stone knives used in harvesting, the *boonal* plank for threshing and the appearance of specialised grindstones.

[20] Presumably where Neolithic cultivators obtained other wild foods they relied on logistic strategies.

[21] Keeley 1988; Rowley-Conwy 1999:136-7.

[22] Wright 1978:203; Contenson 1982:58; Chang 1986:102,112; Henry 1989:19; Crawford 1992:14; Smith 1995a:137; Underhill 1997:123, 125; Bar-Yosef 1998: 167; Belfer-Cohen and Bar-Yosef 2000:23; Mithen et al. 2000:60-1; Kuhn and Stirner 2001:115; Cauvin 2002: 55.

[23] Nishida 1983 (Jomon); Ames 1991:941-2 (Pacific North West).

[24] See Matthew 4:18; Mark 1:16; Luke 5:2-10; John 21:3-11.

largely aquatic and maritime oriented economies,[25] in which catching fish, eels and marine mammals provided a significant part of the food supply. In the case of the Ertebølle, agriculture began diffusing into the region, either by colonisation of sparsely populated land, cultural transfers or demic expansion. Initially agriculture probably did not have an advantage over the pre-existing local economies because of climatic limitations or incompatibility with the existing economic orientation, or both. Whatever the case, the pre-existing local economies had a "comparative advantage," as some acknowledge,[26] and so the northern Europeans were not necessarily "resisting" agriculture. They were simply making rational choices. Perhaps, as agriculture became better adapted to the coastal and northern climates of Europe, it became more viable in areas suited to agriculture. At that point dual, separate, specialisation occurred and both agriculture and fishing economies appear within the same region. Certainly fishing and sea mammal hunting did not disappear because agriculture became the dominant specialisation.[27] As in south west Asia, fishing continued, the antecedents of the Atlantic and North Sea fishing industries of today.[28] A similar dynamic occurred with the Eastern Jomon, with rice farming eventually taking hold in those areas in eastern Japan where there was a predominantly terrestrial economy.[29] But fishing and sea mammal hunting did not disappear in coastal areas, these activities continued, to the present day.[30] It could further be argued that the beginnings of regional specialisation were also evident in the Pacific North West.

One likely indicator of a food production economy is effective surpluses. Surpluses arise when the food procured is in excess of subsistence requirements. Sometimes there are short-term or transitory abundances, such as during the eeling season in south west Victoria, when thousands of eels were left to rot at Lake Bolac. While there is evidence of at least limited preservation, by smoking, of eels in south west Victoria, presumably they did not have the capacity to deal with situations of super-abundance. Similar limitations have been noted in regard to the Pacific North West, their capacity to preserve the salmon being a productive "bottleneck".[31] Nevertheless, a sufficiency of the surplus food may still be preserved, stored or transformed in some way for future use, that not only meets subsistence needs for a period but exceeds those needs. This is an effective surplus.

Although effective surpluses may be indicative of food production, this should not be taken as being prescriptive, there may be cases where the criteria for food production are not fully met, yet extensive storage is undertaken, on a scale that still yields effective surpluses. Another issue in regard to surpluses relates to the problem of specifically detecting or determining if any group in prehistory or ethnographically recent times was producing an effective surplus. Diagnostically, calculations could be made as to how much was kept in storage and balanced against consumption, but often that is just not possible. There are, however, a range of indicators of effective surpluses:

- Sociopolitical Complexity
- Regular or Systematic Trade in Food
- Wealth (Accumulated or Conspicuously Consumed)[32]
- Socially Mediated Savings and Credit Systems
- Increased Time and Resources Devoted to "Non-Productive" Activities (such as leisure pursuits, ceremonial activity, and carving and other artistic endeavours which lead to art and craft specialisation)[33]

Having defined food production, and identified indicators that effective surpluses were being produced, it is now possible to finally answer the question as to whether food production was taking place in the south west Victoria region. Similarly, the question of whether the *Narrinyeri* were food producers is a relevant one, given their sociopolitical complexity and the possibility they were producing significant surpluses.

Regarding the *Gunditjmara* and their congeners of south west Victoria, they seem to have primarily specialised in trapping eels and a few species of fish, with special equipment such as their eel pots being employed. Considerable infrastructure, in the form of weirs, dams, fish traps, races, ditching, fishing platforms and storage structures, was created as part of their specialised

25 See Gerritsen (2001b) for an explanation of the ecozonally based classification of primary economies used here.

26 Bailey 1981:11; Zvelebil 1986:92.

27 Renouf 1984; Zvelebil and Rowley-Conwy 1986:81; 1999:131-135; Gardiner 2003:103; Smart 2003.
This may only being belatedly realised as much of the evidence is submerged. See for example Rowley-Conwy 1999:131-5.

28 Barrett 1997; Rowley-Conwy 1999:145-50; Smart 2003:2.

29 Kalland 1996:1,11,15,21,101,114,119,164,169,180; Imamura 1996: 122-4,138,201,205-6; Matsui 1996.

30 Similar dynamics can be seen in other regions, such as Mesoamerica (Lamberg-Kalovsky and Sabloff 1995:254), South America (Moseley 1975; Feldman 1987) and China (Underhill and Habu 2006:133).

31 Schalk 1977:240; de Laguna 1990a:190.

32 Wealth being defined as effective surpluses being converted to assets of some form.

33 There also appears to be an association in areas where there are other indicators of effective surpluses between this and rates of polygyny. For example men had up to 13 wives in south west Victoria (Clark 2000:97) and at least 6 in the Corners Region (Gason 1895:169, see also Sturt 1849:2:77,118; McKinlay 1861:36; Wells 1894:515). No details are recorded in this regard for the *Nhanda*. However the *Narrinyeri* had up to 6 wives as well (Berndt, Berndt and Stanton 1993:34,61), while up to 10 wives are recorded for the *Yolngu* of Arnhem Land (Keen 1982:620) and 29 amongst the *Tiwi* of Bathurst and Melville Islands. It has been observed that men in these regions "invested" in women (Hart and Pilling 1960: 15; Altman 1987:99). A similar correspondance has been noted in the Pacific North West (Cole and Darling 1990:130) and *Chumash* as well (Arnold 1992:66-7). It would appear that where there are transitory surpluses they may be translated into obligatory relations (betrothal, alliances), effectively being converted to credit, particularly with the *Tiwi*, as an alternative to storage, or where storage was not practised.
Polygyny may have also produced a positive feedback effect on the level of food production as it raised the size of the productive unit, the "household", and possibly increased productive efficiency to some degree. This is exemplified by the *Tiwi*, although they were not food producers *per se* and traditionally, as far as is known, stockpiled but did not store in a systematic way. (Hart and Pilling 1960:34-6,46)

food procurement system. Collective or collaborative procurement activities encompassed the operation of some of this infrastructure, presumably extending to its construction. The involvement of a significant part of the regional population in activities, such as the kangaroo battue, is another significant manifestation of collective procurement. The extent of storage is unknown, though it was undoubtedly an element in the economy.

Even with the limited information at our disposal it is evident effective surpluses were being produced in south west Victoria. Their sociopolitical complexity, as described in Chapter 7, is one tangible sign. Another was the possible conversion of surpluses into some form of wealth, recalling Sievwright's account of the "hut of the Chief of her [the bride's] tribe which was well stored with provisions which were distributed among the party."[34] Ownership of infrastructure, such as the fishing weir Robinson reported as being owned by an unnamed man,[35] is a further indication of the generation of surpluses, in the form of asset-based wealth. Considered in conjunction with the broader scope of evidence, it would seem reasonable, therefore, to conclude that the inhabitants of south west Victoria were indeed engaged in food production.

The situation with the *Narrinyeri* is not so clear cut. Whilst exhibiting possibly the most complex sociopolitical arrangements in pre-contact Australia, they were limited in their degree of specialisation, and were only seasonally sedentary, albeit highly circumscribed. Even though they appear to have specialised in and extensively used nets, some of complex construction and up to 45 metres long, for fishing, catching waterfowl and other birds, as well as small animals, they did not specialise in the species they sought, apart from pelicans.[36] In that case roosting structures up to 10-11 metres long were constructed in the lakes to attract the pelicans so that they could easily be captured.[37] Collective endeavours in terms of the construction of capital items and collaborative procurement were largely centred on the clan-based manufacture and deployment of nets of various types, and in a variety of procurement activities, also limited to the clan.[38] Labelling them as CHGs, Affluent Foragers, or Sedentary Hunter-Gatherer-Fishers is not viable, partly for the reasons outlined in the preceding chapters, and specifically because they weren't highly sedentary and they didn't focus on any particular staple food, even though they relied on storage. The *Narrinyeri* were not quite food producers, although paradoxically they produced significant surpluses in the form of smoked fish, fish oils, and vegetal foods, and accumulated wealth, principally in the form of skin rugs and cloaks. The alternative I propose is that they were incipient food producers in an environment with rich but seasonally variable resources.[39] The point to be taken is that their example underlines the necessity of not being too prescriptive, that each case ought to be considered within a general trajectory, and allowance made for variance in complex local circumstances.

Before finally proceeding to a consideration of theories on the origins of agriculture and other types of specialist food production, some further discussion of trade and exchange would be timely. This is predicated on a view that the emergence of food production does not happen in isolation but is part of a pattern of development. An appreciation of this facet of that pattern of development is not only helpful diagnostically, it also provides a deeper understanding of the processes involved. It is also relevant to the "social competition" theories on the origins of agriculture.

As noted earlier, an important indicator of the production of effective surpluses is systematic or regular trade and exchange in food. This also appears to be a sign that food production, if not actually taking place, is evolving. The logic is fairly simple, one does not normally consistently trade in quantities of food, particularly unidirectionally, unless there is an effective surplus. This means there must be sufficient food to meet subsistence needs through immediate consumption or in a preserved storable form. Where there are converted surpluses that are excess to requirements, then they may be traded.

In the past, the majority of traditional Aboriginal societies in Australia (i.e. most of those outside the case study areas), only consistently traded physical items, such as weapons, utensils, tools, implements, raw materials (adhesive gums, ochre, prized stone, hair or fibre string), decorative items, and intellectual property (songs, dances, corroborees).[40] Trade in food represents a qualitatively different form of trade. In western Victoria, for example, eels were being traded for roots. Buckley recounts how his group travelled for 14 days from the Barwon River, near Geelong, into western Victoria and directly exchanged their roots (probably *murnong*) for eels:

"*two men in each party delivered the eels and roots, on long sheets of bark, carrying them on their heads, from one side to the other, and so on, until the bargain was concluded. In the evening there was another great Corrobberree* [sic] ... ... *After a short time the tribes separated, making an appointment to meet again for the exchange of food.*"[41]

Another reason trade in food is of such significance is that it heralds the beginnings of regional economic articulation,

[34] Sievwright 13 April 1842.
[35] Clark 1998-2000:2:162 (30 April 1841).
[36] Berndt, Berndt and Stanton 1993:92-94,96-99,554-66.
[37] Berndt, Berndt and Stanton 1993:94.
[38] Berndt, Berndt and Stanton 1993:28,75,77-8,9,94-100,562-6.

[39] If one wished to look for an example of Affluent Foragers in traditional circumstances in Australia the *Tiwi* would be prime candidates, they appear to have had a very rich supplies of food but did not store and reputedly only occupied campsites for a few weeks at a time at most (Hart and Pilling 1960).
[40] McBryde 1987:262-5; Berndt and Berndt 1988:128.
[41] Morgan 1852:49.
See also Dawson 1881:78.

with the people in one region specialising in one product which they exchange for complementary products another region may specialise in. This appears to have been the process that occurred in the Levant, the Atlantic coastal regions of north west Europe and in eastern Japan during the Jomon Period. The *Chumash* of the Channel Islands and adjacent Santa Barbara region in California seemed to have taken that a step further, with at least three integrated regions trading such things as fish, nuts, seeds, seal meat and deer, with one of these (the Channel Islands *Chumash*) incorporating specially produced shell beads (which were the local currency) and chert microliths, manufactured on an industrial scale.[42]

Trade in food is also a sign of its commodification. Not only does it become integrated into trade and exchange, but it is regularly used as a form of direct payment or tribute. Thus, food was not only traded in south west Victoria but gifted and used to pay tribute, as mentioned in Chapter 7.[43] Yams were similarly used by the *Nhanda* in exchange and as gift payments.[44] In the Corners Region numerous examples of food trade in protein flour (caterpillar, witchetty grub, liver, grasshopper), *ngardu*, *panicum* seeds, *yaua* bush onions and fish meal were mentioned in Chapter 5. As a consumable *pitjuri* ought to be included in this. And the *Narrinyeri* not only engaged in extensive external food exchange and trade - animal oils, cakes of dried *manteri* berries, preserved *tuwaiki* and *manakura* tubers, and smoked fish - as detailed in Chapter 8, but relied on inter-clan reciprocity in food items as well.[45]

Of perhaps the greatest significance, however, is how trade is conducted. The appearance of systematic trading systems marks a qualitative shift in social and economic organisation in traditional societies.[46] In some cases these are a systematization of the exchange of physical goods, in other cases, such as where consumables are involved, it has the hallmarks of an integrated, commodified network with market characteristics. In the more mobile generalist hunter-gatherer groups in former times in Australia, inter-group trade transactions tended to be opportunistic and *ad hoc*. There may have been a ceremony or corroboree scheduled for some time in the future, so preparations were made by some to make or obtain a few items to exchange at the time, with the aim of acquiring important raw materials or highly valued objects, such as a prized type of boomerang made by a particular group. The primary purpose of these meetings was not, however, to carry out transactions. Nevertheless, if viewed as a continental network, goods flowed as they would if they were being distributed by a commercially based system, even though all transactions were done by some form of gift-exchange or barter.

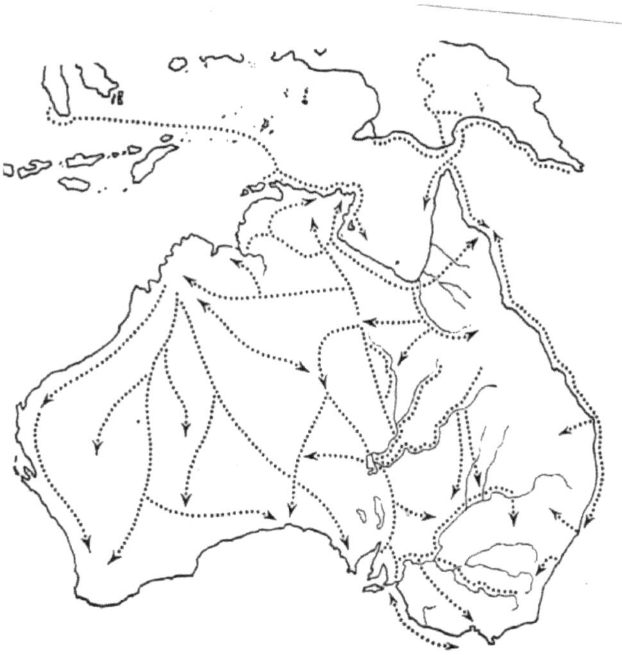

*Figure 26: Trade Routes Within and Beyond Australia (after McCarthy 1938-40:191; Berndt and Berndt 1988:17)*

Where structured trading relations occurred within the continental network, four types of trading systems were evident. These systems usually, but not necessarily, operated within the context of social obligations, socio-religious belief systems, and inter-group relations. They were:

- Partnerships Exchange Networks (e.g. Arnhem Land "Ceremonial Exchange", *Merbok* Exchange System)[47]
- Agents Exchange Systems (e.g. *Narrinyeri "Ngengampi"*)[48]
- Ceremonial Exchange (Ceremonial Meetings with exchange as an important function)
- Market E"xchange (Individual transactions or Gatherings with the specific purpose of transacting)

Except where direct and immediate reciprocation or transactions occurred, the first three of these systems functioned on "credit", perhaps as an unspoken obligation to reciprocate at some unspecified time in the future, perhaps as an expectation to reciprocate in a particular way at a particular time. Social and peer pressure, and threats to an individual's standing, status and reputation were

[42] Arnold 1991; 1992:68-9; Colten and Stewart 1996:238; Arnold 2001: 15; Delaney-Rivera 2001.
As expected this was taking place in the Pacific North West as well, see for example Ames (1991:939-40) and Ames and Maschner (1999:120)
[43] Robinson received gifts of food such as eels on a number of occasions (e.g. Clark 1998-2000:2:122 – 2/4/1841).
[44] Gerritsen 1994a:305n25.
[45] Berndt, Berndt and Stanton 1993:28,117,129,284.
[46] The evidence, limited as it is, indicates that long-distance trade in Australia had commenced by 2200 BP (Smith and Veth 2004).

[47] Stanner 1933-34a,b; Thomson 1949; Warner 1958:95,145-6,148,458-9, 464.
The *Yir Yoront* of west Cape York also appear to have had a Partnerships Exchange Network, though perhaps not as extensive as that found in Arnhem Land (Sharp 1952:75; 1968:345).
[48] Berndt, Berndt and Stanton 1993:116-122,284,533.

the means of enforcing credit obligations.[49] Conversely social standing and status accrued to those who were productive and acquired a surfeit of goods, had high levels of participation (in terms of numbers of partners or agents) or obtained rare items from a long distance.[50] In the Arnhem Land Ceremonial Exchange Cycle one's accomplishments in this regard was measured in terms of increasing one's *marr*, something akin to a spiritual force.[51]

Named partners who carried out transactions in person was the operational mode in Partnerships Exchange. The *ngengampi* system is one example, involving "trade exchange partners"[52] from other dialectal or language groups, often considerable distances away. As noted previously, taboos, avoidance rules and other social restrictions applied in *ngengampi* relations.[53] However, what sets it apart from Partnerships Exchange, was that intermediaries were always used, the *ngengampi* normally never met.[54] Hence, they are characterised here as "Agents". Regarding the *Diyari* of the Corners Region, Gason, in the early Contact Period, noted that "their whole life is spent bartering."[55] As part of their trade and exchange activities they established a relationship with an individual known as their "*yutchin*" [*yutjin*]. The *yutchin* embarked on an expedition in which it was required he "bring back presents".[56] Other members of the group, those who remained behind, according to Howitt, collected "presents" for their returning *yutchin*. As it was an arrangement "used for bartering,"[57] this was probably also an Agents system.

Large volumes of good were also traded through Ceremonial Exchange, particularly where ceremonies were frequently or regularly held.[58] The exchange described by Buckley is a typical example. These were dual purpose "embedded" events, so it is not possible to clearly separate the functions of ceremony and trade. However, at some point, where the scale of trade and exchange is increasing over time, the commercial purpose begins to predominate, and large ceremonial gatherings become "clearing houses" as Altman puts it.[59] It is no surprise, therefore, that elders' recall of Arnhem Land "Ceremonial Exchange" ceremonies focussed on the goods traded (*gerri*) rather than the ceremonies that were conducted.[60] Buckley seems to imply the predominance of commercial imperatives in his account as well, by drawing attention to the fact that some "meetings were not solely for the purpose of exchanging food."[61] While ceremonies and other sacred and ritual performances may have had a significant role, the existence of 12 "trading centres", as documented in Chapter 6, operating in the Corners Region nevertheless suggests that exchange was the primary purpose for many gatherings in those localities. This is reinforced by the observation that the main Australian north-south trade trunk route passed through this region, which seems to have been pivotal in the whole continental network.

Market Exchange was something distinct from simple, straightforward, opportunistic, *ad hoc* barter in that it was planned or structured with the specific purpose of effecting transactions, even if the inevitable ritual, ceremony or corroboree accompanied the transaction or set of transactions. Purposive trading expeditions, such as those "small parties" of *Narrinyeri* embarked upon, travelling upriver for 350 kilometres,[62] is one example. Another, noted in Chapter 5, was the occurrence where 500 people waited at Goyder's Lagoon for the arrival of harvested *pitjuri*. This was clearly a market situation, given the demand for *pitjuri* and the extent of the trade.

These trading systems operated in effect as a formalised distribution system. The *ngengampi* system formed, according to the Berndts, a "supra-economic network," a "market" based "trading economy". Blainey points out, that when dissected, Indigenous trade and exchange "no longer appears so different from modern commerce."[63] As with any normal commercial system, goods moved from "places of plenty to places of scarcity."[64] Although these systems were based on non-monetary transactions the evidence consistently indicates that matching exchange values applied in accordance with supply and demand. This can be seen in such things as the price of stingray barbed fighting spears and stone axes. In times gone by the *Yir Yoront* of west Cape York produced the spears but the axes came from 640 kilometres away. Twelve spears were needed to exchange for a stone axe in *Yir Yoront* country, but closer to the source of the axes, only one spear was required.[65] Similarly the Arnhem Land Ceremonial Exchange network acted as a market system. Though there were a plethora of types of exchanges and transactions with varying degrees of social and ritual import the people of Arnhem Land still recognised basic concepts such as "value" and "wealth", and a set of purely economic transactions, *malli*, to buy or purchase.[66] In latter years "highly prized" modern market goods were incorporated, as they were in the *merbok* system

[49] See for example Howitt 1891:78-9; Stanner 1933-34b:166-72; Thomson 1949:34-7,39,60.
[50] See for example Stanner 1933-34b:162-3,172; Thomson 1949:36,39, 78,80.
Of course modern society is no different in this regard, kudos being attached to possession of special brand name items, items that are rare or those that have some special quality.
[51] Thomson 1949:37,41,57,64,78-80.
[52] Berndt, Berndt and Stanton 1993:533.
[53] Berndt, Berndt and Stanton 1993:116,118-21.
[54] Berndt, Berndt and Stanton 1993:120.
[55] Gason 1879:259.
[56] Howitt 1891:78.
[57] Howitt 1891:79.
[58] See Duncan-Kemp (1934:208-10,211,214-5) for details about such events, how they proceeded, what was traded, and so forth.
[59] Altman 1987:197. See also Berndt 1951:157.
[60] Altman 1987:197.
[61] Morgan 1852:52.
That there were certain ceremonies, such as cult ritual and mortuary ceremonies in Arnhem Land, where no trade took place (Altman 1987:198), supports this and indicates Aboriginal people saw a clear distinction between ceremonial and commercial activities.
[62] Berndt, Berndt and Stanton 1993:19-20.
[63] Blainey 1975:212.
[64] Blainey 1975:212.
[65] Sharp 1952:75; 1968:345.
[66] Thomson 1949:6,46-51.

of the Daly River region in the north west of the Northern Territory, and individual transactions were seen to be matched in dollar terms.[67]

Viewed in broader temporal and spatial terms, trade and exchange in Australia appears to exhibit two developmental trajectories – endogenous growth and exogenous stimulation. The peoples of the Corners Region, south west Victoria, the *Narrinyeri*, and we presume, the *Nhanda*, appear to have been in the midst of an endogenous "take off" when the colonial invasion shattered their societies. Individually, each of these exhibited high levels of sedentism (except for the *Narrinyeri*), all generated some level of surplus and all appeared to manifest some degree of sociopolitical complexity. Economic articulation, in which associated regions become drawn in, articulated, in a formal and systemic manner, hypothetically assumed a form involving dedicated production and "finance". As discussed earlier, trade in food, particularly if mutual, is a strong sign of developing productive integration. *Ngengampi* type arrangements are significant in this regard because they are evidence of extra-territorial linkages and formalised trade relations. Financial articulation is more apparent when effective surpluses are generated and are transformed into wealth, as high value physical items (such as skin cloaks) or surpluses of commoner objects and everyday items. More people from the wider region are drawn into transactions that circulate this wealth, and in which the participants endeavour to increase and then express their wealth, as status and social standing. Perhaps this explains why Gason was drawn to make the observation that "their [the *Diyari*] whole life is spent bartering",[68] and why the *Narrinyeri* used various processes and techniques to "establish a viable production line".[69] This may also have been the fundamental dynamic driving other indigenous wealth circulation systems such as the *potlatch* and the *Kula* ring of the Trobriand Islands.[70]

The clearest case of exogenous stimulation in Australia is apparent in the influence of the *Macassans* from Sulawesi, who began regularly visiting northern Australia from around 1700. The *Macassans*' main interest was in obtaining trepang, also known as the sea slug. They employed Aboriginal people such as the *Yolngu* in coastal Arnhem Land, paying them for their labour and other services, as well as giving tribute for the right to fish. The Arnhem Landers in turn sold the *Macassans* tortoise-shell, pearls and pearl-shell. As a result the Arnhem Landers obtained dugout canoes with sails and stone anchors, bronze fish hooks, harpoons with detachable heads, iron, iron knives and axes, glass, pipes, tobacco, cloth in the form of calico and wool, belts, beads, string, alcohol and some foods.[71] Some of these things then entered the traditional Ceremonial Exchange Cycle.

It has been argued that the trade and engagement with the *Macassans* resulted in the stimulation of the economy in scale and extent, for hundreds of kilometres inland.[72] The introduction of dugouts with sails and anchors (previously there had only been bark canoes), bronze fish hooks and detachable harpoons heads greatly improved the effectiveness and efficiency of shark fishing, dugong and turtle hunting,[73] complementing the pre-existing usage of fish traps and nets.[74] To obtain some of the desired *Macassan* goods, individuals, both on the coast and inland, reputedly intensified their efforts to produce tradeable items.[75] Accordingly, those involved in the supply and exchange of *gerri* engaged in "constant work,"[76] and placed greater emphasis, according to Berndt, on "material values" than was the norm in the traditional generalist Indigenous Australian societies.[77] Increasing craft specialisation in the cold forging of iron "shovel-nosed" spear heads and the production of special detachable flint spear heads (*ngambi lirra*) may have been another outcome.[78] While much of this is inferred from synchronic ethnographic observation it is supported by archaeological evidence indicating subsistence intensification and re-orientation coincident with the commencement of interaction with the *Macassans*.[79] In this instance a part of northern Australia was articulating with a part of what is now Indonesia, resulting in an increase in volume and/or the velocity of goods in circulation in those parts of northern Australia.

Thus far a considerable body of evidence has been put forward detailing the argument that food production was in effect in three areas in Australia at the time of British colonisation. It has been further argued that, as the culmination of five millennia of "intensification", a period

[67] Stanner 1933-34b:161-2; Altman 1987:197,199-200.

[68] Interestingly explorers and other outsiders who made early contact with the *Chumash* commented on their "marked proclivity ... to carry on trade with them, and with one another" (Arnold 2001: 15).

[69] Berndt, Berndt and Stanton 1993:129.

[70] Malinowski 1922.

[71] Thomson 1949:51-2,86; Berndt and Berndt 1954:17-18; Warner 1958: 457; Macknight 1972:285-6,304-6; 1976:84,90-1.
Of course the *Macassan* influence went far beyond material culture transfers, a raft of cultural and linguistic features in north Australia arose from contact with them.
It could be argued that the *Macassans* in effect introduced the "Iron Age" to northern Australia.

[72] Thomson 1949:82-94; Berndt 1951:159; Macknight 1976:84-5.
The Arnhem Land Ceremonial Exchange Cycle seems to have had some degree of articulation with the *merbok* system (Thomson 1949:63).

[73] Warner 1958:460.

[74] Thomson 1949:18-22; Warner 1958:460.
There were signs that at least some Arnhem Landers were beginning to specialise in fishing, marine mammal hunting and shellfish collection. There were also signs of the beginnings of regional subsistence specialisation reflected in the distinction between "Beach People" and "Forest People" (Meehan 1977:502-5,523,526). Limited storage of *mundjutj* fruit (*Buchanania meulleri*) was reportedly another traditional practice (Thomson 1949:23). Such matters point to the possibility that food production may have developed in Arnhem Land as well, if this trajectory had continued.

[75] Thomson 1949; Berndt 1951:159.

[76] Thomson 1949:5,36,39.

[77] Berndt 1951:159,171.

[78] Thomson 1949:9,65,70-2,87,Plate 1; Berndt 1951:159; Macknight 1976:91.

[79] Schrire 1972; Clarke 2000:328-33.

labelled as the Early Australian Neolithic had come into being in those nominated food production areas. Lastly, various aspects of these food producing cultures have been explored in the context of a range of theoretical and definitional issues in an endeavour to arrive at an understanding of the origins of agriculture. At this point it is now time to turn our attention to the central question, why did agriculture arise, when and where it did?

# Chapter 10.
# On the Origins of Agriculture

It was suggested earlier that the question of the origins of agriculture is little misguided, that it ought to be considered as a question of the origins of primary economic specialisation, with agriculture being the first instance of that. Be that as it may, to address the question of the origins of agriculture, it would seem necessary in the first instance to identify when and where agriculture actually developed as an independent, endogenous, localised process. Attempts to systematically identify such places commenced with Russian biogeographer Vavilov in the 1920s,[1] and an immense amount of research into these questions has taken place since then. Table 1[2] lists those regions now identified as such, and their dating.

In view of the contention that agriculture is one form of primary economic specialisation, it should be noted in this context that other economic specialisations have similar timeframes. For example, the origins of the Balkans fishing settlement of Lepenski Vir date from 9000 BP, with a sophisticated culture well established by 7300 BP.[3] In the Pacific North West the type of economy, infrastructure and settlements observed ethnographically began to emerge from 5000 BP, becoming fully formed by 3500 BP.[4] Similarly, Da But, Con Co Ngua, Nong Nor and Khok Phanom Di in Vietnam and Thailand had their origins from 7000 BP, with mature settlement in evidence from 4000 BP.[5] The Ertebølle and Jomon cultures, discussed at length previously, are further examples within this timeframe.

It has been argued in earlier chapters that agriculture had also commenced in Australia, although technically the *Nhanda* example was not a purely endogenous development as it would appear it to have been initiated by a brief cultural transfer of the cultivar and knowledge about it. This transfer can be specifically dated from historical records to late 1629. Thereafter, it followed an exclusively or predominantly endogenous trajectory until 1850,[6] when the forceful intrusion of British settlers began displacing the *Nhanda* and destroying their economy. In regard to the Corners Region, the East Central Australian Non-Centre, there is no specific dating to draw on. The requisite archaeological investigation of the settlements identified in earlier chapters, specific sites such as trade centres, and features such as fish traps and dams have simply not been carried out. However, one could hazard a guess and suggest that the development that reached the form observed in the 19th century only came about in the preceding millennium. For the third Australian example of food production, the aquacultural economy of south west Victoria, there is at least some indicative dating of the developmental trajectory there. As mentioned before, the mounds which were a distinctive feature of the settlement pattern, along with the earliest endeavours to modify the landscape to enhance subsistence, first appear around 2500 BP. Permanent settlements may have been established as early as 600 BP, but this can only be inferred from the dating of Gorrie Swamp Hut, the only identified pre-contact dwelling to be dated. Based on this extremely limited information it can be tentatively concluded that developments in Australia paralleling those in other parts of the world most likely only came into being in the last millennium. The Australian examples must, however, be added to the list, and whatever theory is proposed to explain the origins of agriculture must account for them as well.

Many theories purporting to explain the origins of agriculture have been promulgated over the last 150 years. One of the first people to offer some insight into the question was the venerable Charles Darwin. He noticed that plant growth was richer around the abandoned camps of the Tierra del Fuegans, and suggested that it would "require but little forethought" for them to sow the "seeds of useful plants" around their habitations.[7] A little later de Candolle similarly suggested that the first cultivators "may scatter weeds" around their dwellings, motivated by "pressing necessity, due to insufficient resources in fishing, hunting or ... ... nutritious plants."[8] This foreshadowed a number of "weed theories" of the 20th century, such as the "dump heap hypothesis" and the "floodplain weed theory" which suggested humans had inadvertently created habitats for weedy food plants which were then taken into cultivation.[9]

1 Vavilov 1997.
2 Callen 1967; Coursey 1967:9-14; Harlan 1992a:22; Smith 1995a; Diamond 1997:98-101; Barnett 1999; Lebot 1999; Matthews 2003; Bellwood 2005; Denham 2005 and other sources cited in this work.
3 Srejovic 1972.
4 Ames 1994.
5 Higham 1987:55; Higham and Thosarat 1998:40-63; Higham 2002:37-8, 54-81.
6 With the possibility that there may have been further exogenous influence in 1656 and 1712 if other shipwrecked Dutch sailors had taken up permanent residence in the region.

7 Darwin 1868:1:309-10.
8 De Candolle 1884:1-2.
9 Smith 1995a:20-1,194-6; 1995b

| REGION | PLANTS | DATES |
|---|---|---|
| South West Asia | emmer and einkorn wheats, barley, rye, legumes | 11.5 KBP > |
| Central China | Rice | 11.2 KBP > |
| Northern China | broomcorn and foxtail millets | 10.8 KBP > |
| Mesoamerica | corn, beans, squashes, cassava, millet | 5.5 KBP > |
| Southern Andes | potato, quinoa, beans, squashes, 3 tubers | 4.5 KBP > |
| Lowland South America | cassava, sweet potato, yams | 4.0 KBP > |
| Eastern United States | sunflower, marsh elder, goosefoot, squashes | 4.5 KBP > |
| Sahel - Africa | sorghum, pearl millet | 4.0 KBP > |
| West Africa | yams, kola nuts, rice, cowpea, oil palm | 5.0 KBP > |
| Ethiopia | teff, finger millet, coffee | 5.0 KBP?> |
| SE Asia (incl. PNG) | yams, taro, bananas | 7.0 KBP> |

*Table 1: Earliest Systematic Cultivation ("Domestication") of Food Crops*

Another explanation was offered by Pumpelly in 1908, who argued that agriculture arose in arid environments in which the inhabitants had become tethered to water sources and so "learned to plant the seeds".[10] Childe put forward a comparable hypothesis two decades later, with desiccation causing an "enforced concentration" of people around water sources leading to a "more intensive search for the means of nourishment."[11] As archaeology blossomed after World War II and information and evidence began to accumulate, alternative theories began to proliferate. During the 1950s Braidwood developed his "Hilly Flanks" theory in which it was argued agriculture was triggered by a concentration of population in a particularly favourable locale, in this instance the hilly flanks of Zagros Mountains in the northern Fertile Crescent.[12] In contrast Binford's "Marginality Theory" of 1968 (or Flannery's "Edge Theory") hypothesised that agriculture originated in south west Asia not in the most favourable areas but on their margins, the result of population pressure arising from "fissioning" in favoured areas.[13]

Population pressure, the "development of population density in excess of what could be supported in any region by hunting and gathering,"[14] subsequently emerged as a major theme, beginning with Boserup, and culminating with Cohen, its strongest advocate, as embodied in his 1977 work, *The Food Crisis in Prehistory*.[15] Shortly after Barbara Bender pioneered the innovative "social competition" class of explanations,[16] an idea taken up by many researchers, particularly by Brian Hayden with his "competitive feasting" model,[17] and Harry Lourandos applying it south west Victoria and Australia in general.[18] Co-evolution, based on incidental planting, came to the fore in 1984 when proposed by David Rindos, as discussed in Chapter 3. From the 1980s onwards Ofer Bar-Yosef and his collaborators, and others, have been formulating what are known as Levantine Primacy Models, in various forms.[19] These attempt to explain, within a cultural ecology framework, the occurrence of agriculture in south west Asia as a complex interaction of population growth and responses to climate change. More recently Jared Diamond has argued that agriculture occurred first in south west Asia because it was particularly favoured with domesticable plants and animals.[20] Around the same time Cauvin introduced the idea of some form of innovative ideology driving the transformation leading to the emergence of agriculture.[21]

Climate change, either as a sole driver, in conjunction with population pressure, or as part of a sequence leading to agriculture, has been a significant theme in explanations of agricultural origins. Pumpelly and Childe, along with Peake,[22] were early proponents, and some form of climate change is incorporated in almost all cultural ecology perspectives. Specific climatic effects that reputedly stimulated agriculture included climatic deterioration, periodic drought, increasingly long dry seasons or greater "seasonality".[23]

Perhaps the most all embracing explanation invoking climate induced agricultural development is based on the concept of "synchronicity" – the observation that agriculture transformations "in a number of places" coincided with the transition from the Pleistocene to the Holocene, a transition marked by a significant and enduring change in global climate.[24] This hypothesis has not attracted wider support for a number of reasons. Perhaps the most obvious is

10 Pumpelly 1908:66.
11 Childe 1934:44.
12 Braidwood and Howe 1960.
13 Flannery 1965; Binford 1968.
14 Cohen 1977:6.
15 Boserup 1965; Cohen 1977.
16 Bender 1978,1981.
17 Hayden 1990,1992c.
18 e.g. Lourandos 1985,1997.
19 e.g. Bar-Yosef and Meadows 1995.
20 Diamond 1997:132-141,162-175,405-9.
21 Cauvin 2002.
22 Peake 1928:75-6.
23 e.g. Blumler and Byrne 1991:23: McCorriston and Hole 1991:47,49; Bellwood 2005:21,52.
"Seasonality" is frequently cited as a common factor in agricultural origins. But this may be of little explanatory value as all non-tropical regions of the world experience some degree of seasonality. Qualifying this as "high seasonality" is also questionable given the difficulty in quantifying that and the fact that in numerous instances cultivation commenced in tropical and sub-tropical areas which would not appear to fit that criteria.
24 Byrne 1987; McCorriston and Hole 1991:47; Blumler 1992:106; Cavalli-Sforza 1996:51.

evident in Table 1, the dating of most agricultural transitions does not actually coincide with the advent of the Holocene, the differential being thousands of years in most cases. Furthermore, it has been asserted that in their attempts to make the theory viable, the advocates of synchronicity appear have been distorting concepts and dates for the beginnings of agriculture in different parts of the world.[25]

While synchronicity may not be supportable, many other objections have been raised regarding the putative role of climate change in agricultural transitions. In a number of the cases listed in Table 1 there was no discernable climate change at all in the critical period. Even in cases where some climate change took place there is disagreement about the timing of the agricultural transition and the climate change, whether the climate change was significant enough to have any impact and whether it was climatic deterioration or improvement that was the trigger. Moreover, it has been pointed out that there have been frequent changes in climate in the course of human history but most are not associated with developments such as the advent of agriculture. In south west Asia, from the inception of the Natufian around 14,500 BP to the commencement of the PPNA around 11,500 BP, the region experienced the warmer and wetter Bølling Interstadial, then a brief return to cooler drier conditions during the Older Dryas, followed by the Allerød Interstadial and then another extended cooler drier period, the Younger Dryas, and finally the beginning of the Holocene Interglacial, the current climate regime.[26] The change to a cooler, drier climate during the Younger Dryas witnessed very different changes in the same region. In that period the people of the Harifian culture in the southern Levant reverted to more mobile hunting and gathering, while in the northern Levant the Late Natufian, the pre-cursor of the PPNA, flourished and evolved into the PPNA.[27] While climate change is repeatedly proffered as an explanation for agricultural origins the impact may be overstated and the causal link not evident or proven.[28]

In Australia the climate change at the beginning of the Holocene did not produce any marked reduction in mobility although rain increased and was more reliable. While western New South Wales did appear to experience some deterioration in conditions at that time, this was only in one part of the Corners Region.[29] There was a steady increase in population in Australia during the Holocene, despite slightly drier conditions in the last 3000 years. "Intensification" was well under way by then, but with no attendant climate change correlated with it.[30] Again, the Australian evidence shows there is no clear correlation between climate change and socioeconomic change. Despite this incongruency, climate change is still regularly included in explanatory models for the origins of agriculture.

Regarding population pressure, there has been much criticism of it as a causal factor in the origins of food production, either as a prime mover or in combination with other factors such as climate change. The concept of "population pressure" itself is one that has often been challenged, with some justification. It is a relatively intangible concept, reflecting a supposed imbalance between numbers in the population and availability of food.[31] However, while it may be possible to make estimates of population levels through proxy measurements it is very difficult to determine archaeologically the supply of food and its sufficiency. Population pressure arguments are therefore usually based on evidence of growth in absolute numbers in a discrete area or region, producing an increase in population density, and the presumption that this means that population pressure was being experienced. At some point there are supposedly too many mouths to feed, and according to this logic, the prehistoric populations respond behaviourally to this imbalance by lifting production through adaptation, by developing and implementing extensive cultivation for example.

Of course there is no direct evidence of the state of mind of those people at the time in question, nothing archaeologically to directly demonstrate that population pressure was being experienced, nor to link behavioural change to "population pressure". In 1984 Cohen and Armelagos did report paleopathological evidence of nutritional stress in the early Neolithic, interpreting this as a population pressure marker.[32] However it has since been realised that this simply reflected a change from a varied hunter-gather diet to a starch dominated one, and that Neolithic farmers were fundamentally healthy.[33] A more telling counter-argument is the fact that there is no evidence of any "population pressure" in the Epipaleolithic of south west Asia, or the late Archaic in Mesoamerica, the period when population pressure ought to have been acting to stimulate the subsequent changes.[34] Osteological evidence indicates that these populations were not experiencing the sort of nutritional stress seen as being consistent with population pressure.[35] Furthermore, it has been noted that despite relatively rapid population growth beginning in the Natufian, absolute numbers and population densities were still quite low.[36]

Another strong argument against population pressure relies on the analogous behaviour of hunter-gatherers. Up to

25 Sherratt 1997:275.
26 Yasuda 2003b:22,24,27,Figs. 4,9
27 Bar-Yosef 1998a.
28 Ladizinsky (1998:11) observes, for example, that in the upper Jordan Valley "no significant change of vegetation had occurred" during the Pleistocene/ Holocene transition.
29 O'Connell and Allen 1995:856.
30 Lourandos 1997:224-6,298-301; Williamson 1998:144-5.
31 Hassan 1981:161-2.
32 Cohen and Armelagos 1984:594-5.
33 Bellwood 2005:280n3.
It is quite likely early cultivators and subsistence farmers experienced at least one period of food shortage each year, just before the harvest when stores ran low or ran out, which still happens in subsistence agriculture.
34 Henry 1989:20-4; Hayden 1995a:280; Lamberg-Karlovsky and Sabloff 1995:250-1.
35 Roosevelt 1984.
36 Flannery 1986:11.

the Epipaleolithic all human populations were generalist hunter-gatherers. The point has repeatedly been made that all ethnographically know hunter-gatherers, in fact all human populations, maintain control of their populations, they maintain homeostasis.[37] Thus cultural controls and biological limitations ensure that in normal circumstances an imbalance between the level of population and the available food supply does not normally arise. Human populations usually limit their numbers to the "carrying capacity" of their homelands in the leanest times, in accordance with what is known as Liebig's Law, and so do not experience boom and bust cycling in population size.[38] Common strategies to ensure control of numbers in hunter-gatherer societies include delayed marriage and pregnancy, abortion, infanticide and prolonged lactation.[39] Furthermore, simple logic suggests that if "population density in excess of what could be supported" did develop it would only be in the very short-term. In such situations hunter-gatherers would simply die, either immediately from starvation or within a short period time from malnutrition or susceptibility to diseases as a result of malnourishment. Population pressure requires that this "excess of what could be supported" to implausibly persist for the hundreds of years in which food production took to develop, and assumes no other form of corrective action would be taken. Buckley provided an illustration of how such situations were handled, showing the people he lived with had an acute awareness of the need to limit their numbers, and would make critical decisions to ensure that:

*"If again, a family increases too rapidly, for instance if a woman had a child within twelve months of a previous one, they hold a consultation ... as to whether it shall live or not."*[40]

There are other flaws in the conception of the population pressure as a factor impelling people toward the adoption of extensive cultivation. If there are too many mouths to feed then any population/resource imbalance would become critical should famine conditions develop. In traditional Australia at least, Indigenous populations in arid areas had a range of adaptive responses to drought-induced famine conditions. They relied, for example, on increased mobility, simply moving, sometimes considerable distances, to areas where conditions were better, and this included crossing of territorial boundaries where friendly relations and loose alliances existed.[41] Thus populations can adapt, as they do, so that even if population pressure was affecting them it does not necessarily impel them to move toward food production.

The procurable food supply is a significant limit on population growth in generalist, non-sedentary, hunter-gatherer communities, with "population pressure" supposedly coming into effect if numbers exceeded that limit. However, other factors may restrict population size to a level well below that limit. For example, because of the frequency of relocation and the distances involved, it was very difficult for mobile hunter-gatherers to have more than one baby, or young child who was unable to walk any distance, as part of the family at any given time. This may well have been an important consideration in the situation Buckley referred to. Consequently the logistics involved in following a mobile lifestyle may have been the primary constraint on population size prior to the advent of sedentary food production, not the availability of food necessarily. The removal of that constraint may explain in part why populations grew rapidly in south west Asia, and elsewhere, once they became more sedentary. From the Natufian to the PPNB the population grew by an estimated 1600%.[42]

Theories invoking population pressure and/or climate change have another common weakness, they are unable to explain why food production commenced when it did. What made populations grow in the first place? There have been many climate changes since modern humans first appeared, why didn't they trigger food production? A further weakness is the model of behaviour implicit in these theories. By and large they are "stress models" which presume that people, at least prior to the commencement of food production, only developed food production because they were in difficult circumstances. Some see this as a form of "environmental determinism", where the trajectory toward food production was driven by such things as a supposed "decline in the availability of wild foods."[43] Whatever the case, stress models are based on an adaptationist ecological perspective, people are seen as just being reactive, only making changes because they are "forced" to do so, and in so doing eventually bring about the transformation leading to food production. But the evolution of food production requires a series of innovations in technique, technology, social arrangements and ideas over time. Such innovation is a creative act, it does not occur on demand. Nevertheless, as a social process, significant innovation as a manifestation of that creativity occurs randomly but persistently over time. Seen in broader contexts the evidence of this is all around. New inventions, techniques ideas and social experiments now appear on an almost daily basis, not because people are "forced" to do so, but because they see opportunities, they want to improve things, they want to make things work better. Go back say three centuries to the beginnings of the Industrial Revolution, when the scale of society was considerably smaller and the means of communication far more limited, and see the ferment of ideas, the new mechanical contraptions and "improvements" and the like. Go back to the late Medieval and Renaissance of

37 Deevey 1960:198-200; Cowgill 1975:505-6; Harris 1977a:180.
38 Flannery 1986:11-12; Testart 1982:525; Strassmann and Gillespie 2002.
39 Cowgill 1975:512; Hassan 1981:147-155.
40 Morgan 1852:52.
41 Strehlow 1965:128; Yengoyan 1976.
42 Hayden 1995a:295.
It has been noted in this regard that modern hunter-gatherer populations, such as in Australia, have tended to grow very rapidly where they have been forced to become sedentary.
43 Diamond 1997:110.

Europe, and the contemporaneous Song, Yuan and Ming Dynasties of China, which were marked by the appearance of windmills, gunpowder, cannons, moveable print, eyeglasses, ocean-going fleets and banking, illuminated by the polymath talents of Leonardo, Michelangelo and Galileo. Or to Rome at its peak, with its innumerable and incredible feats of engineering. Go back to the flowering of Greece, the wonders of Egypt, the profound inventions of writing and the wheel from Sumeria. The point is that innovation is not driven by extenuating circumstances, it is an inherent human attribute, that flourishes particularly when there is stability, security and opportunity, not when starvation is lurking or changing climate is undermining subsistence.[44] And, as the work of Solow, Swan and their successors shows, innovation is in fact the primary driver of economic growth and development.[45] This, I would argue, is equally true in prehistoric contexts.

Another theory for the origins of agriculture that has endeavoured to explain its initial timing and location relates to the availability of domesticable species. Diamond argued in *Guns, Germs and Steel* that agriculture occurred first in south west Asia because that region was blessed with the most suitable, domesticable plant and animals species, particularly large-seed species.[46] As can be seen from Table 1, the plants initially cultivated (within the first 1500 years) and animals kept were emmer and einkorn wheats, barley, rye, legumes (pea, lentil, chickpea), sheep and goats. However, the flaw in this theory is immediately evident when the example of China is considered. In central China there was only one founder crop, rice, and no indication that any domesticated animal species were kept. Pigs (*Sus scrofa*) were not domesticated for another 2000 years, chicken perhaps about the same time.[47] Yet agriculture commenced there at almost the same time as in south west Asia. In fact, recent evidence indicates that cultivation of rice may have actually preceded the cultivation of the south west Asian cereals, with contestable evidence from Hunan and Jiangxi indicating rice cultivation may have commenced between 14,000 and 12,000 BP.[48] A further illustration of the weakness of this theory can be found in Australia. As can be seen from Table 2, if the availability of domesticable species was a predictor of agriculture, then perhaps Australia ought to have been among the first, not the last. In all 29 "domesticable species", species cultivated elsewhere or their close relatives, can be identified, 20 definitely endemic to Australia, the remainder probably being introduced in the last millennium.

The last theory proposing a specific model for the origins of agriculture to be considered is one based on social competition. Bender originally proposed that a notional "disequilibrium between escalating demand and productive capacity" may have been the driver of intensification, leading to food production.[49] The origin of escalating demand in her view lay in an increase in the number and scale of ceremonial activities as a mechanism for asserting status differentials and increasingly unequal sociopolitical relations. In this scenario the more powerful beneficiaries, such as elders, encouraged greater production, and used this as a means of signifying, asserting and extending their prestige and influence in relation to other groups.[50]

Hayden developed this conception further with his "competitive feasting" model, based on the social dynamics of the *potlatch* ceremonies common in the Pacific North West, particularly the more spectacular "conspicuous consumption" manifestations of the *potlatch*. A prediction emerging from his analysis was that the plants that were delicacies, intoxicants, or conferred prestige in some way would be the first to be cultivated and domesticated. He pointed to the introduction of the potato in parts of the Pacific North West and the prestige associated with it as validation of the model.[51] A closer examination of the role of the *potlatch*, however, suggests the model is misleading and that there are other explanations for what took place with the ethnographically observed *potlatches*.

From the perspective of the people of the Pacific North West, there were two conceptions embodied in the meaning of the term *potlatch*, "sharing", and "to give [things away]", with the latter having another layer of meaning, translating as "a great thing".[52] The first form of *potlatch* resulted in goods being redistributed, while the other functioned as an affirmation or claim to higher social and political status by the giver of the *potlatch*. It was in effect a "prestige economy".

Traditionally *potlatches* had been modest affairs, conducted on a much more limited scale than in the "golden age" of the Pacific North West. When contact with outsiders first took place in the latter part of the 18th century, it was only the northern groups who conducted *potlatches*.[53] Following contact, the scale and frequency of the traditional *potlatch* began to increase to the levels observed ethnographically. According to the competitive feasting model the demand generated by such feasting or similar sociopolitical dynamics is what led pre-agricultural peoples to intensify their productive capabilities, ultimately to plant crops which eventually became domesticated.

Considered more closely this exemplification of the theory is found wanting. Substantial contact and engagement with outsiders (principally Russians, Europeans and Americans),

44 See for example Reynolds 1986.
45 Solow 1956; Swan 1956; Solow 1957
46 Diamond 1997:132-141,162-175,405-9.
47 Yuan and Flad 2004; Bellwood 2005:117.
48 Toyama 2003:264-5; Yan 2003:152; Yasuda 2003a:134; C. Zhang 2003: 189-90; Bellwood 2005:116.

49 Bender 1981:150.
50 Bender 1981:154.
51 Hayden 1990,1992c.
52 Smithyman and Kalman 2004:18; Sutton and Jonaitis 1990:84; Codere 1990:368
53 Codere 1990:369; Cole and Darling 1990:132.
First recorded contacts appear to have begun in 1774 (Arima and Dewhurst 1990:400; Blackman 1990:255; Cole and Darling 1990:119).

| Plant | Name | Area | Source |
|---|---|---|---|
| Wild Rices | *Oryza sativa**<br>*Oryza meridionalis*<br>*Oryza rufipogon*<br>*Oryza australiensis*** | Arnhem Land, North Queensland, Central Northern Territory | Stuart 1864:506; Palmer 1884b:97; Jones 1975:25; Jones and Meehan 1989:122, 126-30; Yen 1995:841 ABRS 2002:235,305 |
| Millets | *Panicum decompositum*<br>*Panicum laevinode*<br>*Panicum effusum*<br>*Echinochloa crus-galli*<br>*Echinochloa turneriana* | NSW, Queensland, Northern Territory, WA, SA | Maiden 1898:39; ABRS 2002:43:138,213,221-2, 340-1,1,349-50; O'Neill 2006:13; Latz 2007:143. |
| Yams | *Dioscorea alata**<br>*Dioscorea elongata**<br>*Dioscorea bulbifera**<br>*Dioscorea transversa**<br>*Dioscorea hastifolia** | North Coasts, Northern East Coast, Central West Coast | BFF 1986:46:198-202,219; Yen 1989:68. See Chapter 3 |
| Taro | *Colocasia esculentum* | Northern Australia | Thomson 1949:21; Crowley 1983:424; Jones and Meehan 1989:126-7; Yen 1989:68; Matthews 2003,2006 |
| Purslane | *Portulaca oleracea* | Central Australia | See Chapter 5 |
| Coconuts | *Cocos nucifera* | North Coast, North East Queensland | Worsley 1961:165-6 Hynes and Chase 1982:40 |
| Bush Potatoes | *Ipomea costata*<br>*Ipomea polpha*<br>*Ipomea yardiensis*<br>*Ipomea batatas** | Northern and Central Australia | Isaacs 1987:94 Jones and Meehan 1989:128 Yen 1989:60 |
| Native Tobacco | *Nicotiana* spp. | Inland Australia | See Chapter 3 |
| Wild Bananas | *Musa acuminate*<br>*Musa yackeyi*<br>*Musa fitzalanii* | North Queensland | Yen 1995:837-8 |
| Palmgrass (Highlands Pitpit) | *Setaria palmifolia* | North Australia, East Australia | Yen 1995:841 ABRS 2002:43:239,351 |
| Pandanus spp. | *Pandanus spiralis* | Northern Australia | Jones and Meehan 1989:126 |
| Chinese Cabbage | *Ipomoea aquatica** | Northern Territory | Holtze 1892:3 |
| Tamarind | *Tamarindus indica** | Arnhem Land | Worsley 1961:165; Macknight 1976:59 |

* Species possibly or definitely introduced in the millennium prior to British colonisation (*E. crus-galli* is native but also a later introduction in some areas)
** Reputedly "grown" by Aboriginal people (ABRS 2002:43:235).

*Table 2 - Species Found in Continental Australia Cultivated Elsewhere or Closely Related to Those Cultivated Elsewhere*

commencing in the late 18th century, and the establishment of the hegemony of the Canadian and U.S. governments, brought about more peaceful relations, facilitated exchange between tribes and gave rise to the fur trade. Considerable wealth was generated by that trade and the market for tourist artefacts. With access to a wide range of European goods significant importation occurred, including items such as iron, copper, alcohol, tobacco, muskets and molasses. This increase in the scale and value of trade and exchange coincided with an increase in the scale, extent and frequency of *potlatches*. As *potlatches* became "the foundation of the economic system,"[54] of the Pacific North West cultures, the increase in wealth allowed more ostentatious displays and fostered increased inter-group redistribution.[55] It is not surprising, therefore, that the establishment of a trading post at Fort Rupert on Vancouver Island in 1849 coincided with a dramatic increase in the size and frequency of *Kwakwaka'wakw potlatches*.[56] Consequently, one really need look no further than a simple economic explanation of the developments in the Pacific North West. The demands of chiefs and other influential leaders was not driving those developments, it was the economic stimulus arising from a large increase in the scale of exchange, and a significant injection of wealth largely derived from the fur trade. This is what led to the increase in the size and frequency of *potlatches*. Effectively greater economic articulation developed with external economic systems in this period, as well as between different groups in the Pacific North West. In many ways this is comparable to the stimulus to

[54] Garfield quoted in Suttles and Jonaitis (1990:85).
[55] Cole and Darling 1990:119; Mauss 1990:46; Suttles and Jonaitis 1990:85
[56] Cole and Darling 1990:132.

the Ceremonial Exchange Cycle in Arnhem Land arising from contact with the *Macassans*.[57]

Regarding the role of plant foods, in particular the potato, again there are shortcomings in the competitive feasting model. The potato was an introduced cultivar, a novel food with limited availability initially, and this is probably why it was considered a highly sought after prestige item.[58] Moreover, it has been pointed out that the first cultivars "throughout the world" were not delicacies but staples.[59] Generally this is true, but nevertheless plants other than staples, having some valued property, may still have been among the suite of plants first cultivated. Tobacco, as mentioned earlier, is one, gourds another. Similarly, the sowing of the seeds of *Acacia salicina* documented in Chapter 5 was being carried out because the burnt leaves provided the alkali for chewing *pitjuri*.

Since promoting this formulation of social competition concept Hayden has modified his view. Rather than social competition in the form of competitive feasting being a prime mover in agricultural origins, he now accepts that this only occurs "where adequate basic subsistence is guaranteed for all".[60] By extension he argues competitive feasting is actually a mechanism for converting surpluses into storable wealth in those situations where surpluses are already being produced. Furthermore, he notes that this form of socioeconomic competition only takes place when the participants are prosperous, what he calls his "Prosperity Model of Competition". In Hayden's view such competition, if it operated in situations where resource are limited or vulnerable to over-exploitation, as they are for traditional hunter-gatherers, would be self-defeating.[61]

Although conceptually competitive feasting does not appear to be a viable model for explaining agricultural origins, social competition cannot be dismissed on that basis. Broadly speaking social competition is a proposition that suggests production is intensified by the demands associated with individual and sociopolitical prestige. As such this would contribute to aggregate demand. But, as Hayden realised, the additional demand associated with social competition is only realisable when people are producing surpluses, and is not meaningful or significant if people are struggling to assure their basic subsistence. Moreover, even when there is unmet demand, such as when people are not able to assure their subsistence needs are fully met, or additional demand arising from social competition, new forms of production, such as food production, do not necessarily result. This is equally true of the other theories on agriculture. They imply additional demand pressure, from excessive population or worsening climate for example. But demand pressure, additional or otherwise, does not bring about invention and innovation, that is an independent process. If it did then food production would have been devised long before it actually was. In either situation, pre- or post-food production, people may seek to satisfy unmet demand by increasing their "production". This they could do by intensifying their efforts - by working harder or having more people engaged in a particular activity - but this strategy is limited by the productive capacity of that form of production. Production cannot be raised further unless new ways of doing things, new forms of production, are devised.

To put this in perspective it may be helpful to consider some analogous situations. The argument that population pressure or climate change operating at critical periods in prehistory were the factors bringing about the agricultural revolution could equally be applied to the Industrial Revolution. In England, the Industrial Revolution was preceded by the Little Ice Age and strong population growth, numbers rising from 2.7 to 5.7 million between 1541 and 1750.[62] Could a credible argument be sustained that either or both of the factors brought about the Industrial Revolution? Similarly, is it reasonable to argue that the Space Race and the Cold War of recent times, as powerful and expensive forms of social competition, were the causes of the Technological Revolution, the accelerating economic transformation that has been taking place over the last fifty years? Probably not.

**Key Questions**

We now know with some confidence when and where people of various cultures started the pristine planting of crops in a systematic way, deriving a significant part of their subsistence from this. By definition they did this of their own volition and with no direct diffusional transfers from any other people engaged in growing crops. Most of the theories formulated to explain this include components which take account of the fact that these developments took place over an extended period of time and on a geographical scale ranging from local to global. For example, climate change theories usually focus on a period of 1200 years embracing the Younger Dryas and the transition from the Pleistocene to the Holocene. These theories are consequently reliant on medium-term explanations. This is also the case with population pressure and social competition models such as competitive feasting. General social competition models do not usually specify a timeframe, but appear to be couched in terms that imply the causative dynamic operated over longer timescales than those assumed in competitive feasting scenarios. However, the validity and justification for selecting isolated periods is questionable and a case must be put to explain why it is located in that timeframe. Furthermore, it begs the question, what led to that situation

[57] A similar observation has been made about *Nuaulu* intensification in the 19th century (Ellen 1988:123).
[58] A similar dynamic may have occurred with the initial introduction of the yam amongst the *Nhanda*.
[59] Smith 1995b. See also Keeley 1995:266.
[60] Hayden 1994:223.
This has some archaeological support, with large roasting-pits appearing in Late Woodland II communities in eastern United States, interpreted as evidence of feasting (Gallivan 2002: 551).
[61] Hayden 1994.

[62] Hinde 2003:80.

in the first place? To obviate this impasse the approach used here will consider the emergence of agriculture within a framework that takes into consideration all timescales, from long- to short-term. The logical basis for this view is that each phase, as critical developments took shape, cannot be isolated from the preceding period of time in explanatory terms.

Various explanations for the emergence of agriculture also endeavour to account for the location where systematic cropping commenced. In some cases this is couched in terms of a specific locality or region in relation to the other parts of the region, or the rest of the world. Consequently some theories simply identify the type characteristics of localities where cropping first became established, what are considered to be "favourable locations". Alternatively others, such as Diamond's thesis that agriculture occurred first in south west Asia because of its superior endowment of domesticable species, provide global explanations. Implicit in Diamond's argument is the notion that the quantity of domesticable species in particular regions around the world was the critical determinant for the inception of agriculture and decided the order of precedence in the timing of agriculture for each of those regions. While one may contest the validity of either of these explanations the differing approaches are not in irreconcilable, they simply consider the question in terms of different spatial scales. Consequently, any cogent and consistent theory for agricultural origins ought to have the capacity of being able provide an explanation that is valid on any geospatial scale. This condition should also apply in regard to the temporal scale, and so the validity of any explanation should ideally be temporally and spatially scale invariant.

Lastly, in arriving at a consistent explanation three domains need to be taken into account, the environmental context (climate, change in climate, local and regional ecology), social context (population size, social structure) and individual psychology. To this end the processual and systems theories of agricultural origins outlined in the beginning of this chapter often employ complex formulations which identify common or causative factors in the environmental and the social context. While these approaches rely on paradigms such as cultural ecology for their frame of reference, they do not normally appear to draw on any paradigm of individual psychology. This may partially account for the dubious and simplistic behavioural assumptions, alluded to earlier, that have been made regarding the 'motivation to change' on the part of incipient agriculturalists. While strategic motivations, such as "risk-reduction", or "least effort", are often proposed to account for hunter-gatherer behaviour, these are usually limited to single facets of behaviour, and it is not always apparent what body of evidence or psychological theory informs this. Consequently, a complete explanation ideally should incorporate a psychological dimension validated by an accepted psychological theory or paradigm.

**In the Beginning**

According to current thinking modern humans, *Homo sapiens*, first appeared in Africa perhaps around 120,000 to 150,000 years ago. Between 70,000 and 90,000 years ago they had made their way out of Africa to south west Asia before spreading across the world in succeeding millennia. Since then we have increased hugely in numbers and occupied all but the harshest and most inhospitable locations. In the process we have created elaborate and complex tools, equipment, technology, infrastructure, devices and systems for travel and communication, and so forth. These developments have been supported by equally complex social and economic systems with global reach. Such things have never previously been seen on the planet in its 4.5 billion year existence. As a consequence of these activities humans have altered large portions of the planet and its ecosystems, intentionally and unintentionally.

While this is a remarkable occurrence it is nevertheless difficult to identify what qualities humans possess that set them apart from other creatures and has enabled them to do this. One could point to language as a complex form of communication. But we know we are not unique in this regard, many animals, whales, dolphins, birds, chimpanzees, even mice, have well-developed communicative capacities, which we are still struggling to unravel and understand. We are not unique either in our ability to communicate symbolically, chimpanzees have been taught to do that. The ability to use tools is not the exclusive preserve of humans either, chimpanzees, for example, make and use simple tools. There are even differences between various populations of chimpanzees in the type of tools they use (to crack nuts for examples), suggesting there are "cultural" differences between those groups. Many animals build structures such as nests, dams and mounds. Others store food, squirrels being the obvious example. Animals are even capable of being creative, such as elephants and chimpanzees being able to "paint". However, in terms of all our specific cognitive capacities, such as memory, learning, abstract thought, problem solving, planning, initiative, language and creativity we far exceed any comparable capacity in any other animal. In combination these capacities make us clearly unique. Added to this is our ability to consistently collaborate in large numbers, unlike any other mammal, and pass on the benefit of our accumulated experience and learning from generation to generation as "cultural learning".

It would seem these human capacities and capabilities began to strongly manifest themselves from 40,000 to 50,000 years ago.[63] While Neanderthals (*Homo neanderthalensis*) and archaic *Homo sapiens* showed some early signs, it was not until the Upper Paleolithic that human culture began to flourish in different parts of the world. Perhaps this was the result of positive evolutionary feedback reaching a threshold, manifesting as increasing technological

[63] Klein 2000; Wadley 2001; Bar-Yosef 2002.

sophistication, cultural complexity and consistent use of symbolism. Whatever the case, striking examples of symbolic representation, such as early Australian rock art and ceremonial objects,[64] the Lascaux and Altamira cave paintings from France and Spain, and the "Venus" figurines from Dolni Vestonice in the Czech Republic, made their appearance in the succeeding 20,000 or so years. Following this, a series of key technical and sociocultural innovations emerged. These included:

- greater standardisation of tools
- development of grindstones, basketry, clothing, fishing techniques and boiling
- refinement in hunting equipment with barbed bone points, harpoons and spear throwers
- exchange links in the order of several hundred kilometres
- campsites organised into functional areas
- considerate burial as a common custom
- first signs of storage[65]

In the course of the preceding million years hominin populations, particularly *Homo sapiens*, had been slowly increasing. The rate of growth varied from time to time, slowing when adverse climatic conditions, such as the Ice Age, dominated and accelerating when more benign conditions returned. Though the overall growth rate was small, it nevertheless increased over time by a factor of perhaps 10.[66] Spurts appear to have taken place in the Middle Paleolithic, Upper Paleolithic and Epipaleolithic.[67] This would not seem possible without the means of exploiting the environment more effectively and efficiently, through observation and experiment, through innovations in technology and technique, and from accumulated experience translated into cultural learning. Australia provides a clear illustration of this process. When the first humans arrived they would probably have been completely unfamiliar with almost all its native flora and fauna. The haunts and habits of the animals had to be learnt and understood, and methods devised to catch them. What plants were edible and what were not had to be determined. Numerous plants required special preparation, such a cycad nuts or *ngardu*, before they could form part of the diet. And, as elsewhere, grindstones, basketry, clothing, fishing techniques utilising barbed bone points, fish hooks and nets, boomerangs, boiling, spear throwers, storage and considerate burials, such as the famous Lake Mungo interment (40,000 ± 2,000 BP),[68] gradually made their appearance, either as independently invented items or cultural borrowings.[69]

So why were humans doing these things? Population levels were low, climate changes came and went, but humans continued throughout on a trajectory of increasing in numbers in concert with slowly increasing technological and cultural complexity and sophistication, underwritten by increasing productivity in the form of greater yields from hunting and gathering. This trajectory ultimately led to broad-spectrum procurement, intensification and finally food production. Population pressure has been invoked as an explanation in this instance as well, but inexplicably its supposed influence was only felt in the Upper Paleolithic and Epipaleolithic, not the Middle Paleolithic.[70] Conversely, the constant application of evolved or evolving human cognitive capacities, the only constant operating throughout these three periods, provides at a more consistent and viable explanation for the observed trajectory.

In explaining the final transition, to agriculture and other specialised forms of food production, and the adoption of a sedentary life, one difficulty has frequently been advanced, the motivation of hunter-gatherers to change their lifestyle. This issue arose as a result of recent studies of hunter-gatherers, who did not appear to be living in straightened circumstances, as many earlier theorists had assumed was the case. Indeed, most of the time their lives seemed to have been relatively comfortable and carefree. Sahlins consequently suggested that these people formed what he termed the "original affluent society",[71] because only a few hours work each day was sufficient to provide for basic needs. The implication was - why change? But, if the hunter-gatherer lifestyle was so agreeable and there was little incentive to change, how does one explain that in the last 12,000 years virtually every population on Earth has become sedentary and begun to engage in food production? Clearly change did occur, the issue is to identify the motivation for that change and thus resolve the contradiction. Stress model theorists might argue that hunter-gatherers were simply forced to change when adverse circumstances were experienced for extended periods, but the shortcomings of such approaches have already been noted. Consequently alternative approaches need to be explored.

While for much of the time hunter-gatherers may have easily provided for their subsistence needs, whether they were "affluent" hunter-gatherers or not, there were times when this was not the case. The *Montagnais* of Labrador in Canada, for example, went "sometimes two days, sometimes three, without eating for lack of food".[72] Similarly, for the *Watjandi*, the northern *Nhanda* group, Oldfield observed that in the winter time:

"*... should it rain continuously for a lengthened period, the natives suffer severely from hunger, for as at such times all game lies close, hunting is out of the question, and at that season there are no vegetable substances fit for food,*"[73]

[64] Flood 1997:8-16; Gerritsen submitted.
[65] Bar-Yosef 2002:365-9,376.
[66] Deevey 1960; Hassan 1981:199-200.
[67] Stiner et al. 1999.
[68] Bowler et al. 2003.
[69] Flood 2004:171,236,282.
Boiling was not very common in Australia, and in most instances where it was employed may have been the result of foreign contact (Gerritsen 1994a:202-3,208).

[70] Stiner et al. 1999:193.
[71] Sahlins 1972:1-39.
[72] LeJeune quoted in Sahlins 1972:32.
[73] Oldfield 1865:286.

The vagaries of nature and the constraints of the hunter-gatherer lifestyle exposed people such as the *Watjandi* and others to many trying circumstances. Indigenous Australians at times faced extremely hot temperatures, making effort almost impossible, incessant rain, or severe drought eliminating most water supplies. In northern Australia the extreme winds of cyclones posed a significant threat to life and limb. In Central Australia it was normal to sleep on the ground, in the open, with no covering in freezing temperatures. Aboriginal people made light of such exigencies of course, in the knowledge that these were usually just temporary inconveniences. More revealing is how they would really have liked things to be, if they could have brought about permanent change. This insight is provided by cultural expressions of their ideal, their utopia, found in conceptions of the afterlife. In the afterlife, their spirits would, for example, go to where "food is abundant and shade trees and water is everywhere."[74] Along the lower west coast of Western Australia it was a similar story, the spirits of the dead *(djanga)* went to *Kooranup*, the land beyond the sea, "a country where there was plenty of game for men to hunt and *joolal* [edible bulbs] and all other roots for the women to gather."[75] Many increase ceremonies also incorporate a comparable desire, for a reliable and abundant supply of an important, special or preferred food. So, ultimately, Indigenous Australians wished for seemingly basic things such as abundant and reliable food supplies, and suitable shelter. While they may have had that for much of the time, it was not always the case. Perhaps, when the opportunity arose, or when someone with influence in the group had an inspired idea, something new would be tried, to make life just a little better, to negate those aggravating circumstances when hunger struck and so forth.

One revealing example of just such an initiative was the commencement of the planting of yams by the *Kaurareg* of Muralag [Prince of Wales Island just off the tip of Cape York]. This draws on the remarkable account of Barbara Thompson, a young woman who had been the sole survivor of a shipwreck in Torres Strait. She was adopted by the *Kaurareg* and lived with them for 5 years, from 1844 to 1849. The crew from the hydrographic survey vessel, "HMS Rattlesnake" stumbled upon her on 16 October 1849 and within days of her "rescue" the ship's artist, Oswald Walters Brierly, began recording her detailed observations and experiences. According to Thompson, whose veracity was borne out in subsequent ethnographic accounts and investigations,[76] the *Kaurareg*:

"*... carry wood and spread it over the land and burn it. Two days after, when it is cold, they plant setts from the koti yam. This is the time of year* [October] *they cultivate. My father and mother (native) had just sowed some before I left the island.*"[77]

Thompson further reported that the yams were only planted as an insurance against a shortage of the wild yams, which grew profusely on the rocky ridges of the island:

"*They will not touch the kotis when they are ready for taking up this year or cultivate the same ground but let them remain in the ground as a standby if the kotis should get scanty in the rocks.*"[78]

Barbara Thompson's evidence also indicates that yam cultivation on Muralag was a new innovation, the *Kaurareg* had been literally "caught in the act" as cultivation was diffusing across Torres Strait to continental Australia's doorstep - "this was the first year they had taken it [yam cultivation] up,"[79] confirmed by another statement:

"*There was no cultivated pieces of ground on the island I belonged to until they planted the garricoop* [yam garden] *but a few of the men had a garden on Nooroopie* [nearby Nurupai – Horn Island]."[80]

In this account Thompson was describing a typical "risk-reduction" strategy; the planting of yams was initiated as insurance against failure of the wild yams, with the intention of ensuring that they did not go hungry should that happen. It was a small step, the harbinger, perhaps, of a move to reliance on cultivated yams. Of course many such small steps would have been required in other circumstances to develop a fully-fledged subsistence regime based on extensive and systematic cultivation, or to build such things as fish trapping systems, learn to husband flocks and herds of animals, or to move toward a sedentary existence.

Many of the innovations that appeared in the period preceding the south west Asian Neolithic, and the inception of specialised food production elsewhere, were the result of incremental changes such as the planting of yams by the *Kaurareg*. In this instance they had benefited from the experience of other Torres Strait Island groups to the north, as the limited planting of yams, taro and other cultivars apparently spread slowly southward in Torres Strait, despite the predominantly maritime economic orientation.[81] In pristine situations, however, people had to rely principally on their own efforts and not, as the *Kaurareg* did, on the accumulated experimental knowledge of others. It has often been pointed out hunter-gatherers constantly experiment, and that thousands of experiments with plants must have been carried out before the "best" were found and means devised to make them edible. As pointed out in Chapter 3, as knowledge of plants and plant communities increases so does the degree of intervention in their life cycle. Planting can be seen in this context as just one further experiment with plants, to suit human purposes. Equally, for technological innovations there may have been many

74 Palmer 1903:221.
75 Bates n.d.h:20
76 Moore 1979:10.
77 Brierly 1849:148-9.

78 Brierly 1849:178-9.
79 Brierly 1849:179.
80 Brierly 1849:184.
81 Barham 2000; McNiven 2006.

experiments generated by many individuals before the utility of a particular one was recognised, it worked, and that innovation adopted by others. This is what I would call an effective innovation.[82]

One of the effective innovations associated with the advent of agriculture is the edge-grinding of stone cutting and chopping tools, such as stone axes. At the time the British colonised Australia edge-grinding was employed in most of the continent.[83] Edge-grinding first appeared in northern Australia as early as 23,000 BP (Kakadu National Park), but did not come into use in southern and south eastern Australia until 4500 - 4300 BP.[84] The advantage of edge-grinding lay in the stronger working edge that is produced, so that axes and the like stay sharper for longer and the edge is less likely to break or chip.[85] The steel axe is the ultimate axe in this regard. When they were introduced into Australia by the *Macassans* and the British colonists, Aboriginal people immediately recognised their merit. Along the Murray River, for example, the introduction of such axes caused a "sensation", "when it was produced to their astonished gaze, much ejaculation and clucking of the tongue ensued."[86] Just how quickly the value and benefit of this innovation was recognised is well illustrated by an anecdote from Alfred Giles, a member of the Overland Telegraph survey party in Central Australia in 1870. In describing an encounter between members of the survey party and an old Aboriginal man he recounts:

"*Mr Ross made the old man a present of a brand-new and bright tomahawk, which he examined very carefully, feeling the bright steel edge, wondering what sort of stone it could be. After he examined it, he returned it to Mr Ross, seemingly unaware of its use. Mr Ross told me to go and lop off a small limb of a tree near by, which I did with one blow, to the astonishment of the old savage, who examined the limb and eagerly took back the tomahawk.*"[87]

Of course many innovations were generated endogenously and did not derive from direct contact or diffusion from outside Australia. One interesting example which could only have been of local provenance was an automated fish catching device used on the Murray River. Its design and operation was described in detail by Victorian "pioneer" James Kirby:

"*the blacks used to make a fence across these channels* [anabranches of the Murray River which periodically flooded] ... ... *leaving a space at one side ... ... A black would sit near the opening and just behind him a tough stick about ten feet long* [3.0 m] *stuck in the ground with the thick end down. To the thin end of this rod was attached a line with a noose at the other end; a wooden peg was fixed under the water at the opening of the fence to which the noose was caught, and when the fish made a dart to go through the opening he was caught by the gills, his force undid the loop from the peg, and the spring of the stick threw the fish over the head of the black; who would then in a most lazy manner reach back his hand, undo the fish and set the loop again on the peg.*"[88]

Undoubtedly many modern recreational fishers would be highly envious of this fishing apparatus. More seriously, the point here is that innovation amongst hunter-gatherers is based on rational choices, with successful, effective innovations being based on appropriate technology and practices that could readily be integrated into existing lifestyles. The changes detailed above came about not because the individuals and their groups were "forced" to change, they were seizing opportunities that occurred, or presented themselves, to them. The overall intent appeared to be to improve their life in some way. The *Kaurareg* planting yams was to avert hunger should the wild yams fail, the old man from Central Australia, who had obviously never seen a steel axe before, immediately grasped that the axe he was shown was a very useful and effective implement. Whoever devised the automated fish catching system had presumably found an easy and efficient method to catch fish. Broadly speaking we are all constantly looking for ways to make our life better, to solve day to day problems and make the most of our opportunities. Why would people in Paleolithic contexts, in prehistoric Australia, in Australia in the Contact Period, or in the modern world be any different in this regard?

Every day we have little ideas, sometimes big ideas, of ways we can improve our lives, at home, at work, as a society. Sometimes we may face major problems, climate change being one, and need to solve those problems through innovation, adjustment and adaptation. Occasionally we have an inspired idea that produces a major advance or advantage for ourselves and/or others. One need only consider the massive and revolutionary technological changes that have taken place in the last 50 to 100 or so years, the result of innumerable inventions, innovations and discoveries built on knowledge that had been accumulating over preceding centuries. The recent transformation has not occurred because of some exogenous factor, it has been driven by normal human propensities. It is quite apparent from such experiences that the predominant mode for innovation is not crisis or pressure driven. It involves creative attempts, on a daily basis, at improvement, with many failed experiments, unsatisfactory solutions and unsuccessful inventions occurring along the way. While we may now adopt strategic solutions to particular problems, or try to harness that creativity in systematic ways with research, design and development facilities, historically our

82 See for example Kneckt 1994.
83 Davidson and McCarthy 1957:425-8.
The exceptions were Tasmania, the Nullabor Plain and Great Victoria Desert of central southern Australia, and the south west of Western Australia
84 Mulvaney and Kamminga 1999:220-1,253.
85 Underhill 1997:116.
86 Beveridge 1889:69,70.
87 Giles 1926:29.
This encounter took place on Giles Creek, 90 km east of Alice Springs.
88 Kirby 1894:35-6.

efforts at improvement have not been directed but have occurred on an *ad hoc* basis. Unless one would contend that humans were not the same, that they were fundamentally different prior to the Neolithic Revolution, then it must be admitted that this has always been the case. The only difference is one of scale and the pace of change.

In considering the psychological dimension, as an alternative to the stress driven behavioural change models implicit in climate change and population pressure theories, a Hierarchy of Needs approach, as conceived by Maslow,[89] may furnish a more viable means of understanding the motivation behind the trajectory leading to food production and sedentism. It may also provide insight into social competition, as well as much of the motivational underpinnings of the changes, innovations and fashions we create or adopt in the present. Of the needs categories identified by Maslow - Physiological, Safety, Love/Belonging, Esteem and Self-Actualisation - the two basic ones, Physiological and Safety needs, are most applicable to generalist hunter-gatherers. These entail provision for what are considered to be primary needs, including food, water, biological functions, personal safety, bodily homeostasis and maintenance of health. For generalist hunter-gatherers providing food and water for themselves was a daily problem, they were never sure where their next meal was coming from or, particularly in Australia, if their water supplies would last. Most days they managed without much difficulty, but occasionally they went days without food, as attested by the *Montagnais* and *Watjandi* examples. In severe drought they had to abandon their country and find refuge elsewhere. Sleeping in the open may have been acceptable most of the time, but again, if it was cold, if it was raining, it was searingly hot and there was little shade or natural shelter, it was not only very uncomfortable, it compromised health.[90] While hunter-gatherers were no doubt conscious of the limitations of their life, and desirous of change, it was not obvious how this might be done. Generalist hunter-gatherers faced many complexities in their supposedly simple existence, on a day to day basis a man may face a litany of competing demands and scheduling conflicts - where will I find food tomorrow, will it be enough, how long will it last, what if my brother comes with his family, would it be better to go somewhere else, what if it hasn't rained in that other place, what if the ducks unexpectedly fail to turn up there, will there be enemies lurking in that direction, do I need to tell my wife to make a shelter because it might rain tonight, how can I obtain the right stone for the broken point on my best spear, or the right wood to replace my wife's broken digging stick, should I attend the corroboree we have been invited to, will others be offended if I fail to attend, how can I fulfil the other ceremonial duties expected of me, what shall I do if my father's illness worsens and he is unable to move, should I go and see my prospective father-in-law about taking his daughter as my second wife before she is promised to someone else, will I be able to support my in-laws until my second wife is ready to come to me?

In reality the solution of producing food and having permanent shelter was not easily achievable. Hunter-gatherers clearly knew, if Indigenous Australians are any guide, that plants grew from seeds. But what would be the point of planting if you didn't know where you would be tomorrow or next week, let alone in six months when the crop had ripened, presuming it had survived at all while unattended. If such a crop survived or there was an immediate surplus of some other commodity, of tubers, edible gum, eggs or fish for example, without the means to preserve it, it too was of little use. Being mobile also severely limited what could be carried, so even if it could be preserved it needed to be stored somewhere and monitored, otherwise it may well be eaten by insects, birds or mice while you were away. Consequently, mobile hunter-gatherers were faced with a series of chicken and egg problems requiring the discovery of techniques that could effectively increase the quantity and reliability of the food supply and the means of preserving it for periods when it was needed. A series of social adjustments, in mobility, in labour invested in building shelters and storage structures, in food distribution rules for example, were also required in concert to accommodate these other changes. It is no surprise, therefore, that despite the desire of hunter-gatherers for a better life, one in which their basic needs were securely met, the "solution" of food production and permanent sedentism was not an obvious one and was only arrived at by many millennia of learning and innovation through trial and error experimentation.

The nature of the problem faced by hunter-gatherers is well illustrated by an example often cited in the literature, questioning the impact of technical innovation in "development" of hunter-gatherer societies. In that particular case, the researcher Sharp pointed out that the adoption of steel axes by the *Yir Yoront* of Cape York did not produce any "improvement", as might be expected, they simply spent more time sleeping.[91] But this is not surprising really. Using a steel axe could have either resulted in higher output, or a given task being carried out more quickly, leaving time for other productive activities and additional output. However, if preservational techniques and storage infrastructure had not been developed, then producing more may not have been a particularly rational decision, the additional output would simply go to waste. In those circumstances, investing the extra time created by the steel axe in a leisure activity was a more rational choice

As an extension of the Hierarchy of Needs interpretation it is worth noting that once basic needs were met and effective surpluses produced, then the higher order needs, Esteem and Self-Actualisation come to the fore. Esteem in Australian

[89] Maslow 1943,1970.

[90] One method of coping with sleeping in the open, for example, was to sleep between two fires that had burned down to the coals. Unfortunately some individuals received severe burns when they rolled on to one of the fires during the night, often causing crippling disabilities or, at times, fatal injuries.

[91] Sharp 1968:349.

generalist hunter-gatherers, as mentioned in Chapter 7, depended on the age, personal qualities and attributes of the individual and the respect generated by those. When surpluses are produced, allowing a commensurate elaboration of demand, and sociopolitical hierarchies begin to develop, then another dimension is added, esteem may be increased by the accumulation of goods and control of resources, and elevation to higher status positions. Similarly, with hierarchisation, the potential to wield other forms of influence and control through ascension to formal positions increases commensurately, so that the scope for Self-Actualisation in those contexts is greatly increased.

Returning to the developments seen in the Upper Paleolithic, it should not be imagined that the trajectory that eventually led to the appearance of food production was intentionally directed in any fashion, being more a trend or tendency guided by the "unseen hand" of typical human motivations. And, as with much modern social, cultural and economic change, the trend wasn't necessarily smoothly unilinear or pre-determined but probably followed a fitful, meandering, course. Nevertheless, the antecedents of food production began to emerge at this time.

Using south west Asia as the exemplar, the switch to broad-spectrum procurement began there during the Kebaran [23,500 – 18,000 BP]. The first sign of this was the heavy utilisation of small grained grasses at Ohala II, around 23,000 BP, well before the Last Ice Age had reached its zenith.[92] In the succeeding millennia increasing specialisation based on seeds and other "r-selected" species was also in evidence. r-selected species, which includes bacteria, plants and animals, are those that are smaller in size but more numerous and with a high reproduction rate, in contrast to K-selected species, which are usually larger, less numerous and with low reproduction rates. Because they usually consist of innumerable small "packages" of food, r-selected species often require new procurement strategies which may rely on more concentrated, collaborative efforts, such as battues, more sophisticated devices, such as nets or the automatic fish catcher, or the development of procurement infrastructure, such as fish traps.

Trends established in the Kebaran and Geometric Kebaran continued into the Early Natufian when the first signs of significant sedentism appeared. The succeeding periods, the Early Natufian and Late Natufian, were followed by systematic cultivation and the commencement of the Neolithic. In this context the trajectory spanned the Last Ice Age, the warmer and wetter Bølling Interstadial, the brief return to cooler drier conditions of the Older Dryas, then the Allerød Interstadial, the extended cooler, drier Younger Dryas and finally the current warm, moist period which began with the commencement of the Holocene. So, while the climate underwent a series of changes the overall trajectory followed a consistent direction. But it should not be imagined that the sequence followed in south west Asia was replicated elsewhere. In China for example, significant levels of sedentism, if pottery is taken as an indicator, may have preceded broad-spectrum procurement.

[92] Weiss et al. 2004.

### The Setting

There are several fundamental requirements for agriculture - a suitable climate, environment and geographical location for the growing of plants. These fundamental requirements in global perspective are represented schematically below:

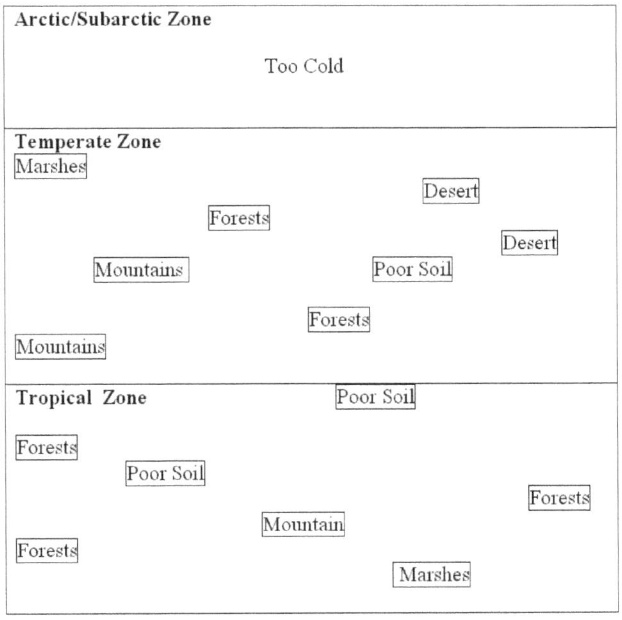


Clearly the pristine development of agriculture was only possible in the Temperate and Tropical Zones. Within these zones many other areas were ruled out because they were unsuitable for the cultivation of crops. Obviously one cannot easily cultivate plants if there is little water, or if there is too much water, it is too cold, the soil is not suitable, the terrain is inhospitable, or there are dense forests. In some instances some of these constraints could be overcome through water storage, irrigation, drainage, enhancement of soil fertility, landscape modification and cutting down of the forests, but these were later development that were not feasible until cultivation, storage and sedentism were already well established. Thus, even in zones with suitable climates there were limits on the locations agriculture could have arisen. It should be noted in this context, that where the climate was unsuitable for agriculture, in the Arctic/Subarctic Zone for example, alternative forms of subsistence specialisation arose, and so we see that fishing and sea mammal hunting predominating in northern latitudes.

While a suitable climate is essential for cultivation, and by extension agriculture, the role of climate change cannot be ignored. Prior to a commitment to large-scale cultivation, humans were largely dependent on the natural productivity of their environment. Colder, drier, periods suppressed

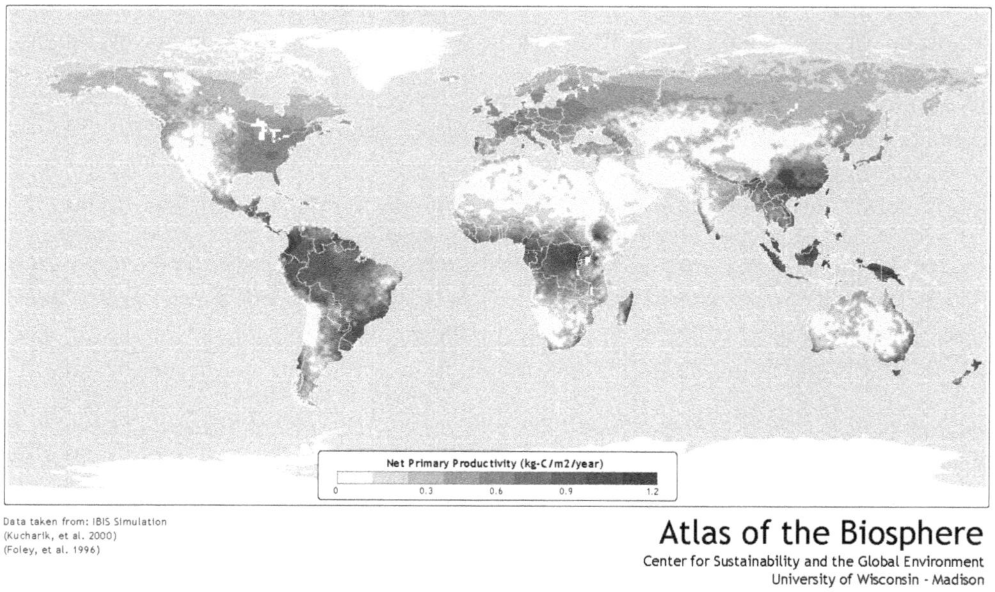


*Figure 27: Global Net Primary Productivity Map*
*(Atlas of the Biosphere, Center for Sustainability and the Global Environments, University of Wisconsin - Madison)*

natural productivity and slowed population growth. The Australian experience during the last Ice Age is an example. It is apparent, although absolute numbers were small, that population stagnated during the 2-3 millennia of the Last Glacial Maximum (see Chapter 2, Figure 1). Conditions were such that some parts of the arid zone were either abandoned completely or barely utilised. Conversely, if the climate changed so that a preferred food source increased then population growth accelerated. In the Natufian the population growth rate slowed when the Younger Dryas began to exert its influence,[93] and it is no surprise that the Neolithic "take off" occurred as conditions improved. Rather than being the trigger for agriculture the Younger Dryas in fact restrained it. Similar observations have been made regarding the rice cultivation. The Interstadials saw the extension of the range of wild rice while the Younger Dryas slowed the trajectory that saw the eventual domestication of rice.[94] Thus climate change cannot be assigned a primary causal role although it can be seen as a significant variable which affected the pace of change. How it affected a specific trajectory depended on whether the change was beneficial or not. In the past, climate change altered the areas where agriculture was possible, with what were formerly suitable localities disappearing or shifting. Where climates changed permanently to a type not suitable for agriculture (or pastoralism) the local trajectory effectively went into reverse. As mentioned in Chapter 6, the *San* people of the Kalahari had adopted pastoralism, metallurgy, and "extensive grain agriculture" by 800 AD but de-intensified as the climate worsened and their homelands turned to desert.

Another highly relevant factor in the trajectory leading to food production is population. As noted earlier, population grew at an increasing rate in almost all inhabited areas of the world from the Middle Paleolithic to the Epipaleolithic. It was argued that this due to greater yields being produced, essentially from increased extractive efficiency, the result of the application of human cognitive capacities and cultural learning. When prehistoric population levels were low, and before people began to extensively manipulate and exert a significant level of control of the local environment and ecology, in line with the scale of intervention outlined in Chapter 3, yields were largely dependent on natural productivity. Consequently the population size in any given area, the population density, reflected the relative richness of that area, its net natural productivity.[95] This factor applied to all geographical scales, from a local area, a region, multi-regional agglomeration, whole continents. A map of global "Net Primary Productivity" (NPP) appears below.

It should be noted, however, that this correlation did not always hold. The central Congo, for example, has high natural productivity but probably had a low prehistoric population because the dense tropical rainforest limited habitability. Consequently, the concept of Net Effective Primary Productivity (NEPP), the primary productivity that can be accessed and exploited by humans, may be more applicable.

Also highly relevant to the consideration of population is the principle that has been enunciated a number of times, that people have an equal capacity to innovate. The logical extension of this is, the greater the number of people, the greater the number of innovations that will occur over

93 Bellwood 2005:51.
94 Zhao 1998:894; Yasuda 2003a:141.
95 Hayden 1981:523; Keeley 1988; Rowley-Conwy 1999:136-7.

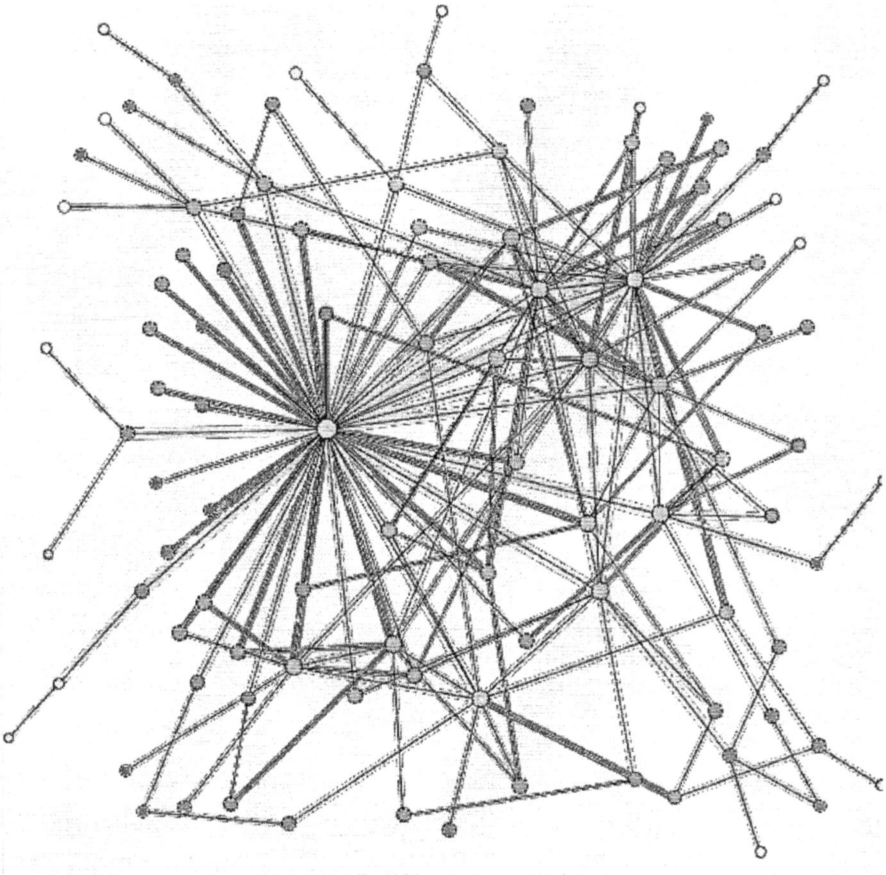

*Figure 28: A Scale-Free Social Network*

time. That is, the overall innovation rate is dependent on population size. But, as has been pointed out, innovations lead to increased yields arising from increased extractive efficiency, and this results in population growth. So in effect population and innovation are dependent variables, with the strength of the feedback between these variables determined by the per capita effective innovation rate. One of the implications of this proposition is that areas or regions with higher Net Effective Primary Productivity than other areas or regions will have higher populations and population density. This would confer an advantage as the number of innovations occurring in a discrete region or area, the innovation density, will be greater than for areas and regions less well endowed.[96]

One further factor relating to innovation that has a significant bearing on the long-term trajectory leading to food production is the communication of information, in the form of the experience, knowledge, skills, ideas, and technical and social innovations, which arise in any locality. While deliberations on the origins of agriculture consider the question in terms of the concept of pristine development this is a misleading construct as such developments do not happen in isolation. Although systematic cultivation may appear to begin in a discrete area, it is the culmination of a protracted trajectory, in the course of which the cultivators, in all likelihood, benefited at some point from the experience, knowledge, skills, ideas and innovations of others. We know that in Upper Paleolithic physical items did not move by trade and exchange any great distances, no more than a few hundred kilometres at most,[97] but that does not mean information flows were similarly limited. Where ever humans are in regular contact information is exchanged. With humans spread over vast areas of the globe a network of communication is formed which spans those distances. So it is possible for some information, even in Paleolithic contexts, to eventually be transmitted over considerable distances.

Such information networks, a form of "self-organising scale-free social networks",[98] would have had physical limits, the geographical boundaries formed by the edges of continents or islands, and uninhabitable regions, so that "bounded networks" are formed. Some features may have inhibited information flows and distorted the lines of communication within the bounded networks, such as deserts and rugged mountain ranges, while other features

[96] A point also made by Flannery (1994:267).
Recent research of comparable modern situations show that the innovation rate (as measured by the number of patents for example) increases disproportionately (in accord with a quarter power law) as population aggregations (large cities) increase in size (Bettencourt et al. 2007).

[97] Bar-Yosef 2002:367.

[98] Barabasi and Albert 1999; Barabasi 2002:55-92.

such as rivers or the spines of mountain ranges may have facilitated communication in a particular direction. Inter-group relations and their perceptions of other groups may have also affected information flows, though probably only in the short-term. The congruence of the potential innovation transfer with the economic orientation of the recipients may have influenced such flows as well.[99] Within a bounded network some locations, typically nodal points in the network ("highly connected hubs"), would have been advantaged by having more linkages to other groups, while those at the margins would have been disadvantaged in relative terms, having fewer linkages. This model can be applied at a regional, multi-regional, continental and multi-continental level.

The real world is a little more complex and complicated than shown here, particularly as there are many more linkages, innovations continue beyond the first linkage and they may be transmitted as multiples or in groups (a complex of innovations), not just singly. Consequently the nodal points with the greatest number of linkages or connections to multiple networks may not be as dramatically advantaged as the model suggests. Also, these bounded networks would naturally have an irregular shape, have barriers within them, or be highly distorted, such as where they follow extended lines of natural communication, further altering the pattern of communication, and transmission and accumulation of innovations. It may be more realistic to conceptualise the networks as information "catchment" zones with groups in areas ("connectors") near the centre of the catchment or line of communication, or linked to more than one catchment zone, deriving most benefit from the flow of innovations, what I would call "Innovation Information Gain". An example of what such a scale-free social network would resemble is shown in Figure 28 below.

In consequence of these arguments, and taking account of the constraints of the possible areas where agriculture was viable, the two factors identified above, Net Effective Primary Productivity and Innovation Information Gain are proposed as the determinants for the timing and location for each example of pristine agriculture. These factors operated in all of those circumstances, the relative strength of each factor being unique to each situation. In some situations the NEPP was highly predominant, in other situations it was Innovation Information Gain (IIG), while elsewhere there was an interplay in which the relative strength of each factor was more balanced.

Applying these factors to what has taken place in prehistory, the first two cases to consider are China and south west Asia. As can be seen in Figure 27, South west Asia is part of a large segment of quite low NEPP running from north Africa to central Asia. However, south west Asia, the Levant, lies at the approximate geographical centre of the inhabitable parts of Europe, Asia and Africa. It had the greatest "catchment" globally and therefore the highest Innovation Information Gain. This is why it was the first area for agriculture to arise. At the regional level the Natufian zone of the northern Levant, the portion of south west Asia where agriculture commenced had higher NEPP than surrounding areas. At the local level, the area with highest NEPP, the Jordan Valley, was also the main line of communication,[100] providing a regionally advantageous IIG. The other part of this region with higher NEPP and forming an important line of communication was the Tigris and Euphrates Rivers, where agriculture arose almost contemporaneously. It is not surprising, therefore, that the Jordan and Tigris/Euphrates should form the famous "Fertile Crescent."

Turning to China, examination of Figure 27 indicates that is had a large area of relatively high NEPP. While Equatorial Africa and Asia (Indonesia, Borneo, New Guinea) may have had higher NPP, as argued earlier, their dense tropical forests meant they generally had lower NEPP. China, therefore had the highest NEPP in the Old World. Consequently China had a relatively high population, while also being relatively central to the habitable parts of east Asia and mainland south east Asia. At the regional level rice cultivation appears to have started in the central part of the Yangtze River Valley, an area with even higher NEPP and a natural line of communication.[101] Millet farming made its appearance a little to the north, again in a rich river basin.

Examples from the New World can also be explained within this framework. Mesoamerica has slightly advantageous NEPP compared to lands to the north but the critical factor in this instance is its position, similar to south west Asia, of being a multi-continental nodal point in communications, particularly as it lies on the northern border of the biogeographic region known as the Neotropics, covering all of Central and South America. In South America, the Andes appear to have been the principal line of communication,[102] and the Southern Andes had somewhat higher NEPP in relative terms, accounting for agriculture there. There is great uncertainty regarding the locations where crops assigned to lowland South America, cassava (*Manihot esculenta*), sweet potatoes and yams, originated. Conflicting claims have been made by different researchers, with the cultivation of cassava reputed to have occurred first in Venezuela, coastal Peru, the Andes, northern Brazil, southern Brazil and southern Surinam.[103] Similarly there is great uncertainty as to where sweet potatoes first came into cultivation,[104] as it is with cultivated yams which are thought to derive from the north of South America, Central America or the West Indies.[105] Consequently it is difficult to identify precisely the locales where cultivation of these species

[99] These limitations would influence the power-law scaling of the networks.

[100] Sherratt (1997:277) also noted that early PPN sites were at nodal points on local communication routes.

[101] Pei 2003:180.

[102] Based on the observation that the diffusion of potatoes, and later maize, followed Cordilleran spine (Carr 1977:666; Perry et al. 2006)

[103] Piperno and Pearsall 1998:123-4; Allem 2002; Perry 2002.

[104] Piperno and Holst 1998; Piperno and Pearsall 1998:126-7; Srisuwan et al. 2006.

[105] Coursey 1967:9-11; Piperno and Pearsall 1998:117.

commenced. However, within the NEPP/IIG framework, it is noteworthy that the highest NEPP in lowland South America occurs in the Andean region of Columbia and its extension into north western Venezuela, southern Brazil, the north western Amazon Basin (north west Brazil, western Columbia and far south of Venezuela) and the contiguous region covering most of Surinam, Guyana and French Guiana. And pockets of quite high NEPP are also evident in parts of Central American "corridor". This extensive and diverse distribution may in fact be indicative of a Non-Centre, with diffuse origins to these crops to be expected, perhaps explaining the difficulties encountered by the investigators.

Regarding North America it can be seen that the highest NEPP north of Mesoamerica is in the south east of the United States. This is the region where systematic cultivation first appears to have arisen in North America. It is through this region that the Mississippi River and important tributaries, such as the Ohio River, run. It is also noteworthy that another small area of higher NEPP can be found in the northwest of the Unites States and the adjacent part of southern Canada, this being the area encompassed by the southern part of the Pacific North West.

In the case of Africa, there are two very large areas of low NEPP, which fragment the continent in terms of habitability, the Sahara and the central Congo. There are two centres of agricultural origins in western Africa, West Africa and Sahel, which lie adjacent to each other. In west Africa, the coastal regions on the north side of the Gulf of Guinea (Liberia to Nigeria) are highly tropical and have the highest Net Primary Productivity but are less habitable because of the forests, swamps and marshes. Consequently the less tropical region inland, particularly along the Niger River, probably has higher NEPP. Further north is the vast Sahara Desert, for the most part of very low NEPP. In communication terms the Sahara and the Congo pose major impediments so that the main flow of innovations in prehistory in the northern half of Africa would have largely been east-west. Based on the innovation model it would be expected that an agricultural centre would be found in west Africa, in this instance where the east-west communication intersects with an area of higher NEPP which has another shorter north-south line of communication along rivers such as the Niger. And this is what is evident in west Africa, except there is not a single locus but a split centre, the northern lobe, the Sahel centre exploiting semi-arid seed plants such as pearl millet and sorghum and the southern lobe where plants more suited to wetter conditions - yams, kola nuts, rice, cowpea, oil palm – have been taken into cultivation.

On the eastern side of Africa is a discrete area of high NPP which also has high NEPP, contiguous with most of Ethiopia. This area also lies at the eastern end of the east-west communication pathway of north Africa as well as lying close to the main lines of communication with southern Africa, around the Congo. Furthermore, the headwaters of the Nile River, presumably a channel of communication with north Africa and the beyond, are found in this region. And this is where the Ethiopian Centre is located.

The only other location where agriculture is believed to have originated, leaving aside Australia, is New Guinea, exemplified by the Kuk site in the central Highlands. This claim has been controversial and the original dating has been revised.[106] But even the revised date is quite a bit earlier than many of the other examples. Be that as it may, the explanation for this occurrence can be found in the high NEPP in the Highlands, relative to the New Guinea lowlands, where agriculture began much later,[107] as a result of diffusion. The Highlands formed a 2000 kilometre line of communication,[108] and in combination with the higher NEPP explains the New Guinea centre.

Finally, Australia as the last inhabited continent in which the growing of crops and other forms of primary specialisation occurred, can be considered from two perspectives, global and continental. In terms of NEPP Australia generally, being the driest inhabited continent with large areas of poor soils, had low NEPP compared to the other continents. It was comparable to Central Asia, Arabia and the Sahara. In terms of its communication "catchment" Australia was significantly disadvantaged as well. Apart from contacts early in the Contact Period with the *Macassans* and spasmodic interaction with marooned Dutch sailors, the only contact with the outside world was through Torres Strait. The main lines of communication leading to Torres Strait came through the Highlands of New Guinea and from limited contact with sea-faring Melanesians. The Highlands were a communications cul de sac however, as the seemed to have been little interaction with the lowlands of New Guinea, which it appear to have had very low populations and were not colonised by agriculturalists until about 3000 BP.[109] Torres Strait itself as a line of communication was disrupted by the rising sea levels in the early Holocene, so that a land bridge was turned into a series of islands that required watercraft to make communication and occupation viable. Consequently the Torres Strait Islands were not re-occupied until 2500 BP. Cultivars such as taro weren't introduced into the northern Islands until 1200 BP.[110] The diffusion of innovations through Torres Strait also centred on maritime subsistence, as evidenced by specialised

[106] Denham 2005.
The original claim was based on anthropogenic landscape modification but this modification may have been a form of General Selective Encouragement as outlined in Chapter 3, aimed at increasing the productivity of an ecosystem. Some corroboration for PNG being the first, but probably not the only independent centre in the SE Asia/New Guinea region can be found in Lebot (1999).

[107] Bellwood 2005:143-4.

[108] Allen 1998:90; Bellwood 2005:143.

[109] Pawley 1998; 2002; O'Connell et al. 2002:301-2; Bellwood 2005: 143-4.

[110] Barham 2000; Carter 2001.
McNiven (2006:9-10) suggests communication and trade relations were also disrupted 7-800 years ago, possibly by widespread climate change in the region coincident with the Little Ice Age.

dugong and turtle hunting and the double-outriggers in north Queensland, cultivating crops was a secondary activity.[111]

An illustrative example of the disadvantageous situation of Australia in regard to IIG can be seen in the introduction of the dingo. Dingoes, related to wild dogs from south and south east Asia, weren't introduced into Australia until about 5000 BP at the very earliest.[112] In North America the remains of domesticated dogs, thought to originate from east Asia, have been found from around 9500 BP (Danger Cave, Utah) and 8500 BP (Koster, Illinois).[113] Although the distances from the area of origin to the entry points of respective continents were similar, dogs reached Australia about 4000 years later than North America.

It is for these reasons, its poor Innovation Information Gain and low NEPP, that agricultural developments and other forms of subsistence specialisation took place much later in Australia than in other parts of the world. In essence Aboriginal Australia had to rely largely on the endogenous generation of innovation based on a relatively low population base. Considering what took place internally in Australia and applying the innovation model, the same pattern as observed globally emerges. Western Victoria had quite a high NEPP relative to other parts of Australia. The volcanic soils were rich and supported high natural productivity. Other areas, such as eastern Victoria, had high NPP, but lower NEPP because of dense forests and rugged mountains. The Corners Region was in general an area of low NEPP although, as pointed out in Chapter 5, there were rich alluvial deposits along parts of some of the rivers, such as the Darling, Diamantina and Coopers Creek. But the main factor operating in this region was IIG. The region was the hub for Australia, set off to the east to some extent because it was more habitable than the actual geographical centre of Australia, with the rivers acting as the main lines of communication.[114] The Australian trade routes, as shown in Figure 26, are a good proxy for those information flows and the degree of connectivity within different parts of Australia, and shows exactly that, the Corners Region as a communication hub.

Elsewhere in Australia, where agricultural or maritime specialisation was emerging, contact with outsiders provided significant IIG. The introduction of the yam to the *Nhanda* was particularly advantageous because, as I have argued, it was congruent with an economy oriented toward the exploitation of terrestrial resources, with the integration of the Dutch sailors fostering the initial change by influencing internal group dynamics. In Arnhem Land and north Queensland the developing specialisations were most pronounced among coastal groups, intensifying their exploitation of maritime resources. A good example, mentioned previously, was the *Yolngu* and other coastal groups from Arnhem Land, who borrowed such things as dugout canoes and bronze fish hooks from the *Macassans*, but not their cultivars.

In summary, it should firstly be noted that in preceding chapters a number of theoretical reorientations were proposed which were considered necessary to arrive at a sounder understanding of how humanity undertook the transition from generalist hunter-gatherers, largely at the mercy of their environment, to sedentary populations of primary specialists, able to manipulate their environment and the evolution of species within it. These theoretical reorientations included initial consideration of the question of the origins of agriculture as a problem in explaining primary economic development rather than the evolution of the biological features of plants under cultivation. The categorisation of the scale of intervention in the lifecycle of plants and a new formulation for characterising differing forms of sedentism and mobility, where populations were not fully sedentary, also contributed to these deliberations.

Lastly, the innovation model approach based on Net Effective Primary Productivity and Innovation Information Gain, has been proposed here as a new explanatory framework for the origins of agriculture and other forms of primary economic specialisation. The model is able to account for all examples of this where they have developed in prehistory and has a predictive capacity. In terms of the timing of agriculture and other specialisations, the larger the area of higher NEPP and/or the higher the connectivity, and thus access to larger information catchments, providing larger Innovation Information Gain, the earlier such developments would take place. Conversely, where these have low values, the later such development would be. In addition, if applied at a regional level this framework may guide future endeavours to identify where specific cultivars originated.

The innovations model approach is capable of resolving a number of debates regarding the origins of agriculture, such as the Centres/Non-Centres conundrum considered in Chapter 5. It would seem that in this case, where there is a broad area of higher NEPP or unfocussed higher IIG, that is, where there are a number of highly connected nodes in the same region, not just a single one, then multiple, apparently synchronous developments would be discerned. Alternatively, if there is a very high NEPP in a limited area, or a focussed hub, then a more discrete development would manifest itself. Ultimately, the value of this approach lies not just in settling esoteric debates, however, but in its potential to provide cogent and consistent explanations for the origins of agriculture. This would not only break the impasse that has existed in this field for some time, it would allow us to move forward in the challenge of understanding of a vital part of human history.

[111] Diamond 1997:315-6; Barham 2000:238-41; McNiven 2006.
[112] Lourandos 1997:274; Dayton 2003:556.
[113] Morey and Wiant 1992; Leonard et al. 2002; Savolainen et al. 2002.
[114] Roth 1897:132; Duncan-Kemp 1934:208; Tindale 1974:81.

# Chapter 11.
# On Australia and the Origins of Agriculture

## Background

The ancestors of Australia's Indigenous population probably arrived in the continent between 43,000 and 50,000 years ago.[1] They were the vanguard of the new hominin, *Homo sapiens*, sweeping across the globe. They were pioneers who manifested all the cognitive and communicative capacities of the new species, as attested by such things as their ability to successfully colonise the continent, their artworks and ceremonial objects, and the Lake Mungo burial. And yet, according to the current view, they did not embrace food production, either as a pristine development or through diffusion. This is the supposed "Australian Problem" alluded to at the very beginning of this work. But it is clear, if the thesis advanced in this work is accepted, that food production subsistence economies had in fact developed endogenously, although it was the last inhabited continent where this occurred. Moreover, agricultural practices had also reached the doorstep of the continent, diffusing through Torres Strait, with the southernmost group, the *Kaurareg*, adopting the cultivation of yams in the historical period.

It has been argued, based on the NEPP and IIG framework, the reason for the belated development of food production in Australia, as well as its delayed diffusion to the islands off the tip of Cape York, lay in Australia's low NEPP and low IIG. Australia's distance from the other continents, and the significant communication barriers it experienced, as evidenced by the delay in innovations reaching Australia from elsewhere, meant that the First Australians had to largely rely upon the endogenous generation of innovations. But, the dryness of the continent, possibly exacerbated by its impoverished soils, meant there was low NEPP. Consequently, the overall size and density of the population generating the innovations was very low in comparison to other parts of the globe, and hence the rate of development was commensurately slower.

## New Insights Into Sedentism and the Beginnings of Agriculture

If one accepts this case, as well as some of the theoretical reorientations that have been proposed, then there is great potential for gleaning valuable insight into the transitional period leading to food production at the local level. For example, researchers have struggled to grasp how people managed the transition from a mobile to a sedentary mode of existence. The Murdoch schema has been partly responsible for this because the categories are too general and imprecise, and so have done little to inform our understandings. Most change is incremental, and the typology presented in Chapter 5 at least offers the possibility of comprehending how people could incrementally increase their levels of sedentism. From this it is possible to conceive, for example, that people who are seasonally sedentary may find a way of identifying, exploiting or developing an additional local food source and so remain for a little longer than a season, extending that to perhaps two full seasons and beyond, and so begin to exhibit multi-seasonal sedentism. Those who followed an embedded round and were sequentially sedentary may find or develop a food source at one of the sites in the round, so extending their length of stay at that location, until they may in the end decide to stay in that location most, if not all, of the time. In situations where people are partially sedentary they may start out with a few old people permanently locating at one settlement, then some of the women and children and so forth. As the capacity to support people at that location grows, so do the numbers permanently residing there. They may then end up being just seasonally mobile, people only leaving to attend ceremonies or to take advantage of an abundant transitory resource elsewhere. Perhaps as people become more sedentary and sociopolitical complexity grows they begin to invest time and effort in the construction of religious structures and shrines which then anchor them further to their settlement.

Another possible route to embracing fixed settlements and a sedentary existence could come through increasing circumscription. The *Narrinyeri* may have been following this trajectory. In this scenario, the clan estates become smaller and smaller as overall extractive efficiency slowly increases and a number of viable food sources become available in at least one part of the estate within reach of a central location, which then develops as the fixed settlement. Their permanent winter huts may well have been such a nucleus. The fact that the *Narrinyeri* were only seasonally sedentary does, however, draw attention to the "problem" of sedentism, why were some groups sedentary and others were not? For example, although some have described the people from the central Murray and its confluence with

[1] Bowler et al. 2003; O'Connell and Allen 2004; Johnson 2006:61.

the lower Darling as Complex Hunter-Gatherers,[2] why weren't they sedentary, despite having high populations, high population densities and in many instances a high level of circumscription?[3] While Sturt noted, "some large huts",[4] just west of the confluence of the Murray and Darling when passing by in January 1829, he later concluded:

"*The natives of the interior ... ... had huts of more solid construction than those of the natives of the Murray or the Darling, although some of their huts were substantially built also ....*"[5]

It would seem that sedentism can arise from situations where there are adequate naturally occurring, multiple food sources that appear sequentially in different seasons of the year, but this would require an unusual set of circumstances. Similarly, natural co-location of sufficient food sources, available in any season is also capable of underwriting a sedentary lifestyle. This is, however, quite unlikely and would also require a special set of circumstances. Lifting production and extractive efficiency could potentially provide enough food but there would still be fluctuations in supply. The most assured pathway in becoming sedentary is to develop the means to preserve an abundant source or sources and be able to store the end product, thus evening out the fluctuations. The *Narrinyeri*, for example, do not appear to have had adequate, naturally co-located or suitable temporally sequential food sources to support full sedentism. Even though they preserved and stored various forms of sustenance, these may not have been sufficient in quantity or longevity to sustain the whole population for extended periods. Fish seemed to keep for the longest, but that was only for a period of 7 months. One gets the impression that there was not the level of abundance of storable foods to sustain full sedentism, even though surpluses were still being produced.

Looking further afield, in the parts of the Pacific North West, where salmon runs formed the economic foundation, there was extensive storage. While there were fixed settlements, a different sort of co-location problem arose. Generally the locations where the salmon were most abundant were not very suitable for settlement owing to the rugged terrain. The settlements were located elsewhere and at least part of the population had to move temporarily to the lower courses of the river for a period of time to catch and preserve the salmon. In effect they were forced to become seasonally mobile, as modern fisherfolk in trawlers often are. Undoubtedly if the lower courses of the rivers of the Pacific North West were suitable for permanent habitation the people in those regions would have been fully sedentary there.

In the East Central Australian Non-Centre, with its mosaic of ecozones and capricious climate, it seems unlikely there were temporally sequential food sources naturally available to support a lifestyle approaching full sedentism in any location. Nevertheless, the Brewarrina fish trap system, once the labour had been invested in building the system and increase extractive efficiency, was one instance where it was possible to support at least a small population on a year round basis. In terms of co-location in the Corners Region, there may well have been instances where foods such as seeds or gum and fish were naturally co-located, but not in sufficient quantities to support sedentism beyond a season or two, even with significant investment in procurement devices such as nets, weirs and fish traps. The planting of *Panicum* spp., *ngardu*, *yaua*, and possibly *munyeroo*, was, however, one way of increasing the possibilities through deliberate co-location. The potential, as outlined in Chapter 9, to increase production and extractive efficiency through planting presumably enhanced the ability of some groups to develop fixed settlements and increase the duration of their stay in such settlements. If one can effectively relocate a food source, as well as augment its productivity, then the possibilities are further enhanced. Perhaps this is part of the reason why agriculture is the most common form of food production. But again, preservation and storage of the more abundant foods - fish, seeds and gum in the case of the Corners Region - appears to have been the critical innovation in supporting a transition to larger, permanently occupied settlements in that region. So storage, often identified as essential to achieving full sedentism, is the commonest and superficially simplest means of achieving that. It is not necessarily the only means, however, and the process cannot be viewed in isolation. The temporal and spatial distribution of foods, and how they can be deliberately manipulated within those dimensions at a local level, shows that there is in reality a complex equation that any group in a pristine situation must struggle to solve, largely through their own efforts and innovations.

As a final example in applying a theoretical perspective to the processes leading to food production, specifically agriculture, the pathways leading to agriculture through transitional planting and sowing will be considered. Others have at times tried to identify some of these pathways, incidental planting, "incidental domestication", for example, being one such pathway identified by Rindos. This begins almost as an accident, an unintended outcome of other behaviour. The broadcasting of seeds by the *Gunditjmara*, mentioned in Chapter 7, may be another form of incidental planting. The hypothetical intention was to attract parrots, but this may have also led to the seeds germinating and stands of plants growing. At other times planting may not be with the intention of producing a crop, the plant may have some other benefit. The planting of *wirra* to provide ash to chew *pitjuri*, *moolyardi* trees for spears or gourds (*Cucurbita* sp.) for containers are examples. But it is likely there are numerous other pathways where small scale planting may begin and then be scaled up, if feasible

[2] e.g. Pardoe 1995:709.
Interestingly, there appear to be strong parallels between the socioeconomic pattern in the central Murray and Watson Brake on the Arkansas River in Louisiana (Pringle 1997).
[3] Kefous 1988.
[4] Sturt 1833:2:115.
[5] Sturt 1849:2:139.

and where there is a comparative advantage over other subsistence activities. The *Kaurareg* example of planting yams as a backup to the supplies of wild yams may be an example. If the watering of the yams for instance, to ensure their survival, could be fitted into the round of subsistence activities then a more assured supply of yams would result. The *Kaurareg* may then have dropped some other marginal subsistence activity to concentrate more on yams and set in motion a positive feedback loop, with the scale of yam planting increasing to the point of becoming the principal form of subsistence.

Another, probably common, transitional pathway may be one where planting begins as a supplementary activity. In the Lake Plains area of New Guinea people clear some jungle, plant bananas and papaya, go off hunting and gathering elsewhere for a few months, return to check on the crop, do minimal maintenance such as weeding, go off again and then return for a period to harvest and consume their produce.[6] Similarly some of the *Apache* groups became seasonally sedentary, planting, growing and then harvesting crops during the growing season before resuming a more mobile subsistence round.[7] In Chapter 3 the *Mardu* of the central northern Western Desert in Western Australia broadcast seeds of "bush tomatoes" around their camp before departing. Again, this is something that could be scaled up, but if there are barriers to that, such as a lack of advantage over other subsistence activities or a limited area for cultivation, then the scale of planting will not increase.

One additional pathway that can be identified may arise from mimicry. Many significant animal and vegetal foods in Aboriginal Australia were traditionally the subject of increase ceremonies, an expression of a desire to maintain or increase the supply of that food. There appears to be a correlation in terms of the Australian evidence between the level of intervention in the life cycle of plants and the specificity in mimicking the germination of the plant. The *katoora* ceremony mentioned in Chapter 5 is a conspicuous instance of this. There appears to be in such cases a convergence of the desire to maintain or increase that food with a realisation expressed in social ideological terms that control of the plant can be exerted. So it comes as no surprise to see that many cultivated crops in traditional societies around the world have fertility rituals associated with their propagation.

Whatever local dynamic led people in the past to begin planting food species, it does not follow that this will necessarily or inevitably lead to agriculture. The conditions that allow planting to be scaled up must also exist. Large-scale planting must offer some advantage as well, over undertaking some other subsistence activity. Finally, conservation of output, through preservation techniques and effective storage, or live storage (such as keeping yams in the ground) must also come about before a fully formed agricultural society begins to emerge.

**Debates and Discourse on Agriculture in Australia**

Given the complex elements required for the development of agriculture, anyone arguing that there is an inevitability in such a transition in any specific circumstance is clearly treading on dangerous ground. Conversely, to argue that it is not possible in other circumstances is equally fraught. Yet this is, in effect, the position that has been taken by some who have considered this question in relation to Australia. Others have not denied the possibility, but, based on the assumption that there had been no agriculture in Australia, insisted that this, the Australian Problem, required explanation.[8] The actual reasons that have been suggested or reflected as to why agriculture couldn't, or didn't, develop in Australia have been:

1. isolation[9]
2. a lack of suitable, domesticable, species, both plant and animal, in Australia[10]
3. that the climate was too dry and capricious[11]
4. that soil types were inappropriate [12]
5. that the population and population densities were too low[13]
6. that Aborigines didn't know plants grew from seeds[14]
7. that Aboriginal societies were inherently conservative and that their belief systems were a sociocultural barrier to the adoption or development of agricultural practices[15]
8. that sociopolitical structures differed from elsewhere[16]
9. that there had been no "crisis of appropriative economy"[17]
10. many of the critical pre-adaptive features, such as sedentism, storage and appropriate tools were not present[18]
11. that there was a lack of comparative advantage of

6 Diamond 1997:106.
7 Basso 1983:468-9; Opler 1983:370; Diamond 2005:140.

8 White 1971:184-5; Harris 1977a:209; Rindos 1984:161-2; Williams 1987:320; Hayden 1995a:297; Price and Gebauer 1995:9.
9 e.g. Duncan-Kemp 1968:xxi; Tindale 1974:36; Tonkinson 1978:14; Yen 1989:68.
10 e.g. de Candolle 1884:4,449; Bennett 1930:13; Shaw 1962:21; Cleland 1966:113; Duncan-Kemp 1968:xxi; MacNeish 1992:321; Butlin 1993:93; Flood 1999:265; Bellwood 2005:36.
Native Australian grasses are actually currently cultivated and sold commercially (O'Neill 2006).
11 e.g. de Candolle 1884:449; Anon. 1907; Jones and Bowler 1980:23; Bailey 1981:11; O'Connell et al. 1983:99; MacNeish 1992:321; Butlin 1993:93; Cribb 1994:382; Flannery 1994:281-2,285; Flood 1999:265; Bellwood 2005:36.
12 e.g. Jones and Bowler 1980:23; Tindale 1974:37.
13 e.g. Flood 1999:265; Diamond 1997:311.
14 e.g. Horne and Aiston 1924:8; Elkin 1954:15.
15 e.g. Meggitt 1964; Duncan-Kemp 1968:xxi; Sharp 1968:352; Tonkinson 1978:14; Chase 1989:49; Jones and Meehan 1989:132-3; Higham 1995:144; David 2002:94.
16 e.g. Anon. 1907; Lourandos 1980a:258-9; Hayden 1990:60.
17 Kabo 1985:602.
18 Anon. 1907.

agriculture over hunter-gatherer lifestyles, what is known as the "original affluence" hypothesis[19]
12. Aboriginal people didn't plant[20]
13. Harvesting method (hand stripping)[21]

Some of these reasons, given the evidence that has been presented in this work, are obviously quite spurious. Some, such as isolation, the climate being too dry, soils types being inappropriate, and population and population densities being too low are based on intuitive reasoning, reflecting the explanation provided by the NEPP and IIG framework for the belated appearance of food production, including its agricultural manifestations, in Australia. It is not necessary to dwell on the reasons advanced for the lack, or impossibility, of agriculture in Australia, but to point out that all assume that there was no agriculturally based food production. Why is this? If one accepts the evidence and arguments that have been presented here, why has this not been realised previously? Contact between people from other parts of the world began over 400 years ago and in-depth documentation of Indigenous cultures commenced when British colonisation began in 1788. Academics studies have been undertaken in Australia for well over a century and have, in recent times, included significant Indigenous perspectives. Yet no-one has argued that food production was being undertaken in any part of Australia. In considering this question more fully it is not the intention to single out or unnecessarily offend other well-intentioned individual researchers, but to draw out the issues. Consequently works which may be quoted or referred to will not necessarily be cited.

Possibly the most insidious reason for the lack of recognition of agriculturally based food production in Australia is paradigm blindness, the assumption that all Indigenous populations in their traditional life were mobile, generalist, hunter-gatherers. It is this filter that time and again has led to "anomalous" evidence, such as presented here, being ignored, dismissed, misinterpreted, reinterpreted or distorted. The large structures at White Lake, near Lake Blanche and in the northern Simpson Desert are quite remarkable in Australian terms, being capable of accommodating 30-50 people. Yet they are almost never mentioned in relevant literature in Australia, let alone elsewhere. The village of 70 dwelling which "appeared to be permanent habitations", each large enough for 12 to 15 people, seen by Sturt on the upper-central Darling in 1829 was recently characterised as being a temporary ceremonial camp with the dwellings simply branch covered structures. In the case of the planting and harvesting of yams by the *Nhanda*, the fact they were planting is constantly ignored in relevant studies, or just seen as a form of selective encouragement. The most frequently cited study in relation to the use of yams on the west coast treated the relationship as small-scale intensive foraging, and consequently failed to recognise the import of the fact that the yam grounds covered many square kilometres of the best alluvial soils in a succession of river valleys in the Victoria District. When the *Nhanda* villages or "towns" reported by Grey are mentioned, it is with no cognisance of what they imply. A recent comprehensive Australian publication on Indigenous plant use missed many of the examples of planting documented earlier, and where planting was acknowledged, treated those instances as just a ritual, plants were only being "ritually planted" or replanted.

Part of the reason for this paradigm blindness lies in the nature of ethnographic and anthropological studies in Australia over the past century. The peoples that have been most intensively studied have been in the Northern Territory, particularly Arnhem Land, and the Western Desert region, where they have still been managing to follow quasi-traditional lifestyles. By default they have become the archetypal Aboriginal societies and are frequently cited as such in the international literature. Consequently their cultural characteristics have been generalised to all Indigenous Australians, as representative of them as mobile, generalist, hunter-gatherers. Conversely, in the areas where food production economies predominated in the past, while there has been a continuity of culture, traditional lifestyles have not been followed in a substantive way for a very long time. Describing, as has been done here, their traditional economy and settlement patterns, and other relevant features, has therefore required the nature of those societies to be reconstructed through historical and reconstructive ethnography, supplemented by memory ethnography and traditions. While some commendable reconstructive studies have been carried out by others in relation to south west Victoria, as well as the *Paakantyi* of the Darling River and people of the Coopers Creek in the Corners Region, they have not been undertaken in the same depth as here. Nor have they sought interpret or present findings informed by theoretical frameworks outside of the hunter-gatherer paradigm, or to explicitly challenge the assumptions of the paradigm.

Another factor in the dominance of the mobile, generalist, hunter-gatherer paradigm relates to the sources. Innumerable 19th century compilers, amateur ethnographers and early academic anthropologists repeatedly pronounced that Indigenous Australians never planted or cultivated, never stored food, and apart from brief periods of immobility, were completely nomadic. This arose from over-generalisation and a lack of recognition of regional differences. If, for example, no planting, storage, permanent dwellings or formal hierarchies were encountered in a locality, or even many localities, the tendency was to pronounce that these things were not part of any Indigenous culture in Australia. This is in spite of the contradictory evidence proffered by their contemporaries, or even by themselves in the case of Horne and Aiston in regard to the broadcast seeding of *wirra* and the staple *yaua*. Curr's rejection of any notion

19 e.g. Berndt and Berndt 1954:38; White 1971:184; Sahlins 1972:14-21; Peterson 1976:274; Bailey 1981:11; Lourandos 1997:76; Flood 1999:264-5; Mithen 2003:338-9.
20 Bellwood 2005:36.
21 Bellwood 2005:36.

of socio-political complexity anywhere in Australia, apart from the *Narrinyeri*, is another example.

Adding to this, the most influential authorities of the late 19th and early 20th centuries, Spencer and Gillen, were at pains to portray in their writings the Aboriginal people of inland Australia in as "primitive" light as possible. This was to validate their position in regard to debates on such things as "group marriage", raging in anthropology at that time. Although most modern researchers are justifiably critical of many of the pronouncements of the 19th century commentaries, few actually question this characterisation or the broader generalisations. In more recent times some highly respected Australian archaeologists and anthropologists have indeed unthinkingly and uncritically echoed the earlier pronouncements, denying that Indigenous Australian planted or cultivated, and asserting they were universally nomadic, that they never had such things as permanent settlements, "built no dwellings" and "accumulated no property".

In addition to the paradigm issue there have been evidentiary problems which have made recognition of high degrees of sedentism and food production in traditional circumstances difficult as well. In general terms, as pointed out in Chapter 6, the colonists had little idea as to what early agricultural and Neolithic societies were like. When they thought of agriculture they most likely thought in terms of "plough agriculture", the well-developed agricultural systems of Europe, Asia and the Americas. And because the local population were engaging in hunting and gathering they assumed they were just hunter-gatherers, unaware that the people of the early Neolithic didn't just grow crops, they also extensively engaged in hunting and gathering. So when pronouncements were made that no cultivation took place, it may well have simply been missed in some cases, or just not recognised because it did not fit the observers' preconceptions in others. Recall Grey conclusion, in trying to convey his impression of the *Nhanda* yam fields, that "these circumstances all combined to give the country an appearance of cultivation ....",[22] and in the Corners Region, McKinlay's comments on an instance he encountered, where the "whole country looks as if it had been carefully ploughed, harrowed and finally rolled."[23] Planting, even in modern agriculture, is usually a very brief activity, so that when some local Indigenous people would, as Smith said, "chuck a bit there ... ... spread it you see - one seed there, one seed there, ... chuck a little bit of dirt on...", it would be over in minutes or hours. The chances of it being commonly observed were slim.[24] The chances of observing a modified landscape, as described by Grey and McKinlay, being aware that planting was being undertaken, and connecting these two things, are even more remote.

Contributing to the evidentiary problems, many of the historical sources that incorporated commentary of Aboriginal cultures contain information, evidence, opinions and judgements that are superficial, confusing, contradictory or plainly wrong. Some of the 19th century amateur ethnographers who claimed intimate knowledge of traditional Aboriginal society and cultural practices were in fact describing situations where profound changes had already taken place, the traditional economy was in ruins, the settlement pattern in disarray and social structures in a chaotic state. These amateur ethnographers may have actually had a role in creating those situations, or, having arrived soon after the initial colonial incursion, thought that what they were observing was the pristine traditional culture. As mentioned in Chapter 5, Hill, despite the fact that he acknowledged the *Maiawali* of the central Diamantina River possessed permanent or near-permanent "dome shaped" dwellings "plastered up with mud," asserted that they were constantly on the move, "rarely staying in one place." The contradiction is resolvable if it is realised that Hill was describing the permanent dwellings often referred to in the Corners Region while reporting the enforced itinerancy of the local population in the post-contact period.

The more sedentary societies engaging in food production in Australia were also possibly more vulnerable to the colonial invasion than more mobile groups. Consequently the traditional economy and settlement pattern broke down more rapidly, so that awareness of them has often had to rely on brief, incidental, fleeting or oblique references by a variety of actors in early contact and frontier situations. When the colonists occupied a locality they invariably appropriated water sources, and the areas with the richest soils, for their cattle and sheep. In the Corners Region the traditional owners were especially vulnerable because water sources, especially permanent water sources, such as waterholes, springs and lagoons were critical to their existence and their way of life. As seen in Chapter 6, much of the population and their settlements were to be found at those locations, major food sources such as fish were derived from them, and they were often where the best soils could be found. In the Victoria District, several colonists noted that the alluvial soils, where the yams grew most profusely, was highly attractive land. Wollaston, referring to the yams, reported that:

*"This root flourishes where the best feed for stock is found. Hence the usurpation of the ground and the secret destruction of the Aborigines."*[25]

Having permanent dwellings and settlements left their Indigenous residents particularly exposed. The unprovoked and wanton destruction of the Scrubby Creek Village is a

[22] Grey 1841b:44.

[23] McKinlay 1861:50.
As mentioned in Chapter 5, Mitchell (1848:274) also came across patches of up to "two acres" which "resembled ground broken up by the hoe" on the Belyando River in 1846, but he too was unable to ascertain for "what purpose" this was done.

[24] In most cases it was only when sympathetic individuals and anthropologists became involved and engaged with Indigenous communities that they were informed of, or observed, planting and cultivation.

[25] Burton and Henn 1954:144.

good illustration of this. The stone houses at Eumeralla Village may also have been deliberately destroyed,[26] not unexpected given the resistance emanating from there during the Eumeralla War, and Fyans' police action against this "Township" in October 1842.

The suppression, or discouragement of public disclosure, of evidence of permanent settlements and more sedentary existence may have been yet another factor contributing to a distortion in historical information about relevant groups, and hence modern understandings. For example, in his original account of his brief reconnaissance in the Victoria District in early December 1841, published at the time in a Perth newspaper, Captain Stokes of the *Beagle* referred to "their [the *Nhanda*'s] winter habitations substantially constructed". But that very line is omitted, the only one to be so, when the substantive part of the newspaper account was reproduced in his published journals in 1846.[27] Similarly, in Victoria, a *Select Committee of the Legislative Council* in 1858-9 took evidence from Thomas and Sievwright in which both discounted that there had been permanency of settlement, even though we know they knew differently.[28] The suspicion is that there was an intent to discredit evidence of permanent settlements because of the implications this may have had for the morality and legality of the colonial dispossession.

Throughout this work there has been a heavy reliance on historical sources. As can be seen from the preceding discussion, many of those sources are problematic. The issue of paradigm blindness has in the past coloured many judgements and interpretations, and by extension much of the evidence. Compounding this, much evidence may well have been either missed or been erroneous for one reason or another. In endeavouring to construct a legitimate case, given these limitations, discerning judgment in assessing the validity of source materials, their consistency, cogency, credibility and sufficiency, has had to be exercised. This is the normal methodology for any historically based study. Nevertheless, many undoubtedly will have difficulty accepting the evidence and arguments that have been presented thus far. While paradigm blindness may limit or distort the understanding and appreciation of honest and dedicated researchers, for some their ideological and moral positions, or unthinking prejudices, may make the propositions advanced here difficult to acceptable. Some, for example, are reluctant to consider the type of historical evidence used here because the assumptions and the interpretations of the authors of those materials were implicitly or explicitly racist, particularly where their views were shaped by the "culture stages" and social Darwinist theories of the time. Others may take a similar position based on the fact that the authors were perpetrators of, or at least complicit in, gross crimes against humanity in the treatment of Indigenous Australians in the Contact Period. Yet others consider the in-depth utilisation of historical sources as a form of "antiquarianism", a throwback to the days of unsophisticated collecting of artefacts with no reference to Aboriginal culture or people.[29]

It is in regard to the mindsets that some adopt, in essence ideological positions, which represent, however, the most insidious barriers to advancing our understanding of traditional cultures in Australia, as well as the origins of agriculture. One of these is the "progressionist trap", that developmental models are inherently flawed because they are deterministic in some way. This is in part a legacy of the culture stage conceptions of earlier days, and in part the result of our inability to frame an effective alternative approach. I would suggest that this is largely because inappropriate paradigms are being relied upon since then. Ecologically based analyses, for example, quite reasonably assume the lives of hunter-gatherers in the past were dominated by their ecological circumstances. However, at some point humans have transcended those circumstances by developing the capacity to manipulate and control their environment, and impose their own selection regimes on plant and animal species, to an extent that negates that assumption. By treating the question as a problem in economic development, as has been done here, new perspectives becomes possible. For one, it frees us from the cultural baggage of "progressionism". To use an analogy, in the modern world we discuss and debate economic development freely, recognising that some societies, some nations, are more developed than others. It is not a moral or cultural failure that a particular nation is not as developed as another, it is the result of history and the advantages and disadvantages that different nations enjoy. While a similar process may be followed by different nations in their development there is no set pathway. Similarly, in the development of agriculture, different cultures have developed in distinctive ways and at distinctive, almost glacial rates, in a process that has taken tens of thousands of years, with some simply lagging others in relative terms not by hundreds of years, but by thousands.

One anthropologist recently commented in regard to Australia that:

*"there are scholars seemingly determined to discover 'villages', so as to prove that there were in parts of Australia sedentary groups – as if nomadism were some kind of moral failure, instead of manifesting sound adaptation to environmental demands"*

[26] See Gerritsen 2000:19,52n134;53n151; also Critchett 1990:30,65, 67.

[27] cf. Stokes 1841:2-3 and Stokes 1846:2:385-93.

[28] Victoria: Legislative Council, Votes and Proceedings 1858-9:61, 74.

[29] An accusation first leveled at one of the pioneers of ethnohistory in Australia, Diane Barwick (Kijas 1997:56), but also recently by a reviewer of one of my papers.
The "Antiquarians" were a group of amateurs active mainly in the first half of the 20th century, who collected Aboriginal surface artifacts in the belief that there was no time depth to the Indigenous presence in Australia. They made classifications of those artifacts based on form or imagined function, with no reference to the people who actually had made and used those artifacts in the past. Some of their papers (e.g. Cobb, Dow, Towle) are cited in this work, as archaeological survey evidence.

This statement encapsulates why, in spite of the evidence, many have closed their minds to the possibility that some Indigenous Australians were living in villages in traditional circumstances and engaging in food production. This rejects the evidence not on the basis of its validity but as a form of apologism for the Indigenous culture. But why should an apology even be necessary? Some people were nomadic, some were not, that was the reality. As mentioned in Chapter 5, palaeontologist Robert Etheridge argued as far back as 1894 that to assert Indigenous Australians were "devoid of any knowledge of husbandry" was a "mistake".[30] Let us not continue to perpetuate that mistake.

[30] Etheridge 1894:110.

# References

NOTES:

1. Battye Library is J. S. Battye Library of West Australian History.
2. WASA is Western Australian State Archives.
3. PR is Printed Record, Battye Library.


Abbott, W. E. 1881 *Notes of a Journey on the Darling*, Sydney: Government Printer.

Aboriginal People of Central Australia 1995 *Desert Bush Tucker Identikit*, P. Latz and G. Wightman (preps.), Darwin: Parks and Wildlife Commission of the Northern Territory.

Adams, K. R. and Bohrer, V. L. 1998 Archaeobotanical indicators of seasonality: Examples from arid southwestern United States. In T. R. Rocek and O. Bar-Yosef (eds) *Seasonality and Sedentism: Archaeological Perspectives From Old and New World Sites*, pp.129-141. Cambridge (Mass.): Peabody Museum of Archaeology and Ethnology.

Aikens, C. M. 1995 First in the world: The Jomon pottery of early Japan. In W. K. Barnett and J. W. Hoopes (eds) *The Emergence of Pottery: Technology and Innovation in Ancient Society*, pp.11-21. Washington: Smithsonian Institution Press.

Aiston, G. 1921 Natives of Central Australia: Fighting With Knives. *The Argus* [Melbourne] 9 July 1921, p.5.

Aiston, G. 1930 *Magic Stones of the Tribes East and North-East of Lake Eyre*, Hobart: Royal Society of Tasmania.

Aiston, G. 1935 The desert Aborigines. *Mankind* 1(12):5-8.

Aiston, G. 1937 The Aboriginal narcotic pitcheri. *Oceania* 7(3):372-377.

Akazawa, T. 1981 Maritime adaptation of prehistoric hunter-gatherers and their transition to agriculture in Japan. In S. Koyama and D. H. Thomas (eds) *Affluent Foragers*, pp. 213-258. Osaka: National Museum of Ethnology, Senri Ethnological Studies No. 9.

Akazawa, T. 1986 Regional variation in procurement systems of Jomon hunter-gatherers. In T. Akazawa and C. M. Aikens (eds) *Prehistoric Hunter-Gatherers in Japan*, pp.73-89. Tokyo: University of Tokyo Press.

Akerman, K. 1975 The double raft or *kalwa* of the West Kimberley. *Mankind* 10(1):20-23.

Akerman, K. 1976 Fishing with stone traps on the Dampierland Peninsula, Western Australia. *Mankind* 10(3):182.

Akerman, K. and Stanton, J. 1994 *Riji and Jakoli: Kimberley Pearlshell in Aboriginal Australia*, Darwin: Northern Territory Museum of Arts and Sciences. *Monograph Series No. 4*.

A Lady [Mrs McPherson] 1860 *My Experiences in Australia. Being Recollections of a Visit to the Australian Colonies 1856-7*. London: J. F. Hope.

Alexander, J. and Coursey, D. G. 1969, The origins of yam cultivation. In P. J. Ucko and G. W. Dimbleby (eds) *Domestication and Exploitation of Plants and Animals*, pp.405-425. London: Duckworth and Co.

Allem, A. C. 2002 The origins and taxonomy of cassava. In R. J. Hillocks, J. M. Thresh and A. C. Belloti (eds) *Cassava: Biology, Production and Utilization*, pp.1-16. Wallingford: CAB International.

Allen, H. 1968 Western plain and eastern hill: A reconstruction of the subsistence activities of the Aboriginal inhabitants of central eastern Australia. Unpublished Bachelor of Arts (Hons) thesis, Department of Anthropology, University of Sydney, Sydney.

Allen, H. 1972 Where the crow flies backwards. Unpublished PhD thesis, Australian National University, Canberra.

Allen, H. 1974 The Bagundji of the Darling River Basin: cereal gatherers in an uncertain environment. *World Archaeology* 5(3):309-322.

Allen, H. 1996 Ethnography and prehistoric archaeology in Australia. *Journal of Anthropological Archaeology* 15:137-159.

Allen, H. 1997 *Autonomy, Mutuality, Hierarchy: Pervasive Qualities in Aboriginal Economic Life*, Darwin: North Australia Research Unit, Australian National University. *Discussion Paper No. 6*.

Allen, H. 1998 Reinterpreting the 1969-72 Willandra Lakes archaeological survey. *Archaeology in Oceania* 33(3):207-220.

Allen, J. 1998 Notions of the Pleistocene in Greater Australia. In T. Murray (ed.) *Archaeology of Aboriginal Australia: A Reader*, pp.86-101. St. Leonards: Allen and Unwin.

Allen, J. and O'Connell, J. F. 1995 Transitions: Pleistocene to Holocene in Australia and New Guinea: Introduction. *Antiquity* 69(265):vii-ix.

Allen, M. W. 1996 Pathways to economic power in Maori chiefdoms: Ecology and warfare in prehistoric Hawkes Bay. *Research in Economic Anthropology* 17:171-225.

Altman, J. C. 1987 *Hunter-Gatherers Today: An Aboriginal Economy in North Australia*, Canberra: Australian Institute of Aboriginal Studies.

Alvard, M. S. and Kuznar, L. 2001 Deferred harvests: The transition from hunting to animal husbandry. *American Anthropologist* 103(2):295-311.

Ames, K. M 1991 The archaeology of the longue duree temporal and spatial scale in the evolution of social complexity on the southern Northwest Coast. *Antiquity* 65(249):935-945.

Ames, K. M. 1994 The Northwest Coast: Complex hunter-gatherers, ecology and social evolution. *Annual Review of Anthropology* 23:209-229.

Ames, K. M. 1995 Chiefly power and household production on the Northwest Coast. In T. D. Price and G. M. Feinman (eds) *Foundations of Social Inequality*, pp.155-187. New York: Plenum Press.

Ames, K. M. and Maschner, H. D. G. 1999 *Peoples of the Northwest Coast: Their Archaeology and Prehistory*, London: Thames and Hudson.

Anderson[-Gerfaud], P. C. 1992 Experimental cultivation, harvest and threshing of wild cereals and their relevance for interpreting the use of Epipaleolithic and Neolithic artefacts. In P. C. Anderson[-Gerfaud] (ed.) *Prehistoire de L'Agriculture: Nouvelles Approches Experimentales et Ethnographiques*, pp.179-209. Paris: Editions du CNRS. *Monographie du CRA No. 6.*

Anderson, P. C. 1998 History of harvesting and threshing techniques for cereals in the prehistoric Near East. In A. B. Damania, J. Valkourn, G. Willcox and C. O. Qualset (eds) *The Origins of Agriculture and Crop Domestication*, pp145-159. Aleppo: ICARDA.

Andresen, J. M., Byrd, B. F., Elson, M. D., McGuire, R. H., Mendoza, R. G., Staski, E. and White, J. P. 1981 The deer hunters: Star Carr reconsidered. *World Archaeology* 13(1):31-46.

Andrews, A. E. J. (ed.) 1986 *Stapylton With Major Mitchell's Australia Felix Expedition*. Hobart: Blubberhead Press.

Andrews, F. W. 1879 Notes on the Aborigines met with on the trip of the exploring party to Lake Eyre in command of Mr J. W. Lewis. In G. Taplin (comp.) *The Folklore, Manners, Customs, and Languages of the South Australian Aborigines*, pp.83-86. Adelaide: Government Printer.

Angas, G. F. 1847 *Savage Life and Scenes in Australia and New Zealand*, 2 volumes, London: Smith Elder.

An Old Hand [G. Hamilton] 1880 *Experiences of a Colonist Forty Years Ago: A Journey From Port Phillip to South Australia in 1839 and a Voyage From Port Phillip to Adelaide in 1846*, Melbourne: J. Williams.

Anonymous [A. Collie] 1834 Anecdotes and remarks relative to the Aborigines at King George's Sound. In *Perth Gazette* 2 August 1834, p.332.

Anonymous 1905 Fish weirs in Australia. *American Antiquarian* 27:226-227.

Anonymous 1907 Why did the blacks not cultivate food. *Science of Man* 9(1):21-24.

Anonymous 1938 Natives of Bentinck Island, Queensland. *Nature* 141:560.

Anonymous 1969 The spinifex Egyptians. *Mimag* 19(8):19-24.

Anonymous 1995 Example of how not to use historical sources. In N. Thieberger (ed.) *Paper and Talk: A Manual for Reconstituting Materials in Australian Indigenous languages from Historical Sources*, p.146. Canberra: Aboriginal Studies Press.

Arima, E. and Dewhurst, J. 1990 Nootkans of Vancouver Island. In W. Suttles, (ed.) *Handbook of North American Indians, Volume 7: Northwest Coast*, pp.391-411. Washington: Smithsonian Institution.

Armstrong, F. F. 1836 The manners and habits of the Aborigines of Western Australia from information collected by Mr F Armstrong, Interpreter 1836. In *Perth Gazette* 29 October 1836, pp.789-90; 5 November 1836, pp.793-4; 12 November 1836, pp.797.

Armytage, G. 1898 Letter In T. F. Bride (ed.) *Letters from Victorian Pioneers*, pp.138-141. Melbourne: Government Printer.

Arnold, J. E. 1991 Transformation of a regional economy: Sociopolitical evolution and the production of valuables in southern California. *Antiquity* 65(249):953-962.

Arnold, J. E. 1992 Complex hunter-gatherer-fishers of prehistoric California: Chiefs, specialists, and maritime adaptations of the Channel Islands. *American Antiquity* 57:60-84.

Arnold, J. E. 1996 The archaeology of Complex Hunter-Gatherers. *Journal of Archaeological Method and Theory* 3(2):77-126.

Arnold, J. E. 2001 The Chumash in world and regional perspective. In J. E. Arnold (ed.) *The Origins of a Pacific Coast Chiefdom: The Chumash of the Channel Islands*, pp.1-19. Salt Lake City: University of Utah Press.

Ashwin, A. C. 1932 From South Australia to Port Darwin with sheep and horses in 1870-71. *Proceedings of the Royal Geographical Society of Australasia, South Australian Branch* 32:47-93.

Asouti, E. and Fairbairn A. 2002, Subsistence economy in Central Anatolia during the Neolithic - the archaeobotanical evidence. In F. Gerard and L. Thissen (eds) The Neolithic of Central Anatolia, Internal Developments and External Relations During the 9th-6th Millennia cal BC, Being the Proceedings of the International CANeW Table Ronde Held in Istanbul on 23-24 November 2001. Pp.181-192. Ege Yanlari. http://www.canew.net/canew-book.htm

Attenbrow, V. 1999 Archaeological research in coastal southeastern Australia: A review. In J. Hall and I. McNiven (eds) *Australian Coastal Archaeology*, pp.195-210. Canberra: ANH Publications, Department of Archaeology and Natural History, Research School of Pacific and Asian Studies.

Attenbrow, V., Fullagar, R. and Szpak, C. 1998 Stone files and shell fish-hooks in southeastern Australia. In R. Fullagar (ed.) *A Closer Look: Recent Australian Studies of Stone Tools*, pp.128-148. Sydney: School of Archaeology, University of Sydney. *Sydney University Archaeological Methods Series 6.*

Attenbrow, V. and Steele, D. 1995 Fishing in Port Jackson, New South Wales – more than meets the eye. *Antiquity* 69(262):47-60.

Austin, R. 1851 Journal of party exploring Peron Peninsula. PR 7728.

Austin, R. 1855 *Journal of Assistant-Surveyor R. Austin Commanding an Expedition Sent by the Government to Explore the Interior of Western Australia*, Perth: Government Printer.

Australian Biological Resource Study/CSIRO 2002 *Flora of Australia: Volume 43 - Poaceae 1: Introduction and Atlas*, Melbourne: ABRS, Canberra/CSIRO.

Bailey, G. N. 1981 Concepts of resource exploitation: Continuity and discontinuity in paleoeconomy. *World Archaeology* 13(1):1-15.

Bakels, C. C. 1997 The beginnings of manuring in western Europe. *Antiquity* 71(272):442-445.

Baker, R. M. 1988 Yanyuwa canoe making. *Records of the South Australian Museum* 22(2):173- 188.

Balme, J. 1991 The antiquity of grinding stones in semi-arid western New South Wales. *Australian Archaeology* 32:3-9.

Balme, J. and Beck, W. 1993 Archaeology and feminism – views on the origins of the division of labour. In H. du Cros and L. Smith (eds) *Women in Archaeology: A Feminist Critique*, pp.61-74. Canberra: Department of Prehistory, Research School of Pacific Studies, Australian National University.

Balter, M. 1998 Why settle down? The mystery of communities. *Science* 282(5393):1442-1445.

Bancroft, J. 1879 *Pituri and Tobacco*, Brisbane: Government Printer.

Banfield, E. J. 1908 *Confessions of a Beachcomber*, London: T. Fisher Unwin.

Banfield, E. J. 1909 Blacks as fishermen: Expedients, devices, stratagems. *Queensland Geographical Journal* 24(10):43-62.

Banfield, E. J. 1918 *Tropic Days*, London: T. Fisher Unwin.

Banks, J. 1768-71 *The Endeavour Journal of Joseph Banks 1768 – 1771*, J. Beaglehole (ed.) 1962, 2 volumes, Sydney: Angus and Robertson.

Barabasi, A-L 2002 *Linked: The New Science of Networks*, Cambridge (Mass.):Perseus Publishing.

Barabasi, A-L and Albert,R. 1999 Emergence of scaling in random networks. *Science* 286:509-512.

Barber, E. J. W. 1991 *Prehistoric Textiles: The Development of Cloth in the Neolithic and Bronze Ages*, Princeton: Princeton University Press.

Barclay, H. V. 1916 Report on exploration of a proportion of Central Australia by the Barclay Macpherson Expedition 1904-5. *Proceedings of the Royal Geographical Society of Australasia, South Australian Branch* 16:106-130.

Barham, A. J. 2000 Late Holocene maritime societies in the Torres Strait Islands, northern Australia. In S. O'Connor and P. Veth (eds) *East of Wallace's Line: Studies of Past and Present Maritime Cultures of the Indo-Pacific Region*, pp.223-314. Rotterdam: A. A. Balkema. *Modern Quaternary Research in Southeast Asia Volume 16.*

Barham, A. J. and Harris, D. R. 1985 Relict field systems in the Torres Strait Region. In I. S. Farrington (ed.) *Prehistoric Intensive Agriculture in the Tropics: Part 1*, pp.247-283. Oxford: British Archaeological Reports. *BAR International Series No. 232.*

Barker, B. 1991 Nara Inlet 1: Coastal resource use and the Holocene marine transgression in the Whitsunday Islands, Central Queensland. *Archaeology in Oceania* 26:102-109.

Barker, B. 1996 Maritime hunter-gatherers on the tropical coast: A social model for change. In S. Ulm, I. Lilley and A. Ross (eds) *Australian Archaeology 95: Proceedings of the Australian Archaeological Association Annual Conference*, pp.31-43. St Lucia: Anthropology Museum, University of Queensland.

Barker, B. 1999 Coastal occupation in the Holocene: Environment, resource use and resource continuity. In J. Hall and I. McNiven (eds) *Australian Coastal Archaeology*, pp.119-127. Canberra: ANH Publications, Department of Archaeology and Natural History, Research School of Pacific and Asian Studies.

Barker, B. 2004 *The Sea People: Late Holocene Maritime Specialisation in the Whitsunday Islands, Central Queensland*, Canberra: Pandanus Books. *Terra Australis 20.*

Barnard, A. and Woodburn, J. 1988 Property, power and ideology in hunting and gathering societies: An introduction. In T. Ingold, D. Riches and J. Woodburn (eds) *Hunters and Gatherers 2: Property, Power and Ideology*, pp.4-31. New York: Berg.

Barnes, G. L. 1993 *China, Korea, Japan: The Rise of Civilization in East Asia*, London: Thames and Hudson.

Barnett, T. 1999 *The Emergence of Food Production in Ethiopia*, Oxford: BAR Publishing. *Cambridge Mono-graphs in African Archaeology No. 45: BAR Interna-tional Series No. 763.*

Barnett, W. K. and Hoopes, J. W. 1995 The shape of early pottery studies. In W. K. Barnett and J. W. Hoopes (eds) *The Emergence of Pottery: Technology and Innovation in Ancient Society*, pp.1-7. Washington: Smithsonian Institution Press.

Barrett, J. A. 1997 Fish trade in Norse Orkney and Caithness: A zooarchaeological approach. *Antiquity* 71(273):616-638.

Barrows, D. P. 1900 *The Ethno-botany of the Coahuilla Indians of Southern California*, Chicago: University of Chicago.

Bar-Yosef, O. 1983 The Natufian in the southern Levant. In T. C. Young, P. Mortensen and P. E. L. Smith (eds) *The Hilly Flanks and Beyond: Essays on the Prehistory of Southwestern Asia*, pp.11-42. Chicago: Oriental Institute of the University of Chicago. *Studies in Ancient Oriental Civilization No. 36.*

Bar-Yosef, O. 1998a The Natufian culture in the Levant, threshold to the origins of agriculture. *Evolutionary Anthropology* 6:159-177.

Bar-Yosef, O. 1998b Introduction. In T. R. Rocek and O. Bar-Yosef (eds) *Seasonality and Sedentism: Archaeological Perspectives From Old and New World Sites*, pp.1-8. Cambridge (Mass.): Peabody Museum of Archaeology and Ethnology.

Bar-Yosef, O. 2002 The Upper Paleolithic revolution. *Annual Review of Anthropology* 31:363-393.

Bar-Yosef, O. 2003 The role of the Younger Dryas in the origin of agriculture in west Asia. In Y. Yasuda (ed.) *The Origins of Pottery and Agriculture*, pp.39-54. New Delhi: Roli Books.

Bar-Yosef, O. and Belfer-Cohen, A. 1989 The origins of sedentism and farming communities in the Levant. *Journal of World Prehistory* 3(4):447-498.

Bar-Yosef, O. and Belfer-Cohen, A. 1992 From foraging to farming in the Mediterranean Levant. In A. B. Gebauer and T. D. Price (eds) *Transitions to Agriculture in Prehistory*, pp. 21-48. Madison: Prehistory Press. *Monographs in World Archaeology No. 4.*

Bar-Yosef, O., Gopher, A., Tchernov, E. and Kislev, M. E. 1991 Netiv Hagdud: An early Neolithic village site in the Jordan valley. *Journal of Field Archaeology* 18(4):405-424.

Bar-Yosef, O. and Kislev, M. E. 1989 Early farming communities in the Jordan Valley. In D. R. Harris and G. C. Hillman (eds) *Foraging and Farming: The Evolution of Plant Exploitation*, pp.632-642. London: Unwin Hyman.

Bar-Yosef, O. and Meadows, R. H. 1995 The origins of agriculture in the Near East. In T. D. Price and A. Gebauer (eds) *Last Hunters – First Farmers: New Perspectives on the Prehistoric Transition to Agriculture*, pp.39-94. Santa Fe: School of American Research Press.

Basedow, H. 1907 Anthropological notes on the western central tribes of the Northern Territory of South Australia. *Transactions and Proceedings of the Royal Society of South Australia* 31: 1-62.

Basedow, H. 1914 Notes on the natives of Bathurst Island. *Journal of the Royal Anthropological Institute of Great Britain and Ireland* 43:291-323.

Basedow, H. 1918 Narrative of an expedition of exploration in north-western Australia. *Proceedings of the Royal Geographical Society of Australasia, South Australian Branch* 18:105-295.

Basedow, H. 1925 *The Australian Aborigines*, Adelaide: F.W. Preece and Sons.

Bassett-Smith, P. W. 1894 The Aborigines of north-west Australia. *Journal of the Anthropological Institute of Great Britain and Ireland* 23:324-331.

Basso, K. H. 1983 Western Apache. In A. Ortiz (ed.) (ed.) *Handbook of North American Indians, Volume 10: Southwest*, pp.462-468. Washington: Smithsonian Institution.

Bates, D. M. n.d.a Notebook 24. Papers of Daisy Bates 1900 – 1940 [manuscript]. National Library of Australia MS 365.

Bates, D. M. n.d.b Miscellaneous Vol. 3 – Notebook 23A. Papers of Daisy Bates 1900 – 1940 [manuscript]. National Library of Australia MS 365.

Bates, D. M. n.d.c Section XII 2D,12. Native Vocabulary compiled from vocabularies submitted by T. Carter *et al.* Net (fishing), Question 8. Papers of Daisy Bates 1900 – 1940 [manuscript]. National Library of Australia MS 365.

Bates, D. M. n.d.d Section IX 3. Weapons etc. Papers of Daisy Bates 1900 – 1940 [manuscript]. National Library of Australia MS 365.

Bates, D. M. n.d.e Section VIII Pt. 2d. Food: Additional notes from various districts of WA. Hunting and fishing: Champion Bay. Papers of Daisy Bates 1900 – 1940 [manuscript]. National Library of Australia MS 365.

Bates, D. M. n.d.f Origin of the Australian Race. Section I. Papers of Daisy Bates 1900 – 1940 [manuscript]. National Library of Australia MS 365.

Bates, D. M. n.d.g Native Beliefs: Cornally. Section VI, Pt. 3a. Papers of Daisy Bates 1900 – 1940 [manuscript]. National Library of Australia MS 365.

Bates, D. M. n.d.h Religion, Superstitions. Section VI Pt. 1. Papers of Daisy Bates 1900 – 1940 [manuscript] National Library of Australia MS 365.

Bates, D. M. n.d.i Songs and dance – Part 1. Section XI, Pt.1a:76a. Papers of Daisy Bates 1900 – 1940 [manuscript]. National Library of Australia MS 365.

Bates, D. M. n.d.j Genealogies. Section III, Pt. 2e. Papers of Daisy Bates 1900 – 1940 [manuscript]. National Library of Australia MS 365.

Bates, D. M. 1938 *The Passing of the Aborigines*, London: John Murray.

Bates, D. M. 1985 *The Native Tribes of Western Australia*, I. White (ed.), Canberra: National Library of Australia.

Baudin, N. 1974 *The Journal of Post Captain Nicolas Baudin*, C. Cornell (trans.), Adelaide: Library Board of South Australia.

Bean, L. J. and Saubel, K. S. 1972 *Temalpakh: Cahuilla Indian Knowledge and Usage of Plants*, Marongo Indian Reservation: Malki Museum Press.

Beard, J. S. 1984 Vegetation: Southwest Botanical Province and South-western Inter-Zone, Western Australia Showing Extent of Kwongan [Map]. Perth: Forests Department, Western Australia.

Beaton, J. M. 1991 Extensification and intensification in central California prehistory. *Antiquity* 65 (249):946-952

Beaton, J. M. 1995 The transition on the coastal fringe of Greater Australia. *Antiquity* 69(265):798- 806.

Becker, L. 1979 *Ludwig Becker: Artist and Naturalist With the Burke and Wills Expedition*, M. Tipping (ed.), Melbourne: Library Council of Victoria.

Beckler, H. 1993 *A Journey to Coopers Creek*, S. Jeffries and M. Kirtesz (trans.), Carlton: Melbourne University Press.

Bedford, C. T. 1887 Reminiscences of a surveying trip from Boulia to the South Australian border. *Proceedings and Transactions of the Queensland Branch of the Royal Geographical Society of Australasia* 2(2):99-113.

Belfer-Cohen, A. and Bar-Yosef, O. 2000 Early sedentism in the Near East. In I. Kuijt (ed.) *Life in Neolithic Farming Communities: Social Organization, Identity and Differentiation*, pp.19-37. New York: Kluwer Academic and Plenum Publishers.

Bellwood, P. 1995 Early agriculture, language history and the archaeological record in China and southeast Asia. In C. T. Yeung and W. B. Li (eds) *Archaeology in Southeast Asia*, pp.11-22. Hong Kong: University Museum and Art Gallery.

Bellwood, P. 2005 *First Farmers: The Origins of Agricultural Societies*, Malden: Blackwell Publishing.

Bender, B. 1975 *Farming in Prehistory: From Hunter-Gatherer to Food Producers*, London: John Baker.

Bender, B. 1978 Gatherer-hunter to farmer: A social perspective. *World Archaeology* 10(1):204- 222.

Bender, B. 1981 Gatherer-hunter intensification. In A. Sheridan and G. Bailey (eds) *Economic Archaeology: Towards an Integration of Ecological and Social Approaches*, pp.149-157. Oxford: British Archaeological Reports. *BAR International Series No. 96.*

Bennett, G. 1834 *Wanderings in New South Wales, Batavia, Pedir Coast ... 1832-34*, London: Richard Bentley.

Bennett, J. W. O. 1879 Vocabulary of the Woolner District dialect, Adelaide River, Northern Territory. In J. D. Wood (comp.) *Native Tribes of South Australia*, pp.309-316. Adelaide: E. S. Wigg.

Bennett, M. M. 1927a The Dalleburra tribe of Queensland. *Journal of the Royal Anthropological Institute of Great Britain and Ireland* 57:399-415.

Bennett, M. M. 1927b *Christison of Lammermoor*, London: Alston Rivers Ltd.

Bennett, M. M. 1930 *The Australian Aboriginal as a Human Being*, London: Alston Rivers.

Bennett, R. 1893 The Aborigines of Australia: Their Habits and Customs. A Paper in Three Parts. National Library of Australia MS 115.

Bennett, S. 1867 *The History of Australian Discovery and Colonisation*, Sydney: Hanson and Bennett.

Benz, B. F. and Long, A. 2000 Prehistoric maize evolution in Teohuacan Valley. *Current Anthropology* 41(3):459-465.

Berndt, C. H. 1974 Digging sticks and spears, or, the two-sex model. In F. Gale (ed.) *Woman's Role in Aboriginal Society*, pp.64-84. Canberra: Australian Institute of Aboriginal Studies.

Berndt, R. M. 1941 The bark-canoe of the Lower Murray, South Australia. *Mankind* 3(1):17-28.

Berndt, R. M. 1951 Ceremonial exchange in western Arnhem Land. *Southwestern Journal of Anthropology*, 7(2):156-176.

Berndt, R. M. 1953 A day in the life of a Dieri man before alien contact. *Anthropos* 48:171-201.

Berndt, R. M. and Berndt, C. H. 1947 Discovery of pottery in north-east Arnhem Land. *Journal of the Royal Anthropological Institute of Great Britain and Ireland* 77:133-138.

Berndt, R. M. and Berndt, C. H. 1954 *Arnhem Land - Its History and People*, Melbourne: Cheshire.

Berndt, R. M. and Berndt, C. H. 1985 *The World of the First Australians*, Adelaide: Rigby.

Berndt, R. M. and Berndt, C. H. 1988 *The World of the First Australians*, Canberra: Aboriginal Studies Press.

Berndt, R. M., Berndt, C. H. and Stanton, J. E. 1993 *A World That Was: The Yaraldi of the Murray River and the Lakes, South Australia*, Carlton: Melbourne University Press.

Berndt, R. M. and Vogelsang, T. 1939 Notes on the Dieri tribe of South Australia. *Transactions of the Royal Society of South Australia* 63:167-171.

Bettencourt, L. M. A., Lobo, J., Helbing, D., Kuhnert, C. and West, G. B. 2007 Growth, innovation, scaling, and the pace of life in cities. *Proceedings of the National Academy of Sciences of the United States of America* 104(17):7301-7306.

Bettinger, R. L. 1978 Alternative adaption strategies in the prehistoric Great Basin. *Journal of Anthropological Research* 34(1):27-46.

Beveridge, P. 1889 *The Aborigines of Victoria and Riverina*, Melbourne: M. L. Hutchinson.

Bicknell, A. C. 1895 *Travel and Adventure in Northern Queensland*, London: Longmans, Green and Co.

Binford, L. R. 1968 Post-Pleistocene adaptations. In S. R. and L. R. Binford (eds) *New Perspectives in Archaeology*, pp.313-341. Chicago: Aldine.

Binford, L. R. 1980 Willow smoke and dog's tails: Hunter-gatherer settlement systems and archaeological site formation. *American Antiquity* 45(1):4-20.

Binford, L. R. 1983 *In Pursuit of the Past*, London: Thames and Hudson.

Bird, C. F. M. and Frankel, D. 1991 Chronology and explanation in western Victoria and south-east South Australia. *Archaeology in Oceania* 26(1):1-16.

Birdsell, J. B. 1953 Some environmental and cultural factors influencing the structure of Australian Aboriginal populations. *The American Naturalist* 87:171-207.

Birtles, F. 1937 Among the Aborigines. *The Australian Museum Magazine* 6(5):152-153.

Black, E. C. 1947 The canoes and canoe trees of Australia. *Mankind* 3(12):351-361.

Black, L. 1943 Here lies ... who? *Wild Life* 5(12):437-438.

Black, L. 1950 *Stone Arrangements*, Perth: Paterson and Brokensha.

Black, R. L. 1949 Notes on the material culture of the Aborigines of the Darling River Valley, western New South Wales. *Mankind* 4(3):102-107.

Blackman, M. B. 1981 *Windows on the Past: The photographic Ethnohistory of the Northern and Kaigani Haida*, Ottawa: National Museum of Canada. *Canadian Ethnology Service Paper No. 74.*

Blackman, M. B. 1990 Traditional Haida culture. In W. Suttles, (ed.) *Handbook of North American Indians, Volume 7: Northwest Coast*, pp.240-260. Washington: Smithsonian Institution.

Blake, B. J. 2003 *The Warrnambool Language: A Consolidated Account of Aboriginal Language of the Warrnambool Area of the Western District of Victoria Based on Nineteenth-Century Sources*, Canberra: Pacific Linguistics, Research School of Pacific and Asian Studies, Australian National University.

Blake, S. T. 1942 *Cyperus rotundas* (Nut grass) and its allies in Australia. *Papers: University of Queensland, Department of Biology* 2(2):1-14.

Blevins, J. 1998 A Dutch Influence on Nhanda? Wanyjidaga innga! *Australian Aboriginal Studies* 1998/1:43-46.

Blevins, J. 2001a *Nhanda: An Aboriginal Language of Western Australia*, Honolulu: University of Hawai'i Press. *Oceanic Linguistics Special Publication No. 30.*

Blevins, J. 2001b A Dutch influence on Nhanda? Malya kanangga! *Australian Aboriginal Studies* 2001/1:74-75.

Blevins, J. 2006 Some notes on Nhanda as spoken by Lucy Ryder (1919 – 2003). *Oceanic Linguistics* 45(1):242-246.

Blount, C. 1937 *Memorandums by James Martin*, Cambridge: Rampant Lion Press.

Blumler, M. A. 1992 Independent inventionism and recent genetic evidence on plant domestication. *Economic Botany* 46(1):98-111.

Blumler, M. A. 1996 Ecology, evolutionary theory and agricultural origins. In D. R. Harris (ed.) *The Origins and Spread of Agriculture and Pastoralism in Eurasia*, pp.25-50. London: UCL Press.

Blumler, M. A. and Byrne, R. 1991 The ecological genetics of domestication and the origins of agriculture. *Current Anthropology* 32(1):23-54.

Blust, R. 1996 Beyond the Austronesian homeland: The Austric hypothesis and its implications for archaeology. In W. H. Goodenough (ed.) *Prehistoric Settlement in the Pacific*, pp.117-140. Philadelphia: American Philosophical Society.

Bogucki, P. and Grygiel, R. 1993 The first farmers of Central Europe. *Journal of Field Archaeology* 20(4):399-426.

Bokonyi, S. 1976 Development of stock rearing in the Near East. *Nature* 264(5581):19-23.

Boldrewood, R. [T. A. Browne] 1884 *Old Melbourne Memories*. Melbourne: George Robertson.

Bollong, C. A., Sampson, C. G. and Smith, A. B. 1997 Khoikhoi and Bushmen pottery in the Cape Colony: Ethnohistory and later Stone Age ceramics of the South African interior. *Journal of Anthropological Archaeology* 16:269-299.

Bonney, F. 1884 On some customs of the aborigines of the River Darling. *Journal of the Anthropological Institute of Great Britain and Ireland* 13:122-137.

Bonney, N. 1995 *Native Plant Uses of Southern South Australia*, Naracoorte:SEBP.

Border, A. 1999 Aboriginal settlement of offshore islands in the Southern Great Barrier Reef Province, central Queensland. In J. Hall and I. McNiven (eds) *Australian Coastal Archaeology*, pp.129-139. Canberra: ANH Publications, Department of Archaeology and Natural History, Research School of Pacific and Asian Studies.

Boserup, E. 1965 *The Conditions of Agricultural Growth*, London: George Allen.

Bowdler, S. 1976 Hook, line and dilly bag: An interpretation of an Australian coastal shell midden. *Mankind* 10(4):248-258.

Bowdler, S. 1990 Before Dirk Hartog: Prehistoric archaeological research in Shark Bay, Western Australia. *Australian Archaeology* 30:46-57.

Bowdler, S. 1995 Offshore islands and maritime explorations in Australian prehistory. *Antiquity* 69(266):944-58.

Bowler, J., Johnson, H., Olley, J. M., Prescott, J. R., Roberts, R. G., Shawcross, W. and Spooner, N. A., 2003: New ages for human occupation and climatic change at Lake Mungo, Australia. *Nature* 421:837-840.

Bowen, G. 1998 Towards a generic technique for dating stone fish traps and weirs. *Australian Archaeology* 47:39-43.

Boyd, A. J. 1896 Narrative of Capt. C. Pennefather's exploration of the Coen, Archer and Batavia Rivers, and the islands in the western coast of the Gulf of Carpentaria in 1880. *Proceedings and Transactions of the Queensland Branch on the Royal Geographical Society of Australasia* 11:46-61.

Boyd, R. T. 1990 Demographic history. In W. Suttles, (ed.) *Handbook of North American Indians, Volume 7: Northwest Coast*, pp.135-148. Washington: Smithsonian Institution.

Brahe, W. 1861-62 W. Brahe's Report. In Report of the Commissioners Appointed to Enquire Into and Report Upon the Circumstances Connected With the Suffering and Death of Robert O'Hara Burke and William John Wills, the Victorian Explorers, pp.88-90 (Appendix I), Melbourne: Government Printers.

Braidwood, R. J. and Howe, B. 1960 *Prehistoric Investigations in Iraqi Kurdistan*, Chicago: University of Chicago Press. *Studies in Ancient Oriental Civilization No. 31.*

Braudel, F. 1980 *On History*, S. Matthews (trans.), London: Weidenfeld and Nicolson.

Bridges, P. S. 1989 Changes in activities with the shift to agriculture in the southeastern United States. *Current Anthropology* 30:385-394.

Brierly, O. W. 1849 Information obtained from Barbara Thompson, 16 October – 12 November 1849. In D. R. Moore 1979 *Islanders and Aborigines at Cape York*, pp.142-212. Canberra: Australian Institute of Aboriginal Studies.

Brierly, O. W. 1862a Canoes in Australia. *The Athenaeum* 1792:304.

Brierly, O. W. 1862b Canoes in Australia. *The Athenaeum* 1795:396-397.

Briggs, J. C. 1987 *Biogeography and Plate Tectonics: Developments in Palaeontology and Stratigraphy*, Amsterdam: Elsevier.

Brock, D. G. 1975 *To the Desert With Sturt: A Diary of the 1844 Expedition*, K. Peake-Jones (ed.), Adelaide: Royal

Geographical Society of Australasia, South Australian Branch.

Brockman, F. S. 1902 *Report of Exploration of North-West Kimberley 1901*, Perth: Government Printer.

Bronson, B. 1977 The earliest farming: Demography as cause and consequence. In C. A. Reed (ed.) *The Origins of Agriculture*, pp.23-48. The Hague: Mouton.

Brookfield, H. 2002 *Exploring Agrodiversity*, Dehra Dun: Bishen Singh Mahendra Pal Singh.

Brown, A. R. 1916 Australian rafts. *Man* 16:8-9.

Brown, J. A. 1985 Long-term trends to sedentism and the emergence of complexity in the American Midwest. In T. D. Price and J. A. Brown (eds) *Prehistoric Hunter-Gatherers: The Emergence of Cultural Complexity*, pp.201-231. Orlando: Academic Press.

Brown, Mrs 1851 Narrative of a journey from York to Champion Bay in the colony of Western Australia, during the months of May and June 1851. *Inquirer and Commercial News*, 10 September 1851, pp.3-4.

Brown, P. L. (ed.) 1986 *Memoirs Recorded at Geelong, Victoria, Australia by Captain Foster Fyans*. Geelong: Geelong Advertiser.

Browne, J. 1856 The Aborigines of Australia. *Nautical Magazine and Naval Chronicle for 1856*, pp.485-493,534-543.

Bruce, R. 1892 *Reminiscences of an Old Squatter*, Adelaide: W. K. Thomas.

Bryant, C. E. 1943 Mallee days. *Wild Life* 5(1):5-8.

Buckley, R. 1981 Alien plants in central Australia. *Botanical Journal of the Linnean Society* 82(4):369-379.

Builth, H. 2000a The connection between the Gunditjmara Aboriginal people and their environment: The case for Complex Hunter-Gatherers in Australia. *People and Physical Environment Research: The Person-Environment Cultural Heritage Journal of Australia andNew Zealand*, 55-56:1-18.

Builth, H. 2000b The connection between the Gunditjmara Aboriginal people and their environment: The case for Complex Hunter-Gatherers in Australia. In G. Moore, J. Hunt and L. Trevillion (eds) *Environment Behaviour Research on the Pacific Rim*, pp.197-210. Sydney: Faculty of Architecture, University of Sydney.

Builth, H. 2002 The archaeology and socioeconomy of the Gunditjmara: A landscape analysis from southwest Victoria, Australia. Unpublished PhD thesis, Department of Archaeology, Flinders University of South Australia.

Bulbeck, D. and Rowley, B. 2001 Macassans and their pots in northern Australia. In C. Frederickson and I. Walters (eds) *Altered States: Material Culture Transformations in the Arafura Region*, pp.55-74. Darwin: Northern Territory University Press.

Bulmer, J. 1994 *Victorian Aborigines: John Bulmer's Recollections 1855 – 1908*, A. Campbell (comp.) and R Vanderwal (ed.), Melbourne: Museum of Victoria. *Occasional Papers in Anthropology and History No. 1.*

Bunbury, H. W. 1930 *Early Days in Western Australia*, Oxford: Oxford University Press.

Bunce, D. 1857 *Australasiatic Reminiscences of Twenty-Three Years Wanderings in Tasmania and the Australias*, Melbourne: J. T. Hendy.

Burch, E. S. and Ellanna, L. J. 1994 Editorial In E. S. Burch and L. J. Ellanna (eds) *Key Issues in Hunter-Gatherer Research*, pp.147-149. Oxford: Berg.

Burchill, E. 1960 *Innamincka* Melbourne: Hodder and Stoughton.

Bureau of Flora and Fauna 1986 *Flora of Australia: Volume 46 - Iridaceae to Dioscoreaceae*, Canberra: Australia Government Publishing Service.

Burges, W. 1851a W. Burges to Acting Colonial Secretary, 9 June 1851. WASA: Colonial Secretary's Records 218/291-3.

Burges, W. 1851b Letter from W Burges, Government resident, Champion Bay. *Inquirer* 17 September 1851, p.2.

Burkill, I. H. 1958-61 The organography and the evolution of Dioscoreae,- the family of yams. *Journal of the Linnean Society,* 56:319-412.

Burkill, I. H. 1966 *A Dictionary of the Economic Products of the Malay Peninsula*, Kuala Lumpur: Ministry of Agriculture and Co-operatives.

Burnham, H. B. 1965 Catal Huyuk – The textiles and twined fabrics. *Anatolian Studies* 15:169-174.

Burton, A. (comp.) and Henn, P. U. (ed.) 1954 *Wollaston's Albany Journals (1848-56): Being Volume 2 of the Journals and Diaries (1841-1856) of Revd. John Ramsden Wollaston*, Perth: Patterson and Brokensha.

Burton, M. L. and White, R. R. 1984 Sexual division of labor in agriculture. *American Anthropologist* 86(3):568-583.

Butlin, N. G. 1983 *Our Original Aggression*, Sydney: George, Allen and Unwin.

Butlin, N. G. 1993 *Economics and the Dreamtime: A Hypothetical History*, Cambridge: Cambridge University Press.

Byrne, R. 1987 Climatic change and the origins of agriculture. In L. Manzanilla (ed.) *Studies in the Neolithic and Urban Revolutions: The V. Gordon Childe Colloquiums, Mexico 1986*, pp. 21-34. Oxford: British Archaeological Reports. *International Series No 349.*

Byrne, R. and McAndrews, J. H. 1975 Pre-Columbian purslane (*Portulaca oleracea* L) in the New World. *Nature* 253:726-727.

Callen, E. O. 1967 First New World cereal. *American Antiquity* 32:535-538.

Cameron, A. L. P. 1885 Notes on some tribes of New South Wales. *Journal of the Anthropological Institute of Great Britain and Ireland* 14:344-370.

Campbell, A. H. 1965 Elementary food production by the Australian Aborigines. *Mankind* 6(5): 206-211.

Campbell, J. B. 1982 Automatic sea-food retrieval systems: Evidence from Hinchinbrook Island and its implications. In S. Bowdler (ed.) *Coastal Archaeology in Eastern Australia*, pp.96- 107. Canberra: Department of Prehistory, Research School of Pacific Studies, Australian National University.

Campbell, V. 1973 Midden archaeology in northern New

South Wales. *Journal of the Anthropological Society of South Australia* 11(2):7-9.
Campbell, V. 1978 Two fish traps on the mid-north coast of New South Wales. In I. McBryde (ed.) *Records of Times Past: Ethnohistorical Essays on the Culture and Ecology of the New England Tribes*, pp.122-134. Canberra: Australian Institute of Aboriginal Studies.
Campbell, W. D. 1914 Description of a rock shelter with Aboriginal markings and an indurated cast of a footprint at the Greenough River, and a cave dwelling at Narandagy Springs, Lockier River. *Journal of the Natural History and Science Society of Western Australia* 5:9-11.
Campbell, W. D. 1921 A description of certain phallic articles of the Australian Aborigines. *Man* 21:145-146.
Cane, S. 1984 Desert camps: A case study of stone artefacts and Aboriginal behaviour in the Western Desert. Unpublished PhD thesis, Department of Prehistory, Australian National University, Canberra.
Cane, S. 1989 Australian Aboriginal seed grinding and its archaeological record: A case study from the Western Desert. In D. R. Harris and G. C. Hillman (eds) *Foraging and Farming: The Evolution of Plant Exploitation*, pp.99-119. London: Unwin Hyman.
Carnegie, D. W. 1898 *Spinifex and Sand*, London: C. Arthur Pearson.
Carr, C. 1977 Why didn't the American Indians domesticate sheep? In C. A. Reed (ed.) *The Origins of Agriculture*, pp.637-691. The Hague: Mouton.
Carstensz, J. 1623 Journal kept by Jan Carstensz on his voyage to New Guinea, 1623. In J. Heeres (comp.) 1899 *The Part Borne By the Dutch in the Discovery of Australia 1606 – 1765*, pp.22-44. London: Luzac and Co.
Carter, M. 2001 New evidence for the earliest human occupation in Torres Strait, northeastern Australia. *Australian Archaeology* 52:51-52.
Cary, J. J. 1904 Canoes of Geelong Aborigines. *The Geelong Naturalist* (2nd series) 1:32-37.
Cashdan, E. 1989 Hunters and gatherers: Economic behaviour in bands. In S. Plattner (ed.) *Economic Anthropology*, pp.21-48. Stanford: Stanford University Press.
Cavalli-Sforza, L. L. 1996 The spread of agriculture and nomadic pastoralism: Insights from genetics, linguistics and archaeology. In D. R. Harris (ed.) *The Origins and Spread of Agriculture and Pastoralism in Eurasia*, pp.51-69. London: UCL Press.
Cauvin, J. 2002 *The Birth of the Gods and the Origins of Agriculture*, T. Watkins (trans.). Cambridge: Cambridge University Press.
Chang, K-C. 1986 *The Archaeology of Ancient China*, New Haven: Yale University Press.
Chase, A. K. 1989 Domestication and domiculture in northern Australia: A social perspective. In D. R. Harris and G. C. Hillman (eds) *Foraging and Farming: The Evolution of Plant Exploitation*, pp.42-54. London: Unwin Hyman.
Chauncy, P. 1878 Notes and anecdotes of the Aborigines of Australia. In R. B. Smyth *The Aborigines of Victoria, Volume 2*, pp.221-284. 2 volumes, Melbourne: Government Printer.
Chikwendu, V. E. and Okezie, C. E. A. 1989 Factors responsible for the enoblement of African yams: Inferences from experiments in yam domestication. In D. R. Harris and G. C. Hillman (eds) *Foraging and Farming: The Evolution of Plant Exploitation*, pp.344-357. London: Unwin Hyman.
Childe, V. G. 1934 *New Light on the Most Ancient East*, London: Kegan Paul, Trench, Trubner and Co.
Childe, V. G. 1937 *Man Makes Himself*, London: Watts and Co.
Claassen, C. 1997 Changing venue: Women's lives in prehistoric North America. In C. Claassen and R. A. Joyce (eds) *Women in Prehistory: North America and Mesoamerica*, pp.65-87. Philadelphia: University of Pennsylvania Press.
Clark, D. and Geering, K. 1986 Aboriginal stone houses on the 'Allambie' property, southwestern Victoria. Unpublished report, Aboriginal Affairs Victoria.
Clark, I. D. 1987 The spatial organisation of the Chap Wurrung – a preliminary analysis. A discussion of the organisation, tribal nomenclature, tribal and clan boundaries. In *Australia Felix: The Chap Wurrung and Major Mitchell*, pp.1-36. Dunkeld and District Historical Museum.
Clark, I. D. 1998 *That's My Country Belonging to Me: Aboriginal Land Tenure and Dispossession in Nineteenth Century Western Victoria*, Beaconsfield: Heritage Matters.
Clark, I. D 1989 A guerrilla base camp in the Eumeralla Aboriginal War in S.W. Victoria (1840s). Unpublished report, Victorian Tourism Commission.
Clark, I. D. (ed.) 1990 A report on an expedition to the Aboriginal tribes of the western interior during the months of March, April, May, June, July and August, 1841 by order of His Honor Charles Joseph Latrobe Esq., Superintendent of Port Phillip. pp.101-121. In "In quest of the tribes: G. A Robinson's unabridged report of his 1841 expedition among the western Victorian Aboriginal tribes; Kenyon's 'condensation' reconsidered." *Memoirs of the Museum of Victoria, Anthropology and History* 1(1):97-130.
Clark, I. D. (ed.) 1998-2000 *The Journals of George Augustus Robinson, Chief Protector, Port Phillip Aboriginal Protectorate*, 6 vols. Melbourne: Heritage Matters.
Clark, I. D. (ed.) 2000 *The Papers of George Augustus Robinson, Chief Protector, Port Phillip Protectorate. Volume 2: Aboriginal Vocabularies – South East Australia*, Ballarat: Heritage Matters.
Clark, I. D. (ed.) 2001 *The Papers of George Augustus Robinson, Chief Protector, Port Phillip Protectorate. Volume 3: Miscellanea*, Ballarat: Heritage Matters.
Clark, I. D. and Heydon, T. 2002 *Dictionary of Aboriginal Placenames of Southwest Victoria*. Melbourne: Victorian Aboriginal Corporation for Languages.
Clark, J. E. and Gosser, D. 1995 Reinventing Mesoamerica's first pottery. In W. K. Barnett and J. W. Hoopes (eds) *The Emergence of Pottery: Technology and Innovation in*

*Ancient Society*, pp.209-221. Washington: Smithsonian Institution Press.

Clarke, A. 1994 Romancing the Stones: The cultural construction of an archaeological landscape in the Western District of Victoria. *Archaeology in Oceania* 29:1-15.

Clarke, A. 2000 The 'Moormans Trowsers': Macassan and Aboriginal interactions. In S. O'Connor and P. Veth (eds) *East of Wallace's Line: Studies of Past and Present Maritime Cultures o the Indo-Pacific Region*, pp.315-335. Rotterdam: A. A. Balkema. *Modern Quaternary Research in Southeast Asia Volume 16.*

Clarke, P. A. 1998 Early Aboriginal food plants in southern South Australia. *Proceedings of the Nutrition Society of Australia* 22:16-20.

Clarke, P. A. 2002 Early Aboriginal fishing technology in the lower Murray, South Australia. *Records of the South Australian Museum* 35(2):147-167.

Clarke, P. A. 2003 *Where the Ancestors Walked*, Crows Nest: Allen and Unwin.

Cleland, J. B. 1966 The ecology of the Aboriginal in south and central Australia. In B. C. Cotton (ed.) *Aboriginal Man in South and Central Australia: Part 1*, pp.111-158. Adelaide: Government Printer.

Cleland, J. B. and Tindale, N. B. 1936 The natives of South Australia. In C. Fenner (ed.) *The Centenary History of South Australia*, pp.16-29. Adelaide: Royal Geographical Society of Australasia, South Australian Branch.

Clemens, H. 1979 Nomad's space: An investigation of the patterning of temporarily occupied settlements and the movements and groupings of the occupants. Unpublished B.A (Hons.) thesis, University of Sydney, Sydney.

Clement, E. 1903 *Ethnographical Notes on the Western-Australian Aborigines*, Leyden: E. J. Brill.

Close, A. E. 1995 Few and far between: Early ceramics in North Africa. In W. K. Barnett and J. W. Hoopes (eds) *The Emergence of Pottery: Technology and Innovation in Ancient Society*, pp.23-37. Washington: Smithsonian Institution Press.

Clutton-Brock, J. 1989 Introduction to domestication. In J. Clutton-Brock (ed.) *The Walking Larder: Patterns of Domestication, Pastoralism and Predation*, pp. 7-9. London: Unwin Hyman.

Clutton-Brock, J. 1999 *A Natural History of Domesticated Mammals*, Cambridge: Cambridge University Press.

Cobb, C. R. 1996 Specialization, exchange, and power in small-scale societies and chiefdoms. *Research in Economic Anthropology* 17:251-294.

Cobb, K. M. 1932 The Paroo River as a collecting ground. *Mankind* 1(4):85-86.

Cobb, K. M. 1933 Burren Junction as a collecting ground. *Mankind* 1(7):163-165.

Codere, H. 1990 Kwakiutl: Traditional culture. In W. Suttles, (ed.) *Handbook of North American Indians, Volume 7: Northwest Coast*, pp.359-377. Washington: Smithsonian Institution.

Coghlan, J. 1898a Vocabulary of Aboriginal dialects of Queensland. *Science of Man* 1(2):48.

Coghlan, J. 1898b Foods of North West Aborigines. *Science of Man* 1(2):48.

Cohen, M. N. 1977 *The Food Crisis in Prehistory: Overpopulation and the Origins of Agriculture*, New Haven: Yale University Press.

Cohen, M. N and Armelagos, G. J. 1984 Editor's summation. In M. N. Cohen and G. J. Armelagos (eds) *Paleopathology at the Origins of Agriculture*, pp. 585-601. Orlando: Academic Press.

Cole, D. and Darling, D. 1990 History of the early period. In W. Suttles, (ed.) *Handbook of North American Indians, Volume 7: Northwest Coast*, pp.119-134. Washington: Smithsonian Institution.

Cole, S. 1967 *The Neolithic Revolution*, London: Trustees of the British Museum.

Coleman, J. 1982 A new look at the north coast: fish traps and villages. In S. Bowdler (ed.) *Coastal Archaeology in Eastern Australia*, pp.1-10. Canberra: Department of Prehistory, Research School of Pacific Studies, Australian National University.

Colledge, S. 1998 Identifying pre-domestication cultivation using multivariate analysis. In A. B. Damania, J. Valkourn, G. Willcox and C. O. Qualset (eds) *The Origins of Agriculture and Crop Domestication*, pp121-131. Aleppo: ICARDA.

Colledge, S. 2001 *Plant Exploitation on Epipaleolithic and Early Neolithic Sites in the Levant*, Oxford: BAR Publishing. *BAR International Series No. 986.*

Colley, S. M. and Jones, R. 1987 New fish bone data from Rocky Cape, north west Tasmania. *Archaeology in Oceania*, 22:67-71.

Collins, P. 2006 *Burn: The Epic Story of Bushfire in Australia.* Crows Nest: Allen and Unwin.

Colten, R. H. and Stewart, A. 1996 An adaptationist model of emergent complexity among hunter gatherers in the Santa Barbara, California region. *Research in Economic Anthropology* 17: 227-250.

Conkey, M. W. and Spector, J. D. 1984 Archaeology and the study of gender. *Advances in Archaeological Method and Theory* 7:1-38.

Conover, D. O. and Munch, S. B. 2002 Sustaining fish yields over evolutionary time scales. *Science* 297(5578):94-96.

Conrick, J. n.d. Conrick, John: Papers. National Library of Australia MS 1317.

Contenson, H. de 1983 Early agriculture in western Asia. In T. C. Young, P. Mortensen and P. E. L. Smith (eds) *The Hilly Flanks and Beyond: Essays on the Prehistory of Southwestern Asia*, pp.57-74. Chicago: Oriental Institute of the University of Chicago. *Studies in Ancient Oriental Civilization No. 36.*

Cook, J. 1768-1771 *The Journal of Captain James Cook on his Voyage of Discovery, Volume 1: The Voyage of the Endeavour 1768 – 1771.* J. Beaghole (ed.) 1968, Cambridge: Hakluyt Society.

Cook, J. 1893 *Captain Cook's Journal During His First*

*Voyage Around the World*, W. L. J. Whaton (ed.), London: Elliot Stock.
Cook, S. 1968 The obsolete 'anti-market' mentality: A critique of the substantivist approach to economic anthropology. In E. E. Clair and H. K. Schneider (eds) *Economic Anthropology: Readings in Theory and Analysis*, pp.208-228. New York: Holt, Rinehart and Winston.
Cooper, D. 1862 Canoes in Australia. *The Athenaeum* 1794:363-364.
Cooper, H. M. 1941 Rock carvings and other Aboriginal relics from near Maree. *The South Australian Naturalist* 21(1):1-4.
Corris, P. 1968 *Aborigines and Europeans in Western Victoria*, Canberra: Australian Institute of Aboriginal Studies. *Occasional Papers in Aboriginal Studies No. 12 Ethnohistory Series No. 1.*
Coursey, D. G. 1967 *Yams: An Account of the Nature, Origins, Cultivation and Utilisation of the Useful Members of the Dioscoreaceae*, London: Longmans.
Coursey, D. G. 1972 The civilization of the yam: Interrelationships of man and yams in Africa and the Indo-Pacific Region. *Archaeology and Physical Anthropology in Oceania* 7:215-233.
Coursey, D. G. 1976a The origins and domestication of yams in Africa. In J. R. Harlan, J. M. J. de Wet and A. B. L. Stemler (eds) *Origins of African Plant Domestication*, pp.383-408. The Hague: Mouton.
Coursey, D. G. 1976b Yams. In N. W. Simmonds (ed.) *Evolution of Crop Plants*, pp.70-74. London: Longmans.
Coutts, P. J. F. 1981 *Readings in Victorian Prehistory*, Melbourne: Ministry for Conservation *Vol. 3: The Victorian Aboriginals 1800 to 1860.*
Coutts, P. J. F., Frank, R. K. and Hughes, P. 1978 *Aboriginal Engineers of the Western District, Victoria.* Melbourne: Ministry of Conservation. *Records of the Victorian Archaeological Survey, No.7.*
Coutts, P. J. F. and Witter, D. C. 1977 New radiocarbon dates for Victorian archaeological sites. *Records of the Victorian Archaeological Survey* 4:59-73.
Coutts, P. J. F., Witter, D., McIlwraith, M. and Frank, R. 1976 *The Mound People of Western Victoria.* Melbourne: Ministry of Conservation. *Records of the Victorian Archaeological Survey, No. 1.*
Coutts, P. J. F., Witter, D. and Parsons, D. M. 1977a Impact of European settlement on Aboriginal society in western Victoria. *Records of the Victorian Archaeological Survey* 4:17-58.
Coutts, P. J. F., Witter, D. and Parsons, D. M. 1977b Impact of European settlement on Aboriginal society in western Victoria. *Search* 8(6):194-205.
Cowgill, G. L. 1975 On the causes and consequences of ancient and modern population changes. *American Anthropologist* 77:505-525.
Cox, J. C. 1880 Notes on some of the habits and customs of Australian natives in Queensland. *Proceedings of the Linnean Society of New South Wales* 5:633-637.
Coxen, C. 1866 The Kommillaroy tribe. *Transactions of the Philosophical Society of Queensland* 1:2/6/1866:1-4.
Crabtree, P. J. 1991 Gender hierarchies and the sexual division of labor in the Natufian culture of the southern Levant. In D. Walde and N. D. Willows (eds) *The Archaeology of Gender: Proceedings of the Twenty-Second Annual Conference of the Archaeological Association of the University of Calgary*, pp.384-391. Calgary: University of Calgary Archaeological Association.
Crawford, G. W. 1992 Prehistoric plant domestication in east Asia. In C. W. Cowan and P. J. Watson (eds) *The Origins of Agriculture: An International Perspective*, pp.7-38. Washington: Smithsonian Institute Press.
Crawford, G. W. 2006 East Asian plant domestication. In M. T. Stark (ed.) *Archaeology of Asia*, pp.77-95. Malden: Blackwell.
Crawford, G. W. and Shen, C. 1998 The origin of rice agriculture: Recent progress in East Asia. *Antiquity* 72(275):34-45.
Crawford, I. M. 1968 *The Art of the Wandjina*, Melbourne: Oxford University Press.
Crawford, I. M. 1982 *Traditional Aboriginal Plant Resources in the Kalumburu Area: Aspects in Ethno-economics*. Perth: Western Australian Museum. *Records of the Western Australian Museum, Supplement No. 15.*
Crawford, J. C. 1965 The diary of James Coutts Crawford: Extracts on Aborigines and Adelaide, 1839 and 1841. D. H. Pike (ed.) *South Australiana* 4(1):3-15.
Craze, B. 1977 The Wiradjuri tribe on the Lachlan and their contact with explorers and settlers. *Armidale and District Historical Society Journal* 20:11-23.
Cribb, R. 1994 Food production. In D. Horton (ed.) *Encyclopaedia of Aboriginal Australia*, 2 volumes, pp. 382-383. Canberra: Aboriginal Studies Press.
Critchett, J. 1981 Introduction to Facsimile Edition in J. Dawson 1981 *Australian Aborigines* Canberra: Australian Institute of Aboriginal Studies.
Critchett, J. 1990 *A 'Distant Field of Murder': Western District Frontiers 1834-1848*. Melbourne: Melbourne University Press.
Crowley, T. 1983 Uradhi. In R. M. W. Dixon and B. J. Blake (eds) *Handbook of Australian Languages, Volume 3*, pp306-428. Amsterdam: John Benjamin.
Curr, E. M. (comp.) 1886 *The Australian Race*, 4 volumes, Melbourne: Government Printer.
Curr, E. M. 1965 *Recollections of Squatting in Victoria*, London: Melbourne University Press.
Currie, J. L. 1858-1859 In *Victoria: Legislative Council, Votes and Proceedings 1858-9 Report of the Select Committee of the Legislative Council on the Aborigines, together with the Proceedings of the Committee, Minutes of Evidence and Appendices*. p.62.
Cusack, D., Moore, T. I. and Ovenden, B. 1965 *Mary Gilmore – A Tribute*, Sydney: Australasian Book Society.
Daley, C. 1931 Food of the Australian Aborigines. *The Victorian Naturalist* 48:23-31.
Dampier, W. 1697 *A New Voyage Round the World*, London: James Knapton.
Dampier, W. 1703 *A Voyage to New Holland... in 1699*, London: James Knapton.

Darbishire, R. D. and Myres, J. L. 1903 On an ornament of unknown use and a quartzite knife from Moreton Bay, Queensland. *Man* 3:33.

Dargin, P. 1976 *Aboriginal Fisheries of the Darling-Barwon Rivers*, Brewarrina: Brewarrina Historical Society.

Darlot, P. F. 1940 James Monckton Darlot. Reminiscences. *The Victorian Naturalist* 18(3):67-76.

Darwin, C. *The Variation of Animals and Plants Under Domestication*, 2 vols. London: John Murray.

David, B. 2002 *Landscapes, Rock-Art and the Dreaming: An Archaeology of Preunderstanding*, London: Leicester University Press.

Davidson, D. S. 1933 Australian netting and basketry techniques. *Journal of the Polynesian Society* 42:257-299.

Davidson, D. S. 1935 The chronology of watercraft in Australia. *Journal of the Polynesian Society* 44:1-16,69-84,137-152,193-207.

Davidson, D. S. 1938 Northwestern Australia and the question of influences from the East Indies. *Journal of the American Oriental Society* 58:61-80.

Davidson, D. S. 1949 Mourning-caps of the Australian Aborigines. *Proceedings of the American Philosophical Society*, 93:57-70.

Davidson, D. S. and McCarthy, F. D. 1957 The distribution and chronology of some important types of stone implements in Western Australia. *Anthropos* 52:390-458.

Davidson, I. 1983 On the edge of the Simpson Desert: Recent additions to understanding of prehistoric artifact distributions in arid Australia. *Australian Archaeology* 17:27-37.

Davidson, I. 1989 Is intensification a condition of the fisher-hunter-gatherer way of life. *Archaeology in Oceania* 24(2):75-78.

Davidson, I., Burke, H., and Mitchell, S. 1989 *Archaeological Sites in Northwest Central Queensland: A Report to the Australian Heritage Commission*, Armidale: Department of Archaeology and Palaeoanthropology, University of New England.

Davies, E. 1881 *The Story of an Earnest Life: A Woman's Adventure in Australia, and in Two Voyages Around the World*, Cincinnati: Central Book Concern.

Davis, H. and Harpending, H. 1977 Some implications for hunter-gatherer ecology derived from the spatial structure of resources. *World Archaeology* 8:275-283.

Davis, J. 1863 *Tracks of McKinlay and Party Across Australia*, London: Sampson Low, Son and Co.

Dawson, J. 1881 *Australian Aborigines*, Melbourne: George Robertson.

Dawson, J. 1981 *Australian Aborigines*, Canberra: Australian Institute of Aboriginal Studies

Dawson, R. 1830 *The Present State of Australia*, London: Smith Elder and Co.

Dayton, L. 2003 On the trail of the first dingo. *Science* 302(5645):555-556.

D'Azevedo, W. L. 1986 Washoe. In W. L. D'Azevedo (ed.) *Handbook of North American Indians, Volume 11: Great Basin*, pp. 466-498. Washington: Smithsonian Institution.

de Candolle, A. 1884 *Origin of Cultivated Plants*, London: Kegan Paul, Tench and Co.

Deetz, J. F. 1968 Discussion, Part VI. In R. B. Lee and I. DeVore (eds) *Man the Hunter*, pp.281-5. Chicago: Aldine.

Deevey, E. S. 1960 The human population. *Scientific American* 203(3):194-204.

Degras, L. 1993 *The Yam: A Tropical Root Crop*, London: Macmillan.

de Haan, G. 1756 Exploratory voyages of the ships Rijder and Buis, Commanded by Lt. Jean Etienne Gonzaal and first mate Lavienne Lodewijk van Asschens, to the Gulf of Carpentaria: Report of the Master Cartographer at Batavia, Gerrit de Haan, to the Governor General and Council September 30, 1756 In J. Heeres (comp.) 1899 *The Part Borne By the Dutch in the Discovery of Australia 1606 – 1765*, pp.92-100. London: Luzac and Co.

de Laguna, Frederica 1990a Eyak. In W. Suttles, (ed.) *Handbook of North American Indians, Volume 7: Northwest Coast*, pp.189-202. Washington: Smithsonian Institution.

de Laguna, Frederica 1990b Tlingit. In W. Suttles, (ed.) *Handbook of North American Indians, Volume 7: Northwest Coast*, pp.203-228. Washington: Smithsonian Institution.

Delaney-Rivera, C. 2001 Groundstone tools as indicators of changing subsistence and exchange patterns in the Coastal Chumash region. In J. E. Arnold (ed.) *The Origins of a Pacific Coast Chiefdom: The Chumash of the Channel Islands*, pp.165-181. Salt Lake City: University of Utah Press.

Denbow, J. R. and Wilmsen, E. N. 1986 Advent and course of pastoralism in the Kalahari. *Science* 234(4783):1509-1515.

Denham, T. 2005 Agricultural origins and the emergence of rectilinear ditch networks in the Highlands of New Guinea. In A. Pawley, R Attenborough, J. Golson and R. Hide (eds) *Papuan Pasts: Cultural, Linguistic and Biological Histories of Papuan-Speaking Peoples*, pp.329-361. Canberra: Pacific Linguistics, Research School of Pacific and Asia Studies, Australia National University.

Dennelll, R. 1985 *European Economic History*, London: Academic Press.

Devitt, J. 1992 Acacias: A traditional food source in Central Australia. In A. P. N. House and C. E. Harwood (eds) *Australian Dry-Zone Acacias for Human Food*, pp.37-53. Canberra: Australian Tree Seed Centre: CSIRO Division of Forestry.

de Tapia, E. M. 1992 The origin of agriculture in Mesoamerica and central America. In C. W Cowan and P. J. Watson (eds) *The Origins of Agriculture: An International Perspective*, pp.143-171. Washington: Smithsonian Institute Press.

Diamond, J. 1997 *Guns, Germs and Steel: The Fates of Human Society*, London: Jonathan Cape.

Diamond, J. 2005 *Collapse: How Societies Choose to Fail or Succeed*, New York: Viking.

Dix, W. C. and Lofgren, M. E. 1974 Kurumi: Possible

Aboriginal incipient agriculture associated with a stone arrangement. *Records of the Western Australian Museum* 3(1):73-77.

Dix, W. C. and Meagher, S. J. 1976 Fish traps in the south-west of Western Australia. *Records of the Western Australian Museum* 4(2):171-187.

Doran, E. 1981 *Wanka Austronesian Canoe Origins*, College Station: Texas A&M University Press.

Dortch, C. E. 1999 Archaeological assessment of Aboriginal estuarine fishing on the Southern Ocean coast of Western Australia. In J. Hall and I. McNiven (eds) *Australian Coastal Archaeology*, pp.25-35. Canberra: ANH Publications, Department of Archaeology and Natural History, Research School of Pacific and Asian Studies.

Dortch, C. E. 2002 Modelling past Aboriginal hunter-gatherer socio-economic and territorial organisation in Western Australia's south-west. *Archaeology in Oceania* 37(1):1-21.

Dortch, C. E. and Gardiner, G. 1976 Archaeological investigations in the Northcliffe District, Western Australia. *Records of the Western Australian Museum* 4(3):257-293.

Douglas, J. 1886 The islands of Torres Strait. *Proceedings of the Queensland Branch of the Geographical Society of Australasia* 1:70-83.

Dow, E. B. 1938a Aboriginal ceremonial cairns near Broken Hill. *Oceania* 9(1):30-36.

Dow, E. B 1938b Aboriginal carvings: West Darling of New South Wales. *Mankind* 2(5):101-120.

Dow, E. B. 1938c Aboriginal stone designs. *Mankind* 2(5):126-133

Downs, J. F. 1966 *The Two Worlds of the Washo*, New York: Holt, Reinhart, Winston.

Draper, P. 1975 !Kung women: Contrasts in sexual egalitarianism in foraging and sedentary contexts. In R. R. Reiter *Towards and Anthropology of Women*, pp.77-109. New York: Monthly Review Press.

Draper, P. 1976 Social and economic constraints on child life among the !Kung. In R. B. Lee and I. DeVore (eds) *Kalahari Hunter-Gatherers: Studies of the Kung San and Their Neighbors*, pp.199-217. Cambridge (Mass.): Harvard University Press.

Drennan, R. D. 1976 Religion and social evolution in Formative Mesoamerica. In K. V. Flannery (ed.) *The Early Mesoamerican Village*, pp.345-368. New York: Academic Press.

Drummond, J. 1842 Extract from a letter addressed by Mr. Drummond to the Editor of the *Inquirer* containing an account of an excursion to the northward of the Moore River, Sept. 7, 1842. In *Exploration Diaries* 4:718-720. Battye Library: PR 5441.

Duff, L. P. n.d. *The History of Caramut*, Ararat: Ararat Advertiser.

Dumont, R. and Vernier, P. 2000 Domestication of yams (*D. cayenensis-D. rotundata*) within the Bariba ethnic group in Benin. *Outlook on Agriculture* 29-2:137-148.

Dunbar, G. K. 1943-44 Notes on the Ngemba tribe of the central Darling River, western New South Wales. *Mankind* 3(5):140-148, 172-180.

Duncan-Kemp, A.M. 1934 *Our Sandhill Country*, Sydney: Angus and Robertson.

Duncan-Kemp, A. M. 1952 *Where Strange Paths Go Down*, Brisbane: W. R. Smith and Paterson.

Duncan-Kemp, A.M. 1961 *Our Channel Country*, Sydney: Angus and Robertson.

Duncan-Kemp, A.M. 1968 *Where Strange Gods Call*, Brisbane: W. R. Smith and Paterson.

Durack, M. 1959 *Kings in Grass Castles*, London: Constable.

Durlacher, J. S. 1900 Account of Aboriginal customs in the north west of Western Australia. [manuscript]. National Library of Australia MS 8039.

d'Urville, D. 1839 *Voyage Pittoresque Autour du Monde*, 2 volumes, Paris: Furne and Co.

Dyall, L. K. 1982 Aboriginal fishing stations on the Newcastle coastline, New South Wales. In S. Bowdler (ed.) *Coastal Archaeology in Eastern Australia*, pp.52-62. Canberra: Department of Prehistory, Research School of Pacific Studies, Australian National University.

Earle, T. K. 1987 Chiefdoms in archaeological and ethnohistorical perspective. *Annual Review of Anthropology* 16:279-308.

Edge-Partington, J. 1903 Notes on the weapons of the Dalleburra tribe, Queensland, lately presented to the British Museum by Mr Robert Christison. *Man* 3:37-38.

Edwards, D. A. and O'Connell, J. F. 1995 Broad spectrum diets in arid Australia. *Antiquity* 69(265):769-783.

Edwards, P. and Hardy-Smith, T. 2004 The garbage crisis in prehistory: Artefact discard patterns at the Early Natufian site of Wadi Hammeh 27 and the origin of household disposal patterns. *Journal of Anthropological Archaeology* 23:253-289.

Edwards, P. C. 1989 Problems of recognizing earliest sedentism: The Natufian example. *Journal of Mediterranean Archaeology* 2(1):5-48.

Edwards, R. 1973 *Aboriginal Bark Canoes of the Murray Valley*, London: Robert Hale and Co.

Eerkens, J. W. 2003 Sedentism, storage, and intensification of small seeds: Prehistoric development in Owens Valley, California. *North American Archaeologist* 24(4):281-309.

Ehrenberg, M. 1989 *Women in Prehistory*, London: British Museum Publications.

Elkin, A. P. 1937 Beliefs and practices concerning death in north-eastern and western South Australia. *Oceania* 7(3):275-299.

Elkin, A. P. 1954 *The Australian Aborigines: How To Understand Them*, Sydney: Angus and Robertson.

Ellen, R. 1988 Foraging, starch extraction and the sedentary lifestyle in the lowland rainforest of central Seram. In T. Ingold, D. Riches and J. Woodburn (eds) *Hunters and Gatherers 1: History, Evolution and Social Change*, pp.117-134. New York: Berg.

Enright, W. J. 1935 An Aboriginal fish trap. *Mankind* 1(12):8-9.

Etheridge, R. 1890 Notes on Australian Aboriginal stone weapons and implements. *Proceedings of the Linnean Society of New South Wales* 15:251-258,289-293,367-372.

Etheridge, R. 1891 Notes on Australian Aboriginal stone weapons and implements. *Proceedings of the Linnean Society of New South Wales* 16:31-43,357-380.

Etheridge, R. 1894 On an aboriginal implement, believe to be undescribed, and supposed to be a hoe. *Journal of the Linnean Society of New South Wales* 9:109-112.

Etheridge, R. 1916 The cylindro-conical and cornute stone implements of western New South Wales and their significance. *Memoirs of the Geological Survey of New South Wales: Ethnological Series* 2:1-41.

Evans, J. 1872 *The Ancient Stone Implements, Weapons and Ornaments of Great Britain*, London: Longmans, Green, Reader and Dyer.

Evans, L. T. 1996 *Crop Evolution, Adaptation and Yield*, Cambridge: Cambridge University Press.

Eylmann, E. 1908 *Die Eingeborenen der Kolonie Sudaustralien*, Berlin: Deitrich Reimer.

Eyre, E. J. 1845 *Journal of Expeditions of Discovery into Central Australia and Overland From Adelaide to King George's Sound in the Years 1840-1*, 2 volumes, London: T. and W. Boone.

Eyre, E. J. 1984 *Autobiographical Narrative of Residence and Exploration in Australia 1832-39*, J. Waterhouse (ed.), London: Caliban Books.

Fagan, B. M. 1995 *People of the Earth: An Introduction to World Prehistory*, New York: Harper Collins.

Fagan, B. M. and Maschner, H. D. G. 1991 The emergence of cultural complexity on the west coast of North America. *Antiquity* 65(249):974-976.

Farwell, G. 1971 *Land of Mirage*, Adelaide: Rigby.

Fawcett, J. W. 1898 Notes on the customs of the Wonnahruah tribe. *Science of Man and Journal of the Royal Anthropological Society of Australasia* 1(7):152-154.

Feinman, G. M. 1995 The emergence of inequality: A focus on strategies and processes. In T. D. Price and G. M. Feinman (eds) *Foundations of Social Inequality*, pp.255-279. New York: Plenum Press

Feldman, R. A. 1987 Architectural evidence for the development of nonegalitarian social systems in coastal Peru. In J. Haas, S. Pozorski and T. Pozorski (eds) *The Origins and Development of the Andean State*, pp.9-14. Cambridge: Cambridge University Press.

Ferguson, W. C. 1987 Mokare's domain. In D. J. Mulvaney and J. P. White (eds) *Australians to 1788*, pp.120-145. Sydney: Fairfax, Syme and Weldon.

Finkel, E. 1997 Native claims muddy waters in fight over Australian lake. *Science* 278:1556- 1557.

Finlayson, H. H. 1943 *Red Centre: Man and Beast in the Heart of Australia*, Sydney: Angus and Robertson.

Fisher, J. 1997 Information exchange, rock art and long run economic growth in Australia. *Australian Archaeology* 44:17-22.

Fitzpatrick, F. A. 1914 *Peeps Into the Past: Pioneering Days on the Manning*, Parramatta: Cumberland Argus.

Flannery, K. V. 1965 The ecology of early food production in Mesopotamia. *Science* 147(3663): 1247-1256.

Flannery, K. V. 1969 Origins and ecological effect of early domestication in Iran and the Near East. In P. J. Ucko and G. W. Dimbleby (eds) *Domestication and Exploitation of Plants and Animals*, pp.73-100. London: Duckworth and Co.

Flannery, K. V. 1972 The origins of the village as a settlement type in Mesoamerica and the Near East: A comparative study. In P. J. Ucko, R. Tringham and G. W. Dimbleby (eds) *Man, Settlement and Urbanism*, pp.23-53. London: Duckworth.

Flannery, K. V. 1976a Linear stream patterns and riverside settlement rules. In K. V. Flannery (ed.) *The Early Mesoamerican Village*, pp.173-180. New York: Academic Press.

Flannery, K. V. 1976b The early Mesoamerican house. In K. V. Flannery (ed.) *The Early Mesoamerican Village*, pp.16-24. New York: Academic Press.

Flannery, K. V. 1986 The Research Problem. In K. V. Flannery (ed.) *Guila Naquitz: Archaic Foraging and Early Agriculture in Oaxaca, Mexico*, pp.3-18. Orlando: Academic Press.

Flannery, T. 1994 *The Future Eaters: An Ecological History of the Australasian Lands and People*, New York: Grove Books.

Flecker, P. O. 1951 Remains of Aboriginal habitation on the Great Barrier Wall. *The North Queensland Naturalist* 19(97):1-3.

Flinders, M 1814 *A Voyage to Terra Australis*, 2 volumes, London: G. and W. Nicol.

Flood, J. 1988 No ethnography, no moth hunters. In B. Meehan and R. Jones (eds) *Archaeology With Ethnography: An Australian Perspective*, pp.270-276. Canberra: Dept. of Prehistory, Research School of Pacific Studies, Australian National University.

Flood, J. 1993 *The Riches of Ancient Australia: An Indispensable Guide to Exploring Prehistoric Australia*, St. Lucia: Queensland University Press.

Flood, J. 1997 *Rock Art of the Dreamtime*, Angus and Robertson. Sydney.

Flood, J. 1999 *Archaeology of the Dreamtime*, Sydney: Angus and Robertson.

Flood, J. 2004 *Archaeology of the Dreamtime*, Marleston: JB Publishing.

Foley, R. 1851 Letter from Dr Foley to E.R. Parker. *Perth Gazette* 14 February 1851.

Foley, R. 1865 Vocabulary of the Champion Bay tribe. *Transactions of the Ethnological Society of London* 3:297-298.

Forrest, A. 1880 *North-West Exploration: Journal of Expedition from De Grey to Port Darwin*, Perth: Government Printer.

Fountain, P. 1907 *Rambles of an Australian Naturalist*, London: John Murray.

Frankel, D. 1982 An account of Aboriginal use of the yam-daisy. *The Artefact* 7 (1-2):43-45.

Fraser, J. 1892 *The Aborigines of New South Wales*, Sydney: Government Printer.
Freeman, W. 1902 Aboriginal burial rights. *Australasian Association for the Advancement of Science* 1902:539-540.
French, M. 1989 *Conflict on the Condamine: Aborigines and the European Invasion*, Toowoomba: Darling Downs Institute Press.
Freslov, J. and Frankel, D. 1999 Abundant fields? A review of coastal archaeology in Victoria. In J. Hall and I. McNiven (eds) *Australian Coastal Archaeology*, pp.239-254. Canberra: ANH Publications, Department of Archaeology and Natural History, Research School of Pacific and Asian Studies.
Fritz, G. J. 1995 New dates and data on early agriculture: The legacy of the Complex Hunter- Gatherers. *Annals of the Missouri Botanical Gardens* 82(1):3-15.
Fry, J. (ed.) 1997 *W. R. Cave Diary: A Journey From Farina to Warburton in 1882*, Crafers: White Bay Nominees.
Fullagar, R. and Field, J. 1997 Pleistocene seed-grinding implements from the Australian arid zone. *Antiquity* 71:300-307.
Fyans, F. 1844 Captain Fyans to Superintendent La Trobe. Eumeralla, 9 Oct. 1842. Enclosure 5 in No.59 (No. 195). In *British Parliamentary Papers (Aborigines) 1844, Volume 3: Despatches Illustrating the condition of the Aboriginal population of the Australian colonies*. p.235.
Fyans, F. 1845 Evidence of Foster Fyans JP, Commissioner of Crown Lands for the Portland District, Port Phillip. In *New South Wales: Votes and Proceedings of the Legislative Council During the Session of the Year 1845 - New South Wales, Aborigines. Report from the Select Committee in the condition of the Aborigines with Appendix, Minutes of Evidence and replies to a circular letter*, pp.43-44.
Fyans, F. 1898 Letter. In T. F. Bride (ed.) *Letters from Victorian Pioneers*, pp.114-129. Melbourne: Government Printer.
Gallivan, M. D. 2002 Measuring sedentariness and settlement population: Accumulations research in the Middle Atlantic region. *American Antiquity* 67(2):535-557.
Garber, P. A. 1987 Foraging strategies among living primates. *Annual Review of Anthropology* 16:339-364.
Gardiner, P. 2003 Caught in the act – where is the transition. In L. Bevan and J. Moore (eds) *Peopling the Mesolithic in a Northern Environment*, pp.103-112. Oxford: BAR Publishing. *BAR International Series No. 1157*.
Gason, S. 1874 *The Dieyerie Tribe of Australian Aborigines*, Adelaide: Government Printer.
Gason, S. 1879 The manners and customs of the Dieyerie tribe. In J. D. Woods (comp.) *The Native Tribes of South Australia*, pp.253-307. Adelaide: E. S. Wigg.
Gason, S. 1895 Of the tribes Dieyerie, Auminie, Yandrawontha, Yarawuarka, Billadapa Lat. $31^0$ 0' S., Long. $138^0$ 55' E. *Journal of the Anthropological Institute of Great Britain and Ireland* 24:167-176.
Gaughwin, D. and Fullagar, R. 1995 Victorian offshore islands in a mainland coastal economy. *Australian Archaeology* 40:38-50.
Geering, K. 1985 Report on VAS fieldtrips to 'Allambie' 12-16 March and 14-18 April 1985. Unpublished report, Aboriginal Affairs Victoria.
Gerritsen, R. 1994a *And Their Ghosts May Be Heard*, South Fremantle: Fremantle Arts Centre Press.
Gerritsen, R. 1994b *And Their Ghosts May Be Heard: Appendix: English - Nanda - Dutch Vocabulary*, South Fremantle: Fremantle Arts Centre Press.
Gerritsen, R. 1997 The note on page 146 of 'Paper and Talk': A Response. *Australian Aboriginal Studies* 1997/2:55-57.
Gerritsen, R. 2000 *The Traditional Settlement Pattern in South West Victoria Reconsidered*, Canberra: Intellectual Property Publications.
Gerritsen, R. 2001a Aboriginal fish hooks in southern Australia: Evidence, arguments and implications. *Australian Archaeology* 52:18-28.
Gerritsen, R. 2001b A Dutch influence on Nhanda?: A Reply to Blevins. *Australian Aboriginal Studies* 2001/2:69-73.
Gerritsen, R. 2001c Factors influencing the development of indigenous Aboriginal agriculture in Australia. Paper presented to the International Conference on Cultural Encounters in the Indian-Pacific Region, 1200 -1800, Perth.
Gerritsen, R. 2002 *Nhanda Villages of the Victoria District, Western Australia*, Canberra: Intellectual Property Publications.
Gerritsen, R. 2004 Historical problems and methodological issues regarding Nhanda, an Aboriginal language of Western Australia. *Anthropological Linguistics* 46(1):84-99.
Gerritsen, R. 2007a Papers of Rupert Gerritsen: Ethnohistory and the Antipodes. [manuscript] National Library of Australia MS 10002.
Gerritsen, R. 2007b The debate over where Australia's first European residents were marooned in 1629 – Part 1. *Hydrographic Journal* 126:20-25.
Gerritsen, R. 2008 The landing site debate: Where were Australia's first European residents marooned in 1629? In P. Hornsby and J. Maschke (eds) *Hydro 2007 Conference Proceedings: Focus on Asia*, Belrose: International Federation of Hydrographic Societies.
Gerritsen, R., Slee, C. and Cramer M. 2005 *The Batavia Legacy*, Geraldton: Batavia Coast Maritime Heritage Association.
Gibson, J. L. 1974 Poverty Point: The first North American chiefdom. *Archaeology* 27(2):97-105.
Giles, A. 1926 *Exploring in the Seventies and the Construction of the Overland Telegraph Line*, Adelaide: W. K. Thomas.
Gill, E. D. 1970 Aboriginal antiquities in Victoria. In F. D. McCarthy (ed.) *Aboriginal Antiquities in Australia: Their Nature and Preservation*, pp.26-33. Canberra: Australian Institute of Aboriginal and Torres Strait Islanders.

*Australian Aboriginal Studies No. 22; Prehistory and Material Culture Series No. 3.*

Gill, E. D. 1977 Constraints imposed by the earth sciences on the interpretation of the Warrnambool 'Mahogany Ship' wreck site. *The Artefact* 2(1):9-17.

Gill, T. 1907-8 Some anthropological notes on the South Australian Aborigines. *Proceedings of the Royal Geographical Society of Australasia, South Australian Branch* 10:220-248.

Gillespie, L. L. 1984 *Aborigines of the Canberra Region*, Canberra: Gillespie.

Gilmore, M. 1932 *Under the Wilgas*, Melbourne: Robertson and Mullens.

Gilmore, M. 1934 *Old Days, Old Ways*, Sydney: Angus and Robertson.

Gilmore, M. 1935 *More Recollections*, Sydney: Angus and Robertson.

Goddard, R. H. 1939 Aboriginal sculpture. *Report of the Twenty-Fourth Meeting of the Australian and New Zealand Association for the Advancement of Science* 24:160-163.

Godelier, M 1991 An unfinished attempt at reconstructing the social process which may have prompted the transformation of great-men societies into big-men societies. In M. Godelier and M. Strathern (eds) *Big Men and Great Men: Personification of Power in Melanesia*, pp.275-304. Cambridge: Cambridge University Press.

Godelier, M. and Strathern, M. (eds) 1991 *Big Men and Great Men: Personification of Power in Melanesia*, Cambridge: Cambridge University Press.

Godfrey, H. 1858-1859 In *Victoria: Legislative Council, Votes and Proceedings 1858-9 Report of the Select Committee of the Legislative Council on the Aborigines, together with the Proceedings of the Committee, Minutes of Evidence and Appendices*. p.62.

Godwin, L. 1988 Around the traps: A reappraisal of stone fishing weirs in northern New South Wales. *Archaeology in Oceania* 23(2):49-59.

Gorecki, P., Grant, M., and O'Connor, S. 1997 The morphology, function and antiquity of Australian grinding implements. *Archaeology in Oceania* 32(2):141-150.

Gott, B. 1982 Ecology of root use by the Aborigines of southern Australia. *Archaeology in Oceania* 17(1):59-67.

Gott, B. 1983 Murnong – *Microseris scapigera*: A study of a staple food of Victorian Aborigines. *Australian Aboriginal Studies* 1983/2:2-18.

Gott, B. 1987 Grampians Aboriginal plants. In *Australia Felix: The Chap Wurrung and Major Mitchell*, pp.37-51. Dunkeld and District Historical Museum.

Gott, B. 1999 Koorie use and management of the Plains. In Roger N. Jones (ed.) *The Great Plains Crash: Proceedings of a Conference on the Grasslands and Grassy Woodlands of Victoria*, pp.41-45. http://home.vicnet.net.au/~iffa/GPCconf/Title.pdf Gott, B. and Conran, J. 1991 *Victorian Koorie Plants*, Hamilton: Yangennanock Women's Group.

Gough, K. 1975 The origin of the family. In R. R. Reiter (ed.) *Toward and Anthropology of Women*, pp.51-76. New York: Monthly Review Press.

Gould, R. A. 1969 Subsistence behaviour among the Western Desert Aborigines of Australia. *Oceania* 39(4):255-274.

Gow, R. G. 1860-71 Diary and journal 1860-1871. National Library of Australia MS 24.

Goyder, G. W. 1857 Northern Exploration. *Parliamentary Papers 1857-8 South Australia*, No. 72 of 1857, pp.1-4.

Granovetter, M. 1985 Economic action and social structure: The problem of embeddedness. *American Journal of Sociology* 91(3):481-510.

Gregory, A. C. 1858 *New South Wales Legislative Council: Expedition in Search of Dr Leichhardt (Report of Proceedings)*, Sydney: Government Printer.

Gregory, A. C. 1884a North Australian expedition 1855-6. In A. C. and F. T. Gregory (comps.) *Journals of Australian Expeditions*, pp.99-194. Brisbane: Government Printers.

Gregory, A. C. 1884b Expedition in search of Dr Leichhardt – Report of Proceedings. In A. C. and F. T. Gregory (comps.) *Journals of Australian Expeditions*, pp.200-210. Brisbane: Government Printers.

Gregory, A. C. 1886 Inaugural address. *Proceedings of the Queensland Branch of the Geographical Society of Australasia* 1:18-25.

Gregory, A. C. 1887 Memorandum on the Aborigines of Australia. In H. L. Roth 1887 On the Origin of Agriculture. *Journal of the Anthropological Institute of Great Britain and Ireland* 16:131-133.

Gregory, E. (comp.) 1896 *Narrative of James Murrell's ('Jemmy Morrill') Seventeen Year Exile Among the Wild Blacks of North Queensland*, Brisbane: Edmund Gregory.

Gregory, F. T. 1884a Gascoyne River 1858. In A. C. and F. T. Gregory (comps.) *Journals of Australian Expeditions*, pp.37-51. Brisbane: Government Printers.

Gregory, F. T. 1884b North-West Coast. In A. C. and F. T. Gregory (comps.) *Journals of Australian Expeditions*, pp.52-97. Brisbane: Government Printers.

Gregory, J. W. 1906 *The Dead Heart of Australia*, London: John Murray.

Gresser, P. J. 1962 New distributional records of stone implements in New South Wales and Queensland. Part 1. *Mankind* 5(12):522-533.

Gresser, P. J. 1963a New distributional records of stone implements in New South Wales and Queensland. Part 2. *Mankind* 6(1):27-40.

Gresser, P. J. 1963b New distributional records of stone implements in New South Wales and Queensland. Part 3. *Mankind* 6(2):80-87.

Gresser, P. J. 1964 New distributional records of stone implements in New South Wales and Queensland. Part 4. *Mankind* 6(3):120-135.

Grey, G. 1841a *A Journal of Two Expeditions in North-West and Western Australia During the Years 1837 – 1839*, 2 volumes, London: T. and W. Boone.

Grey, G. 1841b Notes upon that portion of Western Australia which lies between Gantheaume Bay and the Arrowsmith River. In H.S. Chapman (comp.) *The New Settlement of Australind*, pp.41-54. London: Harvey and Darton.

Griffiths, C. 1845 *The Present State and Prospects of the Port Phillip District of New South Wales*. Dublin: William Curry Jnr.

Grossman, D. 1992 *Rural Process-Pattern Relationships: Nomadization, Sedentarization and Settlement Fixation*. New York: Praeger.

Haaland, R. 1997 Emergence of sedentism: New ways of living, new ways of symbolizing. *Antiquity* 71(272):374-385.

Habu, J. 1996 Jomon sedentism and intersite variability: Collectors of the early Jomon Moroiso phase in Japan. *Arctic Anthropology* 33(2):38-49.

Hahn, S. K. 1995 Yams *Dioscorea* spp. (Dioscoreaceae). In J. Smartt and N. W. Simmonds (eds) *Evolution of Crop Plants*, Harlow: Longman Scientific and Technical.

Hale, H. M. and Tindale, N. B. 1933-36 Aborigines of Princess Charlotte Bay, North Queensland. *Records of the South Australian Museum* 5:63-172.

Hall, C. B. 1898. Letter. In T. F. Bride (ed.) *Letters from Victorian Pioneers*, pp.210-222. Melbourne: Government Printer.

Hall, J. 1982 Sitting on the crop of the bay: An historical and archaeological sketch of Aboriginal settlement and subsistence in Moreton Bay, southeast Queensland. In S. Bowdler (ed.) *Coastal Archaeology in Eastern Australia*, pp.75-95. Canberra: Department of Prehistory, Research School of Pacific Studies, Australian National University.

Hall, J. 1999 The impact of sea level rise on the archaeological record of the Moreton region, southeast Queensland. In J. Hall and I. McNiven (eds) *Australian Coastal Archaeology*, pp.169-184. Canberra: ANH Publications, Department of Archaeology and Natural History, Research School of Pacific and Asian Studies.

Hall, T. 1917 *A Short History of the Downs Blacks Known as the 'Blucher Tribe'*, Warwick, T. Hall.

Halpin, M. M. and Seguin, M. 1990 Tsimshian peoples: Southern Tsimshian, Coast Tsimshian, Nishga, and Gitksan. In W. Suttles, (ed.) *Handbook of North American Indians, Volume 7: Northwest Coast*, pp.267-284. Washington: Smithsonian Institution.

Hammond, J. E. 1933 *Winjan's People*, Perth: Imperial Press.

Hancock, J. F. 1992 *Plant Evolution and the Origin of Crop Species*, Englewood Cliffs: Prentice Hall.

Happ, G. 1977 To catch a fish: The aquatic factor in Aboriginal Australian fishing technology. BA (Hons) thesis, University of Sydney, Sydney.

Hardy, B. 1969 *West of the Darling*, Milton: Jacaranda Press.

Hardy-Smith, T. and Edwards, P. C. 2004 The garbage crisis in prehistory: Artefact discard patterns at the Early Natufian site of Wadi Hammeh 27 and the origins of household refuse disposal strategies. *Journal of Anthropological Archaeology* 23:253-289.

Harlan, J. R. 1967 A wild wheat harvest in Turkey. *Archaeology* 20:197-201.

Harlan, J. R. 1971 Agricultural origins: Centers and noncenters. *Science* 174(4008):468-474.

Harlan, J. R. 1989 Wild-grass seed harvesting in the Sahara and sub-Sahara of Africa. In D. R. Harris and G. C. Hillman (eds) *Foraging and Farming: The Evolution of Plant Exploitation*, pp.79-98. London: Unwin Hyman.

Harlan, J. R. 1992a Wild grass seed harvesting and implications for domestication. In P. C. Anderson[-Gerfaud] (ed.) *Prehistoire de L'Agriculture: Nouvelles Approches Experimentales et Ethnographiques*, pp.21-27. Paris: Editions du CNRS. *Monographie du CRA No. 6*.

Harlan, J. R. 1992b *Crops and Man*, Madison: American Society of Agronomy, Crop Society of America.

Harlan, J. R. 1995 *The Living Fields*, Cambridge: Cambridge University Press.

Harney, W. E. 1950 Roads and trade. *Walkabout* 16(5):42-45.

Harney, W. E. 1961 *North of 23⁰*, Adelaide: Rigby.

Harper, W. R. 1898 A description of certain objects of unknown significance formerly used by some New South Wales tribes. *Proceedings of the Linnean Society of New South Wales* 23(3):420-436.

Harrington, S. P. M. 1997 Earliest agriculture in New World. *Archaeology* 50(4):11.

Harris, D. R. 1977a Alternative pathways toward agriculture. In C. A. Reed (ed.) *The Origins of Agriculture*, pp.179-243. The Hague: Mouton.

Harris, D. R. 1977b Subsistence strategies across Torres Strait. In J. Allen, J. Golson and R. Jones (eds) *Sunda and Sahul: Prehistoric Studies in Southeast Asia, Melanesia and Australia*, pp.421-463. London: Academic Press.

Harris, D. R. 1979 Foragers and farmers in the Western Torres Strait Islands: An historical analysis of economic, demographic, and spatial differentiation. In P. C. Burnham and R. F. Ellen (eds) *Social and Ecological systems*, pp.75-109. London: Academic Press.

Harris, D. R. 1989 An evolutionary continuum of people-plant interaction. In D. R. Harris and G. C. Hillman (eds) *Foraging and Farming: The Evolution of Plant Exploitation*, pp.11-26. London: Unwin Hyman.

Harris, D. R. 1996 Introduction: Themes and concepts in the study of early agriculture. In D. R. Harris (ed.) *The Origins and Spread of Agriculture and Pastoralism in Eurasia*, pp.1-9. London: UCL Press.

Harris, M. 1968 *The Rise of Anthropological Theory: A History of Theories of Culture*. New York: Thomas Y. Crowell and Co.

Hart, C. W. M. and Pilling A. R., 1960 *The Tiwi of Northern Australia*, New York: Holt, Rinehart and Winston.

Harvey, A. 1943 A fishing legend of the Jaralde tribe of Lake Alexandrina, South Australia. *Mankind* 3(4):108-112.

Harvey, A. 1945 Food preservation in Australian tribes. *Mankind* 3(7):191-192.

Hassan, F. 1981 *Demographic Archaeology*, New York: Academic Press.

Hawker, J. C. 1901 *Early Experiences in South Australia*, 2 parts, Adelaide: E. S. Wigg and Son.

Hayden, B. 1981 Research and development in the Stone Age: Technological transitions among hunter-gatherers. *Current Anthropology* 22(5):519-548.

Hayden, B. 1990 Nimrods, piscators, pluckers and planters: The emergence of food production. *Journal of Anthropological Archaeology* 9:31-69.

Hayden, B. 1992a Observing prehistoric women. In C. Claassen (ed.) *Exploring Gender Through Archaeology: Selected Papers From the 1991 Boone Conference*, pp.33-47. Madison: Prehistory Press, *Monographs in World Archaeology No. 11*.

Hayden, B. 1992b Conclusions: Ecology and complex hunter-gatherers. In B. Hayden (ed.) *A Complex Culture of the British Columbia Plateau*, pp.525-559. Vancouver: University of British Columbia Press.

Hayden, B. 1992c Models of domestication. In A. B. Gebauer and T. D. Price (eds) *Transitions to Agriculture in Prehistory*, pp. 11-19. Madison: Prehistory Press. *Monographs in World Archaeology No. 4*.

Hayden, B. 1994 Competition, labor, and Complex Hunter-Gatherers. In E. S. Burch and L. J. Ellanna (eds) *Key Issues in Hunter-Gatherer Research*, pp.223-239. Oxford: Berg.

Hayden, B. 1995a A new overview of domestication. In T. D. Price and A. Gebauer (eds) *Last Hunters – First Farmers: New Perspectives on the Prehistoric Transition to Agriculture*, pp.273-299. Santa Fe: School of American Research Press.

Hayden, B. 1995b Pathways to power: Principles for creating socioeconomic inequalities. In T. D. Price and G. M. Feinman (eds) *Foundations of Social Inequality*, pp.15-86. New York: Plenum Press.

Heizer, R. F. (ed.) 1978 *Handbook of North American Indians, Volume 8: California*, Washington: Smithsonian Institution.

Helbaek, H. 1960 The paleoethnobotany of the Near East and Europe. In R. J. Braidwood and B. Howe (eds) *Prehistoric Investigations in Iraqi Kurdistan*, pp.99-118. Chicago: University of Chicago Press. *Oriental Institute of the University of Chicago Studies in Ancient Oriental Civilization No. 31*.

Helbaek, H. 1963 Textiles from Catal Huyuk. *Archaeology* 16(1):39-46.

Helbaek, H. 1970 The plant husbandry of Hacilar. In J. Mellaart (ed.) *Excavations at Hacilar, Volume 1*, pp.189-244, 2 volumes, Edinburgh: Edinburgh University Press.

Helpman, Lt. R. 1846 Champion Bay to the coal on the Irwin River by Lt. Helpman R.N. In *Exploration Diaries* 4:6-19. Battye Library: PR 5441.

Helpman, Lt. R. 1849 Letter from Lt. Helpman to Government Gazette. *Government Gazette* 18 December 1849, p.3.

Helpman, Lt. R. 1851 Extracts from a report of a voyage to Sharks Bay and Exmouth Gulf, Feb. to June 1851. In *Exploration Diaries* 4:265-276. Battye Library: PR 5441.

Henderson, J. 1851 *Excursions and Adventures in New South Wales: With Pictures of Squatting Life in the Bush*, 2 volumes, London: Saunders and Otley.

Henry, D. O. 1989 *From Foraging to Agriculture: The Levant at the End of the Ice Age*, Philadelphia: University of Pennsylvania Press.

Henry, D. O. 1991 Foraging, sedentism, and adaptive vigor in the Natufian: Rethinking the linkages. In G. A. Clark (ed.) *Perspectives on the Past: Theoretical Biases in Mediterranean Hunter-Gatherer Research*, pp.353-370. Philadelphia: University of Pennsylvania Press.

Hepburn, J. 1853 Letter. In T. F. Bride (ed.) 1969 *Letters from Victorian Pioneers*, Melbourne: Heinemann.

Heritage Concepts 2004 *Final Report: Aboriginal and Historic Cultural Heritage Assessment and Management Guidelines – Opal Prospecting Areas, Lightning Ridge*, www.dpi.nsw.gov.au/__data/.../20040401-OPA1-4-REF-Narran-Warrambool-Reserve-Lightning-Ridge-Appendix-E-Heritage-Report.pdf

Heun, M., Shafer-Pregl, R., Klawan, D., Castagna, R., Accerbi, M., Borghi, B., and Salamini, F. 1997 Site of einkorn wheat domestication identified by DNA fingerprinting. *Science* 278(5341):1312-14.

Hey, N. and Roth, W. E. 1903 *An Elementary Grammar of the Nggerikudi Language*, Brisbane: Government Printer. *North Queensland Ethnography Bulletin No. 6*.

Hiatt, B. 1968 The food quest and the economy of the Tasmanian Aborigines [Part 2]. *Oceania* 38(3):190-219.

Hiatt, B. 1974 Woman the gatherer. In F. Gale (ed.) *Woman's Role in Aboriginal Society*, pp.4-15. Canberra: Australian Institute of Aboriginal Studies. *Australian Aboriginal Studies No. 36*.

Hiatt, L. R. 1996 *Arguments About Aborigines: Australia and the Evolution of Social Anthropology*, Cambridge: Cambridge University Press.

Hic Et Ubique 1877 The explorer: Out Eyre's Creek Way. *The Queenslander* 24 March 1877, p.13

Higgs, E. S. 1968 Archaeology – Where now? *Mankind* 6(12):617-620.

Higgs, E. S. and Jarman, M. R. 1969 The origins of agriculture. *Antiquity* 43(169):31-41.

Higgs, E. S. and Jarman, M. R. 1972 The origins of animal and plant husbandry. In E. S. Higgs (ed.) *Papers in Economic Prehistory*, pp.3-13. London: Cambridge University Press.

Higham, C. 1987 Rice cultivation and its place in Southeast Asian prehistory: Some first indications. In L. Manzanilla (ed.) *Studies in the Neolithic and Urban Revolutions: The V. Gordon Childe Colloquiums, Mexico, 1986*, pp.51-67. Oxford: British Archaeological Reports. *BAR International Series 349*.

Higham, C. 1989 *The Archaeology of Mainland Southeast Asia*, Cambridge: Cambridge University Press.

Higham, C. 1995 The transition to rice cultivation in Southeast Asia. In T. D. Price and A. Gebauer (eds) *Last Hunters – First Farmers: New Perspectives on the Prehistoric Transition to Agriculture*, pp.127-155. Santa Fe: School of American Research Press.

Higham, C. 2002 *Early Cultures of Mainland Southeast Asia*, Bangkok: River Books.

Higham, C. and Lu, T. L-D. 1998 The origins and dispersal of rice cultivation. *Antiquity* 72(278):867-877.

Higham, C. and Thosarat, R. 1998 *Prehistoric Thailand*, Bangkok: River Books.

Hill, S. 1901 Ceremonies, customs and foods of the Myoli tribe. *Science of Man and Journal of the Royal Anthropological Society of Australasia* 4(2):24-25.

Hillman, G. C. 1996 Late Pleistocene changes in wild plant-foods available to hunter-gatherers of the northern Fertile Crescent: Possible preludes to cereal cultivation. In D. R. Harris (ed.) *The Origins and Spread of Agriculture and Pastoralism in Eurasia*, pp.159-203. London: UCL Press.

Hillman, G. C. and Davies, M. S. 1990 Measured domestication rates in wild wheats and barley under primitive cultivation and their archaeological implications. *Journal of World Prehistory* 4(2):157-222

Hillman, G. C. and Davies, M. S. 1992 Domestication rates in wild wheats and barley under primitive cultivation: Preliminary results and archaeological implications of field measurements of selection coefficient. In P. C. Anderson[-Gerfaud] (ed.) *Prehistoire de L'Agriculture: Nouvelles Approches Experimentales et Ethnographiques*, pp.113-158. Paris: Editions du CNRS. *Monographie du CRA No. 6.*

Hinde, A. 2003 *England's Population: A History Since the Doomsday Survey*, London: Hodder Arnold.

Hiscock, P. 1986 Technological change in the Hunter River Valley and the interpretation of late Holocene change in Australia. *Archaeology in Oceania* 21:40-50.

Hiscock, P. 1999 Holocene coastal occupation of western Arnhem Land. In J. Hall and I. J. McNiven (eds) *Australian Coastal Archaeology*, pp.91-103. Canberra: ANH Publications, Dept. of Archaeology and Natural History, Research School of Pacific and Asian Studies, Australian National University.

Hiscock, P. and Wallis, L. A. 2005 Pleistocene settlements of deserts from an Australian perspective. In P. Veth, M. A. Smith and P. Hiscock (eds) *Desert Peoples: Archaeological Perspectives*, pp.34-57. Oxford, Blackwell Publishing.

Ho, P-T. 1977 The indigenous origins of Chinese agriculture. In C. A. Reed (ed.) *The Origins of Agriculture*, pp.413-484. The Hague: Mouton.

Hobhouse, L. T., Wheeler, G. C. and Ginsberg, M. 1915 *The Material Culture and Social Institutions of the Simpler Peoples: An Essay in Correlation*, London: Chapman and Hall.

Hodgkinson, W. O. 1877 Diary of the north-western expedition, continued. *Votes and Proceedings of the Legislative Assembly of Queensland During the Session of 1877, Volume 3*. pp.205- 277.

Holdaway, S. J., Fanning, P. C. and Witter, D. 2000 Prehistoric Aboriginal occupation of the rangelands: Interpreting the surface archaeological record of far western New South Wales, Australia. *The Rangeland Journal* 22(1):44-57.

Holdaway, S. J. and Stern, N. 1996 *Written in Stone: Decoding the Australian Flaked Stone Record*, Melbourne: Museum Victoria.

Hole, F. 1984 A reassessment of the Neolithic Revolution. *Paleorient* 10:49-60.

Hole, F. 2000 Is size important: Function and hierarchy in Neolithic settlements. In I. Kuijt (ed.) *Life in Neolithic Farming Communities: Social Organization, Identity and Differentiation*, pp.191-209. New York: Kluwer Academic and Plenum Publishers.

Holm, B. 1990a Art. In W. Suttles, (ed.) *Handbook of North American Indians, Volume 7: Northwest Coast*, pp.602-32. Washington: Smithsonian Institution.

Holm, B. 1990b. Kwakiutl winter ceremonies. In W. Suttles, (ed.) *Handbook of North American Indians, Volume 7: Northwest Coast*, pp.378-386. Washington: Smithsonian Institution.

Holmes, B. 2004 Manna or millstone. *New Scientist* 183(2465):29-31.

Holtze, M. 1892 Introduced plants in the Northern Territory. *Transactions of the Royal Society of South Australia* 15(1):1-4.

Hoopes, J. W. 1995 Interaction of hunting and gathering societies as a context for the emergence of pottery in the Central American Isthmus. In W. K. Barnett and J. W. Hoopes (eds) *The Emergence of Pottery: Technology and Innovation in Ancient Society*, pp.185-198. Washington: Smithsonian Institution Press.

Horne, G. and Aiston, G. 1924 *Savage Life in Central Australia*, London: Macmillan.

Hornell, J. 1940 The genetic relation of the bark canoe to dug-outs and plank-built boats. *Man* 40:114-119.

Horton, D. R. 1982 The burning question: Aborigines, fire and Australian ecosystems. *Mankind* 13(3):237-251.

Horton, D. R. 2000 *The Pure State of Nature: Sacred Cows, Destructive Myths and the Environment*. St Leonards: Allen and Unwin.

Howitt, A. W. 1861 Howitt's Diary. In *The Victorian Exploring Expedition: Diary of Burke and Wills, Journal of Howitt and Narrative of King*, pp.6-14. Melbourne: The Age.

Howitt, A. W. 1891 The Dieri and other kindred tribes of Central Australia. *Journal of the Anthropological Institute of Great Britain and Ireland* 20:30-104.

Howitt, A. W. 1904 *The Native Tribes of South-East Australia*, London: Macmillan.

Howitt, A. W. 1907 *Inaugural Address: Personal Reminiscences of Central Australia and the Burke and Wills Expedition*, Melbourne: A. W. Howitt.

Howitt, A. W. 1908 The native tribes of southeast Australia. *The American Antiquarian* 30:81-95.

Howitt, A. W. and Siebert, O. 1904 Legends of the Dieri and kindred tribes of Central Australia. *Journal of the*

*Anthropological Society of Great Britain and Ireland* 34:100-129.

H. P. 1897 Information about Australian tribes. *Australasian Anthropological Journal* 1(4):80-82.

Hughes, P. and Hiscock, P. 2005 The archaeology of Lake Eyre South area. In W. J. H. Slaytor (ed.) *Archaeology of the Lake Eyre South Region*, pp.1-20. Adelaide: Royal Geographical Society of South Australia. *Lake Eyre South Monograph Series, Volume 6.*

Hughes, P. J. and Djohadze, V. 1980 *Radiocarbon Dates From Archaeological Sites on the South Coast of New South Wales and the Use of Depth/Age Curves.* Canberra: Department of Prehistory, Australian National University. *Occasional Paper in Prehistory No. 1.*

Hughes, P. J. and Lampert, R. J. 1982 Prehistoric population change in southern coastal New South Wales. In S. Bowdler (ed.) *Coastal Archaeology in Eastern Australia*, pp.16-29. Canberra: Department of Prehistory, Research School of Pacific Studies, Australian National University.

Hughes, S. S. 1991 Division of labor at a Besant hunting camp in eastern Montana. *Plains Anthropologist* 36:25-49.

Hurum, H. J. 1977 *A History of the Fish Hook*, London: Adam and Charles Black.

Hyam, G. N. 1940 The vegetable foods of the Australian Aborigines. *The Victorian Naturalist* 56(6):95-98; 56(7):115-119.

Hynes, R. A. and Chase, A. K. 1982 Plants, sites and domiculture: Aboriginal influence on plant communities in Cape York Peninsula. *Archaeology in Oceania* 17(1):38-50.

Ibanez, J. J., Gonzalez, J. E., Palomo, A. and Ferrer, A. 1998 Pre-pottery Neolithic A and Pre- Pottery Neolithic B lithic agricultural tools on the Middle Euphrates: The sites of Tell Mureybit and Tell Halula. In A. B. Damania, J. Valkourn, G. Willcox and C. O. Qualset (eds) *The Origins of Agriculture and Crop Domestication*, pp.132-144. Aleppo: ICARDA.

Idriess, I. L. 1953 *The Red Chief*, Sydney: Angus and Robertson.

Idriess, I. L. 1964 *Our Stone Age Mystery*, Sydney: Angus and Robertson.

Ilbery, E. S. 1927 The Battle of Pinjarra, 1834: The Passing of the Bibbulmun. *Journal and Proceedings of the Western Australian Historical Society* 1(1):24-30.

Imamura, K. 1996 *Prehistoric Japan: New Perspectives on Insular East Asia*, Honolulu: University of Hawai'i Press.

Imlay, Dr. 1838 Journey of Dr. Imlay to the River Murray. In T. H. James (comp.) *Six Months in South Australia*, pp.229-237. London: J. Cross.

Ingold, T. 1983 The significance of storage in hunting societies. *Man* 18:553-571.

Ingold, T. 1996 Growing plants and raising animals: An anthropological perspective. In D. R. Harris (ed.) *The Origins and Spread of Agriculture and Pastoralism in Eurasia*, pp.12-24. London: UCL Press.

Institute of Aboriginal Development 1992 *Women's Gathering and Hunting in the Pitjantjatjara Homelands*, Alice Springs: IAD Press.

Irwin, F. C. 1835 *The State and Position of Western Australia; Commonly Called the Swan-River Settlement*, London: Simpkin, Marshall and Co.

Isaac, B. L. 1993 Retrospective on the formalist-substantivist debate. *Research in Economic Anthropology* 14:213-233.

Isaac, B. L. 1996 Introduction. *Research in Economic Anthropology* 17:1-10.

Isaacs, J. 1984 *Australia's Living Heritage: Arts of the Dreaming*, Sydney: Lansdowne Press.

Isaacs, J. 1987 Bush Food: Aboriginal Food and Herbal Medicine, McMahon's Point: Weldons.

Isaacs, J. 1996 *A Companion Guide to Bush Food*, Sydney: Lansdowne.

Jackson, C. F. V. 1902 *Report on a Visit to the West Coast of the Cape York Peninsula and Some Islands of the Gulf of Carpentaria Also Reports on Horn Island and Possession Island Gold Fields, and Recent Prospecting of the Cretaceous Coals of the Cook District*, Brisbane: Government Printer. *Queensland Department of Mines Geological Survey Report No. 180.*

Jackson, G. and de Gand, D. 1996 *Report on an Aboriginal Heritage Study of the Nanda Area Central, Kalbarri Region, Western Australia*, Perth: Anthropos Australia.

Jackson, T. L. 1991 Pounding acorns: Women's production as social and economic focus. In J. M. Gero and M. W. Conkey (eds) *Engendering Archaeology: Women and Prehistory*, pp.301- 325. Oxford: Blackwell.

Jarman, Captain 1863 The 'Tien Tsin' track to the Harding River. Our extracts from Captain Jarman's. *Exploration Diaries* 5:55-60. Battye Library: PR 5441.

Jarman, H. N. 1972 The origins of wheat and barley cultivation. In E. S. Higgs (ed.) *Papers in Economic Prehistory*, pp.15-26. London: Cambridge University Press.

Jarman, M. R., Bailey, G. N. and Jarman, H. N. 1982 *Early European Agriculture*, Cambridge: Cambridge University Press.

Jenkin, G. 1985 *Conquest of the Ngarrinjeri*, Point McLeay: Raukkan Publishers.

Jewell, P. 1981 A summing-up. In R. Mercer (ed.) *Farming Practices in British Prehistory*, pp.223- 230. Edinburgh: Edinburgh University Press.

Johnson, C. 2006 *Australia's Mammal Extinctions: A 50 000 Year History*, Cambridge: Cambridge University Press.

Johnson, G. A. 1982 Organisational structure and 'scalar stress'. In A. C. Renfrew, M. J. Rowland and B. A. Segraves (eds) *Theory and Explanation in Archaeology*, pp.389-421. New York: Academic Press.

Johnson, J. E. 1963 Observations on some Aboriginal campsites in South Australia and adjoining states. *Mankind* 6(2):64-79.

Johnson, K. and Burchett, M. (eds) 1996 *Native Australian Plants*, Sydney: University of New South Wales Press.

Johnston, T. H. 1943 Aboriginal names and utilization of

fauna in the Eyrean region. *Transactions of the Royal Society of South Australia* 67:244-311.

Johnston, T. H. and Cleland, J. B. 1943 Native names and uses of plants in the north-eastern corner of South Australia. *Transactions of the Royal Society of South Australia* 67:149-173.

Jones, R. 1969 Firestick farming. *Australian Natural History* 16(7):224-228.

Jones, R. 1970 Tasmanian Aborigines and dogs. *Mankind* 7:256-271.

Jones, R. 1975 The Neolithic, Paleolithic and the hunting gardeners: Man and land in the Antipodes. In R. P. Suggate and M. M. Cresswell (eds) *Quaternary Studies*, pp.21-34. Wellington: *Royal Society of New Zealand Bulletin No. 13*

Jones, R. 1976 Tasmania: Aquatic machines and off-shore islands. In G. de G. Sieveking, I. H. Longworth and K. E. Wilson (eds) *Problems in Economic and Social Archaeology*, pp.235- 263. London: Duckworth.

Jones, R. 1984 Hunters and history: A case study from Tasmania. In C. Schrire (ed.) *Past and Present in Hunter Gatherer Studies*, pp.27-65. Orlando: Academic Press.

Jones, R. 1990 Hunters of the Dreaming: Some ideational, economic and ecological parameters of the Australian Aboriginal productive systems. In D. E. Yen and J. M. J. Mummery (eds) *Pacific Production Systems: Approaches to Economic Prehistory*, pp.25-53. Canberra: Department of Prehistory, Australian National University. *Occasional Papers in Prehistory No. 18.*

Jones, R. and Bowler, J. 1980 Struggle for the savannah: Northern Australia in ecological and prehistoric perspective. In R. Jones (ed.) *Northern Australia: Options and Implications*, pp.3-31. Canberra: Research School of Pacific Studies, Australian National University.

Jones, R. and Meehan, B. 1989 Plant foods of the Gidjingali: Ethnographic and archaeological perspectives from northern Australia on tuber and seed exploitation. In D. R. Harris and G. C. Hillman (eds) *Foraging and Farming: The Evolution of Plant Exploitation*, pp.120-135. London: Unwin Hyman.

Jones, T. L. 1996 Mortars, pestles and the division of labor in prehistoric California: A view from Big Sur. *American Antiquity* 61(2):243-264.

Jones, W. 1979 Up the creek: Hunter-gatherers in the Cooper Basin. Unpublished BA (Hons) thesis, Armidale, University of New England.

Jordan, D. K. 1999 Chronological Table of Mesoamerican Archaeology.http://weber.ucsd.edu/ ~dkjordan/arch/mexchron.html

Jukes, J. B. 1862 Canoes in Australia. *The Athenaeum* 1793:331.

Jung, K. E. 1884 *Australia: The Country and Its Inhabitants*, London: W. Swan Sonnenschein and Co.

Kaberry, P. M. 1935 The Forrest River and Lyne River tribes of north-west Australia. *Oceania* 5(4):408-436.

Kabo, V. R. 1985 The origins of the food producing economy. *Current Anthropology* 26(5):601- 616.

Kalland, A. 1995 *Fishing Villages of Tokugawa Japan*, Honolulu: University of Hawaii Press.

Keegan, W. F. and Diamond, J. M. 1987 Colonization of islands by humans: A biogeographical perspective. *Advances in Archaeological Method and Theory* 10:49-92.

Keeley, L. H. 1988 Hunter-gatherer economic complexity and "Population Pressure": A cross- cultural analysis. *Journal of Anthropological Archaeology* 7:373-411.

Keeley, L. H. 1995 Protoagricultural practices among hunter-gatherers. In T. D. Price and A. Gebauer (eds) *Last Hunters – First Farmers: New Perspectives on the Prehistoric Transition to Agriculture*, pp.243-272. Santa Fe: School of American Research Press.

Keen, I. 1982 How some Murngin men marry ten wives: The marital implications of matrilineal cross-cousin structures. *Man* 17:620-642.

Kefous, K. C. 1988 Butlin's bootstraps: Aboriginal population in the pre-contact Murray-Darling region. In B. Meehan and R. Jones (eds) *Archaeology With Ethnography: An Australian Perspective*, pp.225-237. Canberra: Department of Prehistory, Research School of Pacific Studies, Australian National University.

Keighery, G. J. 1990 The vegetation and flora of Shark Bay. In P. F Berry, S. D. Bradshaw and B. R. Wilson (eds) *Research in Shark Bay: Report of the France-Australe Bicentenary Expedition Committee*, pp.61-87. Perth: Western Australian Museum.

Keighery, G. J. and Gibson, N. 1993 Biogeography and composition of the flora of the Cape Range peninsula, Western Australia. In W. F. Humphreys (ed.) *The Biogeography of the Cape Range, Western Australia*, pp.51-85. Perth: Western Australian Museum. *Records of the Western Australian Museum, Supplement No. 45.*

Kelly, J. D. 1968 Hut sites, rock engravings, stone arrangements and Tjuringa. Mulligan River, Queensland. *Mankind* 6(11):563-566.

Kelly, R. L. 1992 Mobility/sedentism: Concepts, archaeological measures and effects. *Annual Review of Anthropology* 21:43-66.

Kelly, R. L. 1995 *The Foraging Spectrum: Diversity in Hunter-Gatherer Lifeways*, Washington: Smithsonian Institution Press.

Kelly, R. L. 1998 Foraging and sedentism. In T. R. Rocek and O. Bar-Yosef (eds) *Seasonality and Sedentism: Archaeological Perspectives from Old and New Worlds Sites*, pp.9-23. Cambridge (Mass.): Peabody Museum of Archaeology and Ethnology.

Kennedy, D. I. D. and Bouchard, R. T. 1990 North Coast Salish. In W. Suttles, (ed.) *Handbook of North American Indians, Volume 7: Northwest Coast*, pp.441-452. Washington: Smithsonian Institution.

Kennedy, E. B. 1852 Extracts from the Journal of an exploring expedition into Central Australia to determine the course of the River Barcoo (or the Victoria of Sir T. L. Mitchell). *Journal of the Royal Geographical Society of London* 22:228-280.

Kenrick, D. M. 1995 *Jomon Japan: The World's Oldest Pottery*, London and New York: Kegan Paul International.

Kent, S. 1989 Cross-cultural perceptions of farmers as hunters and the value of meat. In S. Kent, (ed.) *Farmers as Hunters : The Implications for Sedentism*, pp.1-17. Cambridge: Cambridge University Press.

Kenyon, A. S. 1912 Camping places of the Aborigines of south-east Australia. *Victorian Historical Magazine* 2(3):99-110.

Kenyon, A. S. 1930-31 Stone structures of the Australian Aborigines. *The Victorian Naturalist* 47: 71-75.

Kenyon, K. M. 1957 *Digging Up Jericho*, London: Ernst Benn.

Kenyon, K. M. 1960-83 *Excavations at Jericho. Volume 3*, London: British School of Archaeology in Jerusalem.

Kerwin, B. and Breen, J. G. 1981 The land of stone chips. *Oceania* 51(4):286-311.

Kijas, J. 1997. An 'Unfashionable concern with the past': The historical anthropology of Diane Barwick. *Australian Aboriginal Studies* 1997/1:48-60.

Kilham, C., Pamulkan, M., Pootchemunka, J. and Wolmby, T. 1986 *Dictionary and Source Book of the Wik-Mungkan Language*, Darwin: Summer Institute of Linguistics.

Kimber, R. G. 1976 Beginnings of farming? Some man-plant-animal relationships in central Australia. *Mankind* 10(3):142-150.

Kimber, R. G. 1984 Resource use and management in central Australia. *Australian Aboriginal Studies* 1984/2:12-23.

Kimber, R. G. 1988 Smallpox in Central Australia: Evidence for epidemics and postulation about the impact. *Australian Archaeology* 27:63-8.

Kimber, R. G. 1996 *Man From Arltunga: Walter Smith, Australian Bushman*, Victoria Park: Arltunga Hotel and Bush Resort and Hesperian Press.

King, J. 1861 John King's Narrative. In The Victorian Exploring Expedition: Diary of Burke and Wills, Journal of Howitt and Narrative of King, pp.2-5. Melbourne: The Age.

King, P. P 1826 *Narrative of a Survey of the Intertropical Coasts of Australia 1818-1822*, 2 volumes, London: John Murray.

Kirby, J. 1894 *Old Times in the Bush of Australia*, Ballarat: James Curtis.

Kirkpatrick, J., McDougall, K. and Hyde, M. 1995 *Australia's Most Threatened Ecosystems: The Southeastern Lowland Native Grasslands*, Chipping Norton: Surrey Beatty and Sons.

Kislev, M. E. 1992 Agriculture in the Near East in the VIIth millennium bc. In P. C. Anderson [-Gerfaud] (ed.) *Prehistoire de L'Agriculture: Nouvelles Approches Experimentales et Ethnographiques*, pp.87-93. Paris: Editions du CNRS. *Monographie du CRA No. 6.*

Kislev, M. E., Hartmann, A. and Bar-Yosef, O. 2006 Early domesticated fig in the Jordan Valley. *Science* 312(5778):1372-1374.

Kislev, M. E., Weiss, E. and Hartmann, A. 2004 Impetus for sowing and the beginning of agriculture: Ground collecting of wild cereals. *Proceedings of the National Academy of Sciences of the United States of America* 101(9):2692-2695.

Klein, R. G. 2000 Archaeology and the evolution of human behaviour. *Evolutionary Anthropology* 17:17-36.

Knapp, A. B. 1992 Preface. In A. B. Knapp (ed.) *Archaeology, Annales, and Ethnohistory*, pp.xv- xvi. Cambridge: Cambridge University Press.

Kneckt, H. 1994 Late Ice Age hunting technology. *Scientific American* 271(1):82-87.

Knight, J. 1990 A broken Juin knife from Yandan Creek: Some implications. *Archaeology in Oceania* 25(2):68-74.

Kotani, Y. 1981 Evidence of plant cultivation in Jomon Japan: Some implications. In S. Koyama and D. H. Thomas (eds) *Affluent Foragers*, pp. 201-212. Osaka: National Museum of Ethnology. *Senri Ethnological Studies No. 9.*

Koyama, S. and Thomas, D.H. 1981 Introduction. In S. Koyama and D. H. Thomas (eds) *Affluent Foragers*, pp.1-12. Osaka: National Museum of Ethnology. *Senri Ethnological Series No. 9.*

Krefft, G. 1862-65 On the manners and customs of the Aborigines of the lower Murray and Darling. *Transactions of the Philosophical Society of New South Wales* n.v.:357-374.

Kuhn, S. L. and Stiner, M. C. 2001 The antiquity of hunter-gatherers. In C. Panter-Brick, R. H. Layton and P. Rowley-Conwy (eds) *Hunter-Gatherers: An Interdisciplinary Perspective*, Cambridge: Cambridge University Press.

Kuijt, I. 2000 Life in Neolithic farming communities. In I. Kuijt (ed.) *Life in Neolithic Farming Communities: Social Organization, Identity and Differentiation*, pp.3-13. New York: Kluwer Academic and Plenum Publishers.

Kuijt, I. and Mahasneh, H. 1998 Dhra': An Early Neolithic village in the southern Jordan valley. *Journal of Field Archaeology* 25:153-161.

Kuzmin, Y. V. and Orlova, L. A. 2000 The neolithization of Siberia and Russian far east: Radiocarbon evidence. *Antiquity* 74:356-364.

Ladizinsky, G. 1987 Pulse domestication before cultivation. *Economic Botany* 41(1):60-65.

Ladizinsky, G. 1998 *Plant Evolution Under Domestication*, Dordrecht: Kluwer Academic Publishers.

Lamb, L. and Barker, B. 2001 Evidence for early Holocene change in the Whitsunday Islands: A new radiocarbon determination from Nara Inlet 1 rockshelter. *Australian Archaeology* 53:42-43.

Lamberg-Karlovsky, C. C. and Sabloff, J. A. 1995 *Ancient Civilizations: The Near East and Mesoamerica*, Prospect Heights: Waveland Press.

Lamond, H. G. 1950 Aboriginal net making. *Mankind* 4(4):168-169.

Lambert, P. M. and Walker, P. L. 1991 Physical anthropological evidence for the evolution of social complexity in coastal Southern California. *Antiquity* 65(249):963-973.

Lampert, R. J. 1966 An excavation at Durras North, New

South Wales. *Archaeology and Physical Anthropology in Oceania* 1(2):83-118.

Lampert, R. J. 1981 *The Great Kartan Mystery*, Canberra: Department of Prehistory, Research School of Pacific Studies, Australian National University.

Lampert, R. J. 1985 Archaeological reconnaissance on a field trip to Dalhousie Springs. *Australian Archaeology* 21:57-62.

Lampert, R. J. 1988 A preliminary report on fieldwork in the Lake Eyre Basin 1988. *Australian Archaeology* 27:69-74.

Lampert, R. J. 1989 Archaeology. In W. Zeidler and W. F. Ponder (eds) *Natural History of Dalhousie Springs*, Adelaide: South Australian Museum.

Lampert, R. J. and Hughes, P. J. 1974 Sea level changes and Aboriginal coastal adaptations in southern New South Wales. *Archaeology and Physical Anthropology in Oceania* 9(3):226- 35.

Landsborough, W. 1862 *Journal of Landsborough's Expedition From Carpentaria in Search of Burke and Wills*, Melbourne: F. F. Baillliere.

Lang, G. S. 1865 *The Aborigines of Australia*, Melbourne: Wilson and Mackinnon.

Larsen, C. S. and Ruff, C. B. 1994 The stresses of conquest in Spanish Florida: Structural adaptation and change before and after contact. In C. S. Larsen and G. R. Milner (eds) *In the Wake of Contact: Biological Responses to Conquest*, pp.21-34. New York, Wiley-Liss.

Latz, P. 1995 *Bushfires and Bushtucker: Aboriginal Plant Use in Central Australia*, Alice Springs: IAD Press.

Latz, P. 2007 *The Flaming Desert: Arid Australia – A Fire Shaped Landscape*, Alice Springs: P. Latz.

Lawson, E. 1989 *Louisa Atkinson: The Distant Sound of Native Voices*, Canberra: English Department, University College, Australian Defence Force Academy. *Occasional Papers No. 15.*

Leach, H. M. 1997 The terminology of agricultural origins and food production systems – a horticultural perspective. *Antiquity* 71:135-148.

Leakey, R. E 1981 *The Making of Mankind*, London: Michael Joseph Limited.

Lebot, V. 1999 Biomolecular evidence for plant domestication in Sahul. *Genetic Resources and Crop Evolution* 46:619-628.

Lee, G-A., Crawford, G. W., Lui, L. and Chen, X. 2007 Plants and people from the Early Neolithic to Shang periods in North China. *Proceedings of the National Academy of Sciences of the United States of America* 104(3):1087-1092.

Lee, R. B. 1968 What hunters do for a living, or how to make out on scarce resources. In R. B. Lee and I. DeVore (eds) *Man the Hunter*, pp.30-43. Chicago: Aldine.

Lee, R. B. 1976 Introduction. In R. B. Lee and I. DeVore (eds) *Kalahari Hunter-Gatherers: Studies of the !Kung San and Their Neighbors*, pp.3-24. Cambridge (Mass.): Harvard University Press.

Legge, T. 1996 The beginning of caprine domestication in southwest Asia. In D. R. Harris (ed.) *The Origins and Spread of Agriculture and Pastoralism in Eurasia*, pp.12-24. Washington DC: Smithsonian Institute Press.

Leichhardt 1847a Letter to Mr Lynd. In J. D. Lang 1847 *Cooksland in North-East Australia*, pp.375-6. London: Longman, Brown, Green and Longmans.

Leichhardt, L. 1847b *Journal of an Overland Expedition in Australia From Moreton Bay to Port Essington*, London: T. and W. Boone.

Leonard, J. A., Wayne, R. K., Wheeler, J., Valadez, R. Guillen, S. and Vilas, C. 2002 Ancient DNA evidence for Old World origins of New World dogs. *Science* 298(5598):1613-1616.

Le Souef, W. H. D. 1907 *Wild Life in Australia*, Christchurch: Whitcombe.

Lewis, B. S. 1996 The role of attached and independent specialisation in the development of sociopolitical complexity. *Research in Economic Anthropology* 17:357-388.

Lewis, J. 1922 *Fought and Won*, Adelaide: W. K. Thomas.

Lewis, J. W. 1975a *Report on the Lake Eyre Expedition*, Adelaide: Government Printer.

Lewis, J. W. 1875b *Journal of Mr Lewis's Lake Eyre Expedition, 1874-5*, South Australian Parliamentary Papers, No. 19 of 1875.

Lieberman, D.E. 1993 The rise and fall of seasonal mobility among hunter-gatherers: The case of the southern Levant. *Current Anthropology* 34(5):599-631.

Lieberman, D. E. 1998 Natufian "sedentism" and the importance of biological data for estimating reduced mobility. In T. R. Rocek and O. Bar-Yosef (eds) *Seasonality and Sedentism: Archaeological Perspectives From Old and New World Sites*, pp.75-92. Cambridge (Mass.): Peabody Museum of Archaeology and Ethnology.

Lindsay, D. 1890 Explorations in the northern territory of South Australia. *Proceedings of the Royal Geographical Society of Australasia, South Australian Branch* 2(3):1-16.

Littleton, J. 2002 Mortuary behaviour on the Hay Plain: Do cemeteries exist. *Archaeology in Oceania* 37(3):105-122.

Lloyd, G. T. 1862 *Thirty-Three Years in Tasmania and Victoria*. London: Houston and Wright.

Lockyer, E. 1826-7 Journal of Major Lockyer, commandant of the expedition sent from Sydney in 1826 to found a settlement at King George's Sound, Western Australia. In Lockyer, E. Papers 1826-1827 [manuscript]. National Library of Australia MS 123.

Lockyer, E. 1888 Journal of an excursion up the River Brisbane in the year 1825 by Edmund Lockyer. In H. S. Russell (comp.) *The Genesis of Queensland*, Sydney: Turner and Henderson.

Londo, J. P, Chiang, Y-C., Hung, K-H., Chiang, T-Y., and Schaal, B. 2006 Phylogeography of Asian wild rice, *Oryza rufipogon*, reveals multiple independent domestications of cultivated rice, *Oryza sativa*. *Proceedings of the National Academy of Sciences of the United States of America* 103(25):9578-9583

Longacre, W. A. 1995 Why did they invent pottery anyway?

In W. K. Barnett and J. W. Hoopes (eds) *The Emergence of Pottery: Technology and Innovation in Ancient Society*, pp.277- 280. Washington: Smithsonian Institution Press.

Lord, W. B. and Baines, T. 1876 *Shifts and Expedients of Camp Life, Travel and Exploration*, London: H. Cox.

L'Oste-Brown, S., Godwin, L. and Morwood, M. 2002 Aboriginal bark burials: 700 years of mortuary tradition in Central Queensland. *Australian Aboriginal Studies* 2002/1:43-50.

Lourandos, H. 1976 Aboriginal settlement and land use in south western Victoria: A report on current field work. *The Artefact* 1(4):174-193.

Lourandos, H. 1977 Aboriginal spatial organization and population: South western Victoria reconsidered. *Archaeology and Physical Anthropology in Oceania* 12:202-225.

Lourandos, H. 1980a Change or stability?: hydraulics, hunter-gatherers, and population in temperate Australia. *World Archaeology* 11(3):245-264.

Lourandos, H. 1980b Forces of Change: Aboriginal technology and population in south western Victoria. Unpublished Ph.D. thesis, University of Sydney, Sydney.

Lourandos, H. 1983 Intensification: A late Pleistocene-Holocene archaeological sequence from southwestern Victoria. *Archaeology in Oceania* 18(2):81-94.

Lourandos, H. 1985 Intensification and Australian prehistory. In T. D. Price and J. A. Brown (eds) *Prehistoric Hunter-Gatherers: The Emergence of Complexity*, pp.385-423. San Diego: Academic Press.

Lourandos, H. 1988 Paleopolitics : Resource intensification in Aboriginal Australia and Papua New Guinea. In T. Ingold, D. Riches and J. Woodburn (eds) *Hunters and Gatherers 1: History, Evolution and Social Change*, pp.148-160. Oxford: Berg.

Lourandos, H. 1993 Hunter-gatherer cultural dynamics: Long- and short-term trends in Australian prehistory. *Journal of Archaeological Research* 1:67-88.

Lourandos, H. 1997 *Continent of Hunter-Gatherers: New Perspectives in Australian Prehistory*, Cambridge: Cambridge University Press.

Lourandos, H. and David, B. 1998 Comparing long-term archaeological and environmental trends: North Queensland and semi-arid Australia. *The Artefact* 21:105-114.

Lourandos, H. and Ross, A. 1994 The great 'Intensification' debate: Its history and place in Australian archaeology. *Australian Archaeology* 39:54-63.

Love, J. R. B. 1917 Notes on the Wororra tribe of north-western Australia. *Transactions and Proceedings of the Royal Society of South Australia* 41:21-38.

Love, J. R. B. 1936 *Stone-Age Bushmen of Today*, London: Blackie and Sons.

Lu, H., Liu, Z., Wu, N., Berne, S., Sato, Y., Liu, B., and Wang, L. 2002 Rice domestication and climatic change: Phytolith evidence from East China. *Boreas* 31:378-385.

Lu, T. L-D. 1998 Some botanical characteristics of green foxtail (*Setaria viridis*) and harvesting experiments on the grass. *Antiquity* 72(278):902-907.

Lu, T. L-D. 1999 *The Transition from Foraging to Farming and the Origins of Agriculture in China*, Oxford: BAR Publishing. *BAR International Series No. 774.*

Lubbock, J. 1865 *Prehistoric Times As Illustrated by Ancient Remains and the Manners and Customs of Modern Savages*, London: Williams and Norgate.

Luebbers, R. A. 1978 Meals and menus: A study of change in prehistoric coastal settlement in South Australia. Unpublished PhD thesis, Australian National University, Canberra.

Lumholtz, C. 1889 *Among Cannibals*, New York: Charles Scribner and Sons.

Macgillivray, J. 1852 *Narrative of the Voyage of the HMS Rattlesnake*, 2 volumes, London: T. and W. Boone.

Macknight, C. C. 1972 Macassans and Aborigines. *Oceania* 42(4):283-321.

Macknight, C. C. 1976 *The Voyage to Marege: Macassan Trepangers in Northern Australia*, Melbourne: Melbourne University Press.

Macknight, C. C. 1986 Macassans and the Aboriginal past. *Archaeology in Oceania* 21(1):69-75.

MacNeish, R. S. 1972 The evolution of community patterns in the Tehuacan Valley of Mexico and speculations about the cultural process. In P. J. Ucko, R. Tringham and G. W. Dimbleby (eds) *Man, Settlement and Urbanism*, pp.67-93. London: Duckworth.

MacNeish, R. S. 1986 The preceramic of Middle America. *Advances in World Archaeology* 5:93- 129.

MacNeish, R. S. 1992 *The Origins of Agriculture and Settled Life*, Norman: University of Oklahoma Press.

MacPherson, P. 1885 The oven-mounds of the Aborigines of Victoria. *Journal and Proceedings of the Royal Society of New South Wales* 18:49-59.

Maddock, K. 1982 *The Australian Aborigines*, Ringwood: Penguin.

Maiden, J.H. 1889 *The Useful Native Plants of Australia.* Sydney: Turner and Henderson.

Maiden, J. H. 1898 *A Manual of the Grasses of New South Wales*, Sydney: Government Printer.

Major, R. H. 1859 *Early Voyages to Terra Australis*, London: Hakluyt Society.

Malcolm, J. 1845 Evidence of James Malcolm Esq. In *New South Wales: Votes and Proceedings of the Legislative Council During the Session of the Year 1845 - New South Wales, Aborigines. Report from the Select Committee in the condition of the Aborigines with Appendix, Minutes of Evidence and replies to a circular letter*, pp.12-15.

Malinowski, B. 1922 *Argonauts of the Western Pacific: An Account of Native Enterprise and Adventure in the Archipelagos of Melanesian New Guinea*, London: George Routledge and Sons.

Mann, J. F. 1893 Canoes and fishing. In *Pamphlets Issued by the New South Wales Commissioners for the World's Columbian Exposition, Chicago, 1893*, volume 1, pp.98-101. 2 volumes, Sydney: Government Printer.

Marchant, L. R. 1988 *An Island Unto Itself: William Dampier and New Holland*, Victoria Park: Hesperian Press.
Marcus, J. 1976 The size of the early Mesoamerican village. In K. V. Flannery (ed.) *The Early Mesoamerican Village*, pp.79-90. New York: Academic Press.
Marcus, J. and Flannery, K. V. 1996 *Zapotec Civilization: How Urban Society Evolved in Mexico's Oaxaca Valley*, London: Thames and Hudson.
Marquardt, W. H. 1988 Politics of production among the Calusa of south Florida. In T. Ingold, D. Riches and J. Woodburn (eds) *Hunters and Gatherers 1: History, Evolution and Social Change*, pp.161-188. New York: Berg.
Marshall, A. 1937 Where the Diamantina flows. *Walkabout* 4(2):57-61.
Martin, J. 1865 Explorations in North-Western Australia. *Journal of the Royal Geographical Society* 35:237-289.
Maschner, H. D. G. 1991 The emergence of cultural complexity on the northern Norwest Coast. *Antiquity* 65(249):924-934.
Maschner, H. D. G. and Fagan, B. M. 1991 Introduction; Hunter-gatherer complexity on the west coast of North America. *Antiquity* 65(249):921-923.
Maslow, A. H. 1943 A theory of human motivation. *Psychological Review* 50:376-396.
Maslow, A. H. 1970 *Motivation and Personality*, New York: Harper and Rowe.
Massola, A. 1956 Australian fish hooks and their distribution. *Memoirs of the National Museum of Victoria* 22(1):1-17.
Massola, A. 1963 Aboriginal relics in Victoria. *Australian Natural History* 14(8):254-258.
Massola, A. 1968 The Aboriginal fish traps at Homerton. *The Victorian Naturalist* 85(7):197-200.
Massola, A. 1969 *Journey to Aboriginal Victoria*, Adelaide: Rigby.
Massola, A. 1970 *Aboriginal Mission Stations in Victoria: Yelta, Ebenezer, Ramahyuck, Lake Condah*. Melbourne: Hawthorn Press.
Mathew, J. 1899 *Eaglehawk and Crow*, London: David Nutt.
Mathew, J. 1910 *Two Representative Tribes of Queensland*, London: T. Fisher Unwin.
Mathews, R. H. 1896 Australian ground and tree drawings. *American Anthropologist* 9(2):33-49.
Mathews, R. H. 1900-01 Ethnological notes on the Aboriginal tribes of the Northern Territory. *Queensland Geographical Journal* 16(2):69-90.
Mathews, R. H. 1903 The Aboriginal fisheries at Brewarrina. *Journal and Proceedings of the Royal Society of New South Wales* 37:146-156.
Mathews, R. H. 1904 Ethnological notes on the Aboriginal tribes of New South Wales and Victoria. *Journal and Proceedings of the Royal Society of New South Wales* 38:203-381.
Mathews, R. H. 1908 Aboriginal navigation. *Queensland Geographical Journal* 23(9):66-81.
Mathews, R. H. 1908-9 Some articles used in burial and other rites by the Australian Aborigines. *Queensland Geographical Journal* 24(10):63-72.
Mathews, R. H. 1909a Some rock pictures and ceremonial stones of the Australian Aborigines. *Australasian Association for the Advancement of Science* 1909:493-498.
Mathews, R. H. 1909b Ceremonial stones used by the Australian Aborigines. *Proceedings of the American Philosophical Society* 48:1-7.
Matsui, A. 1996 Archaeological investigations of anadromous salmonid fishing in Japan. *World Archaeology* 27(3):444-460.
Matsuoka, Y., Vigouroux, Y., Goodman, M. M., Jesus, S. G. Buckler, E. and Doebley, J. 2002 A single domestication for maize shown by multilocus microsatellite genotyping. *Proceedings of the National Academy of Sciences of the United States of America* 99(9): 6080-6084.
Matthews, J. (ed.) 1977 *The Two Worlds of Jimmy Barker*, Canberra: Australian Institute of Aboriginal Studies.
Matthews, P. J. 2003 Taro planthoppers (*Tarophagus* spp.) in Australia and the origins of taro (*Colocasia esculenta*) in Oceania. *Archaeology in Oceania* 38(3):192-202.
Matthews, P. J. 2006 How and when did taro become the most widespread food crop in the world. In R. Torrance and H. Barton (eds) *Ancient Starch Research*, pp.22-23. Walnut Creek: Left Coast Press.
Mauss, M. 1990 *The Gift: The Form and the Reason for Exchange in Archaic Societies*, W. D. Halls (trans.) New York: W. W. Norton.
McArthur, M. 1960 Report of the nutrition unit. In C. P. Mountford (ed.) *Records of the American Australian Scientific Expedition to Arnhem Land, Volume 2, Anthropology and Nutrition*. pp.1-143. Parkville: Melbourne University Press.
McBryde, I. 1984a Exchange in south eastern Australia: An ethnohistorical perspective. *Aboriginal History* 8:132-153.
McBryde, I. 1984b Kulin greenstone quarries: The social contexts of production and distribution for the Mt William site. *World Archaeology* 16(2):267-285.
McBryde, I. 1987 Goods from another country: Exchange networks and the people of the Lake Eyre Basin. In D. J. Mulvaney and J. P. White (eds) *Australians to 1788*, pp.253-273. Sydney: Fairfax, Syme and Weldon.
McCarthy, F. D. 1938-40 Trade in Aboriginal Australia, and trade relationships with Torres Strait, New Guinea and Malaya. *Oceania* 9:405-438, 10:80-104,171-195.
McCarthy, F. D. 1962 Aborigines: Domestic and economic life. In A. H. Chisholm (ed.) *Australian Encyclopaedia, Volume 1*, pp.29-37. Sydney: Grollier.
McCarthy, F. D. 1970 Aboriginal antiquities in New South Wales. In F. D. McCarthy (ed.) *Aboriginal Antiquities in Australia: Their Nature and Preservation*, pp.15-25. Canberra: Australian Institute of Aboriginal and Torres Strait Islanders. *Australian Aboriginal Studies No. 22; Prehistory and Material Culture Series No. 3.*
McCarthy, F. D. 1976 *Australian Aboriginal Stone Implements*, Sydney: Australian Museum Trust.

McCarthy, F. D. and McArthur, M. 1960 The food quest and the time factor in Aboriginal economic life. In C. P Mountford (ed.) *Records of the American Australian Scientific Expedition to Arnhem Land, Volume 2, Anthropology and Nutrition*. pp.145-194. Parkville: Melbourne University Press.

McConnel, U. 1930-1 The Wik-Munkan tribe of Cape York Peninsula: Part 1. *Oceania* 1(1):97- 104; Part 2:1(2):181-205.

McConnel, U. 1957 *Myths of the Munkan*, Melbourne: Melbourne University Press.

McCorriston, J. and Hole, F. 1991 The ecology of seasonal stress and the origins of agriculture in the Near East. *American Anthropologist* 93(1):46-69.

McCourt, T. 1975 *Aboriginal Artefacts*, Adelaide: Rigby.

McGuire, K. R. and Hildebrandt, W. R. 1994 The possibilities of women and men: Gender and the California Milling Stone Horizon. *Journal of California and Great Basin Anthropology* 16(1):49-54.

McKellar, H. 1984 *Matya-Mundu: A History of the Aboriginal People of South West Queensland*, T. Blake (ed.), Cunnamulla: Cunnamulla Native Welfare Association.

McKelson, K. 1989 *Topical Vocabulary in Northern Nyangumarta*, Broome: Nulungu Catholic College.

McKinlay, J. 1861 *McKinlay's Journal of Exploration in the Interior of Australia*, Melbourne: F. F. Balliere.

McLellan, A. S. 1901 The Aboriginals of the Maranoa. *The Queenslander* 9 February 1901, p.258.

McNiven, I. J. 1998 Aboriginal settlement of the saline lake and volcanic landscapes of Corangamite Basin, western Victoria. *The Artefact* 21:63-94.

McNiven, I. J. 2000 Treats or retreats? Aboriginal island use along the Gippsland coast. *The Artefact* 23:22-34.

McNiven, I. J. 2006 Dauan 4 and the emergence of ethnographically-known social arrangements across Torres Strait during the last 600 – 800 years. *Australian Archaeology* 62:1-12.

Meadows, R. 1996 The origins and spread of agriculture and pastoralism in northwestern South Asia. In D. R. Harris (ed.) *The Origins and Spread of Agriculture and Pastoralism in Eurasia*, pp.390-412. London: UCL Press.

Meehan, B. 1977 Man does not live by calories alone: The role of shellfish in a coastal cuisine. In J. Allen, J. Golson and R. Jones (eds) *Sunda and Sahul: Prehistoric Studies in Southeast Asia, Melanesia and Australia*, pp.493-531. London: Academic Press.

Meehan, B. 1988 The 'dinner time camp'. In B. Meehan and R. Jones (eds) *Archaeology With Ethnography: An Australian Perspective*, pp.171-181. Canberra: Dept. of Prehistory, Research School of Pacific Studies, Australian National University.

Meggitt, M. J. 1964 Aboriginal food-gatherers of tropical Australia. In *The Ecology of Man in the Tropical Environment: Proceedings and Papers – International Union for the Conservation of Nature and Natural Resources: 9th Technical Meeting*. pp.30-37. Morges: The Union.

Meggitt, M. J. 1966 Indigenous forms of government among the Australian Aborigines. In I. Hogbin and L. R. Hiatt (eds) *Readings in Australian and Pacific Anthropology*, London: Melbourne University Press.

Mellaart, J. 1975 *The Neolithic of the Near East*, London: Thames and Hudson.

Mellars, P., and Reinhardt, S. C. 1978 Patterns of Mesolithic land-use in southern England: A geological perspective. In P. Mellars (ed.) *The Early Postglacial Settlement of Northern Europe: An Ecological Perspective*, pp.243-293. London: Duckworth.

Memmot, P. 2007 *Gunyah, Goondie and Wurley: The Aboriginal Architecture of Australia*, St Lucia: University of Queensland Press.

Meston, A. 1895 *Geographic History of Queensland*, Brisbane: Government Printer.

Meston, A. 1901 Maranoa Aboriginals. *The Queenslander* 12 January 1901, pp.63-64.

Meyer, H. A. E. 1846 *Manners and Customs of the Aborigines of the Encounter Bay Tribe, South Australia*, Adelaide: Government Printer.

Meyer, H. E. A. 1879 Manners and customs of the Aborigines of the Encounter Bay tribe, South Australia. In J. D. Wood (comp.) *Native Tribes of South Australia*, pp.183-206. Adelaide: E. S. Wigg.

Miller, J. and Seaburg, W. R. 1990 Athapascans of southwestern Oregon. In W. Suttles, (ed.) *Handbook of North American Indians, Volume 7: Northwest Coast*, pp.580-588. Washington: Smithsonian Institution.

Minnis, P. E 1992 Earliest plant cultivation in the Desert Borderlands of North America. In C. W. Cowan and P. J. Watson (eds) *The Origins of Agriculture: An International Perspective*, pp.121-41. Washington: Smithsonian Institute Press.

Mitchell, S. 1996 Dugongs and dugouts, sharptacks and shellbacks: Macassan contact and Aboriginal marine hunting on the Cobourg Peninsula, north western Arnhem Land. In I. C. Glover and P. Bellwood (eds) *Indo-Pacific History: The Chang Mai Papers, Volume 2*, pp. 181-191. *Bulletin of the Indo-Pacific Prehistory Association No. 15*.

Mitchell, S. R. 1949 *Stone-Age Craftsmen: Stone Tools and Camping Places of the Australian Aboriginals*, Melbourne: Tait Book Co.

Mitchell, T. L. 1839 *Three Expeditions into the Interior of Eastern Australia*, 2 volumes, London: T. and W. Boone.

Mitchell, T. L. 1848 *Journal of an Expedition Into the Interior of Tropical Australia*, London: Longman, Brown, Green and Longmans.

Mithen, S. 2003 *After the Ice*, London: Weidenfeld and Nicholson.

Mithen, S., Finlayson, B., Pirie, A., Carruthers, D., and Kennedy, A. 2000 New evidence of economic and technological diversity in the Pre-Pottery Neolithic A: Wadi Faynan 16. *Current Anthropology* 41(4):655-663.

Moore, A. M. T. 1983 The first farmers of the Levant. In T. C. Young, P. Mortensen and P. E. L. Smith (eds) *The Hilly Flanks and Beyond: Essays on the Prehistory of*

*Southwestern Asia*, pp.91-111. Chicago: Oriental Institute of the University of Chicago. *Studies in Ancient Oriental Civilization No. 36.*

Moore, A. M. T. 1995 The inception of potting in Western Asia and its impact on economy and society. In W. K. Barnett and J. W. Hoopes (eds) *The Emergence of Pottery: Technology and Innovation in Ancient Society*, pp.39-53. Washington: Smithsonian Institution Press.

Moore, D. R. 1979 *Islanders and Aborigines at Cape York*, Canberra: Australian Institute of Aboriginal Studies.

Moore, G. F. 1884 *Diary of Ten Years of an Early Settler in Western Australia*, London: M. Walbrook.

Morey, D. F. and Wiant, M. D. 1992 Early Holocene domestic dog burials from the North American Midwest. *Current Anthropology* 33(2):224-229.

Morey, E. 1952 *The Morey Papers*, V. R. de V. Voss (ed.), Emu Park.

Morgan, J. 1852 *The Life and Adventures of William Buckley*, Hobart: Archibald Macdougall.

Morse, K. 1993 Who can see the sea? Prehistoric Aboriginal occupation of the Cape Range Peninsula. In W. F. Humphreys (ed.) *The Biogeography of the Cape Range, Western Australia*, pp.227-242. Perth: Western Australian Museum. *Records of the Western Australian Museum, Supplement No. 45.*

Morse, K. 1999 Coastwatch: Pleistocene resource use on the Cape Range Peninsula. In J. Hall and I. McNiven (eds) *Australian Coastal Archaeology*, pp.73-78. Canberra: ANH Publications, Department of Archaeology and Natural History, Research School of Pacific and Asian Studies.

Morwood, M. J. 1981 Archaeology of the central Queensland highlands: The stone component. *Archaeology in Oceania* 16(1):1-52.

Morwood, M. J. 1984 The prehistory of the central Queensland highlands. In F. Wendorf and A. Close (eds) *Advances in World Archaeology, Volume 3*, pp.325-380. New York: Academic Books.

Morwood, M. J. 1986 The Whitsunday Islands: Initial historical and archaeological observations and implications for future work. *Queensland Archaeological Research* 3:72-87.

Morwood, M. J. and Hobbs, R. R. 1997 The Asian connection: Preliminary report on Indonesian trepang sites on the Kimberley coast, N.W. Australia. *Archaeology in Oceania* 32(3):197- 206.

Moseley, M. E. 1975 *The Maritime Foundations of Andean Civilization*, Menlo Park: Cummings Publishing

Mountford, C. P. 1939 Aboriginal methods of fishing and cooking as used on the southern coast of Eyre's Peninsula, South Australia. *Mankind* 2(7):196-200.

Mountford, C. P. 1976 *Nomads of the Australian Desert*, Rigby: Adelaide.

Mulvaney, D. J. 1964 Prehistory of the basalt plains. *Proceedings of the Royal Society of Victoria* New Series 77(2):427-432.

Mulvaney, D. J. 1975 *The Prehistory of Australia*. Ringwood: Penguin.

Mulvaney, D. J. 1989 *Encounters in Place: Outsiders and Aboriginal Australians 1606 – 1985*, St Lucia: University of Queensland Press.

Mulvaney, D. J. and Green, N. (eds) 1992 *Commandant of Solitude: The Journals of Captain Collet Barker*, Melbourne: Miegunyah Press.

Mulvaney, D. J. and Joyce, E. B. 1965 Archaeological and geomorphological investigations on Mt Moffat Station, Queensland, Australia. *Proceedings of the Prehistoric Society* 31:147-212.

Mulvaney, D. J. and Kamminga, J. 1999 *The Prehistory of Australia*. St. Leonards: Allen and Unwin.

Murdock, G. P. 1967 *Ethnographic Atlas*. Pittsburgh: University of Pittsburgh Press.

Murdock, G. P. and Provost, C. 1973 Measurement of cultural complexity. *Ethnology* 12(4):379- 392.

Murdock, G.P. and Wilson, S.F. 1972 Settlement patterns and community organization: Cross- cultural codes. *Ethnology* 11(3):254-295.

Murray, T. 1988 Ethnoarchaeology or palaeoethnology? In B. Meehan and R. Jones (eds) *Archaeology With Ethnography: An Australian Perspective*, pp.1-16. Canberra: Department of Prehistory, Research School of Pacific Studies, Australian National University.

Neill, J. 1845 Catalogue of reptiles and fish found at King George's Sound. In E. J. Eyre *Journal of Expeditions of Discovery into Central Australia and Overland From Adelaide to King George's Sound in the Years 1840-1, Volume 1*, pp.412-431. London: T. and W. Boone.

Newland, S. 1890 The Parkengees, or Aboriginal tribes on the Darling River. *Proceedings of the Royal Geographical Society of Australasia, South Australian Branch* 2:20-32.

Newland, S. 1920-21 Annual address of the President. *Proceedings of the Royal Geographical Society of Australasia, South Australian Branch* 22:1-64.

Newland, S. 1926 *Memoirs of Simpson Newland*, Adelaide: F. W. Preece and Sons.

Nicholson, A. and Cane, S. 1991 Archaeology on the Anxious Coast. *Australian Archaeology* 33:3- 13.

Nicholson, A. and Cane, S. 1994 Pre-European coastal settlement and use of the sea. *Australian Archaeology* 39:108-117.

Nind, S. 1832 Description of the natives of King George's Sound (Swan River Colony) and adjoining country. *Journal of the Royal Geographical Society of London* 1:21-51.

Nishida, M. 1983 The emergence of food production in Neolithic Japan. *Journal of Anthropological Archaeology* 2:305-322.

Nobbs, C. W. 1992 The inhabitants of Cooper Creek: A part translation of Carl Emil Jung's 'Am Cooper Creek'. *Records of the South Australian Museum* 26(2):129-138.

Normile, D. 1997 Yangtze seen as earliest rice site. *Science* 275(5298):309.

North, A. J. 1916-17 The birds of Coolabah and Brewar-

rina, North-Western New South Wales. *Records of the Australian Museum* 16:121-162.
Northern Territory Museum 1976-83 Site Files – Fish-traps.
Norton, A. 1907 Stray notes about our Aborigines. *Science of Man* 9(5):68-69;9(7):101-102.
Noy, T., Legge, A. J. and Higgs, E. S. 1973 Recent excavations at Nahal Oren, Israel. *Proceedings of the Prehistoric Society* 39:75-99.
Noy, T., Schuldenrein, J., and Tchernov, E. 1980 Gilgal, a Pre-Pottery Neolithic A site in the Lower Jordan Valley. *Israel Exploration Quarterly* 30:63-82.
Oakden, J. 1838 Notes on an excursion to the Murray by Messrs Hill, Wood, Willis and Oakden. In T. H. James (comp.) *Six Months in South Australia*, pp.289-291. London: J. Cross.
O'Connell, J. F. and Allen, J. 1995 Human reactions to the Pleistocene-Holocene transition in Greater Australia: A summary. *Antiquity* 69(265):855-862.
O'Connell, J. F. and Allen, J. 2004 Dating the colonization of Sahul (Pleistocene Australia – New Guinea): A review of recent evidence. *Journal of Archaeological Science* 31:835-853.
O'Connell, J. F. and Hawkes, K. 1981 Alyawara plant use and optimal foraging theory. In B. Winterhalder and E. A. Smith (eds) *Hunter-Gatherer Foraging Strategies: Ethnographic and Archaeological Analyses*, pp.99-125. Chicago: Chicago University Press.
O'Connell, J. F., Hawkes, K, and Jones, N. B. 1988 Hadza hunting, butchering and bone transport and their archaeological implications. *Journal of Anthropological Research* 44(2):113-161.
O'Connell, J. F., Latz, P. K. and Barnett, P. 1983 Traditional and modern plant use among the Alyawara of Central Australia. *Economic Botany* 37(1):80-109.
O'Connor, S. 1987 The stone house structures of High Cliffy Island, North West Kimberley, Western Australia. *Australian Archaeology* 25:30-39.
O'Connor, S. 1992 The timing and nature of prehistoric island use in northern Australia. *Archaeology in Oceania* 27(2):49-60.
O'Connor, S. 1993 Saltwater people of the southwest Kimberley coast. In G. Burenhult (ed.) *People of the Stone Age: Hunter-Gatherers and Early Farmers*, pp.226-227. St. Lucia: University of Queensland Press.
O'Connor, S. 1994 A 6700 bp date for island use in the West Kimberley, Western Australia: New evidence from High Cliffy Island. *Australian Archaeology* 39:102-108.
O'Connor, S. 1999a *30,000 Years of Aboriginal Occupation: Kimberley, North West Australia*, Canberra: ANH Publications and The Centre for Archaeological Research. *Terra Australis 14.*
O'Connor, S. 1999b A diversity of coastal economies: Shell mounds in the Kimberley. In J. Hall and I. McNiven (eds) *Australian Coastal Archaeology*, pp.37-50. Canberra: ANH Publications, Department of Archaeology and Natural History, Research School of Pacific and Asian Studies.
O'Connor, S., Apkin, K., Spriggs, M., Veth P. and Ayliffe, L. K. 2002 From savanna to rainforest: Changing environments and human occupation at Liang Lembudu, Aru Islands, Maluku (Indonesia). In A. P. Kershaw, N. J. Tapper, B. David, P. Bishop and D. Penny (eds) *Bridging Wallace's Line: The Environment and Cultural History and Dynamics of the Southeast Asian - Australasian Region*, pp. 279-306. Reiskirchen: Catena.
O'Connor, S., Veth, P. and Campbell, C. 1998 Serpent's Glen Rockshelter: Report of the Pleistocene-aged occupation sequence from the Western Desert. *Australian Archaeology* 46:12-22.
Officer, G. 1901 Notes on an Aboriginal grave in the Darling River district. *Proceedings of the Linnean Society of New South Wales* 26(2):238-241.
Oldfield, A. 1865 On the Aborigines of Australia. *Transactions of the Ethnological Society of London* 3:215-298.
Olsen, S. J. 1985 *Origins of the Domestic Dog: The Fossil Record*, Tucson: University of Arizona Press.
Olszewski, D. 1991 Comments [on Blumler and Byrne 1991]. *Current Anthropology* 32(1):43.
O'Neill, G. 2006 How green is my valley. *Australian Life Scientist* 3(6):12-13.
Opler, M. E. 1983 The Apachean culture pattern and its origins. In A. Ortiz (ed.) *Handbook of North American Indians, Volume 10: Southwest*, pp.368-392. Washington: Smithsonian Institution.
O'Shea, J. 1981 Coping with scarcity: Exchange and social storage. In A. Sheridan and G. Bailey (eds) *Economic Archaeology: Towards an Integration of Ecological and Social Approaches*, pp.168-183. Oxford: British Archaeological Reports. *BAR International Series No. 96.*
Oxley, J. 1820 *Journal of Two Expeditions into the Interior of New South Wales Undertaken by the Orders of the British Government in the Years 1817-18*, London: John Murray.
Palmer, E. 1884a Notes on some Australian tribes. *Journal of the Anthropological Institute of Great Britain and Ireland* 13:276-334.
Palmer, E. 1884b On plants used by the natives of north Queensland, Flinders and Mitchell Rivers, for food, medicine, etc. *Journal and Proceedings of the Royal Society of New South Wales* 17:93-114.
Palmer, E. 1903 *Early Days in North Queensland*, Sydney: Angus and Robertson.
Palmer, E. G. W. 1893 Aboriginal fisheries. *Pamphlets Issued by the New South Wales Government Commissioners for the World's Columbia Exposition, Chicago, 1893*, volume 1, pp.95-98. 2 volumes, Sydney: Government Printer.
Palsson, G. 1988 Hunters and gatherers of the sea. In T. Ingold, D. Riches and J. Woodburn (eds) *Hunters and Gatherers 1: History, Evolution and Social Change*, pp.189-204. New York: Berg.
Pamphlett, T and Uniacke, J. (comp.) 1825 Narrative of Thomas Pamphlett. In B. Field (ed.) *Geographical Memoirs on New South Wales*, pp.87-129. London: John Murray.

Pardoe, C. 1988 The cemetery as symbol. The distribution of prehistoric Aboriginal burial grounds in southeastern Australia. *Archaeology in Oceania* 23(1):1-16.
Pardoe, C. 1995 Riverine, biological and cultural evolution in southeastern Australia. *Antiquity* 69(265):696-713.
Pardoe, C. 2003 The Menindee Lakes: A regional archaeology. *Australian Archaeology* 57:42-53.
Parker, E. R. 1886 York District – Whajook Tribe. In E. M. Curr (comp.) *The Australian Race*, vol. 1, pp.336-341. Melbourne: Government Printer.
Parker, K. L. 1897 *Australian Legendary Tales: Folklore of the Noongaburrahs as Told to the Piccaninnies*, London: David Nutt.
Parker, K. L. 1898 *More Australian Legendary Tales*, London: David Nutt.
Parker, K. L. 1905 *The Euahlayi Tribe*, London: Archibald, Constable.
Parker, K. L. and Muir, M. 1982 *My Bush Book: K. Langloh Parker's 1890s Story of Outback Station Life*, Adelaide: Rigby.
Parkinson, S. 1773 *Journal of a Voyage to the South Seas in His Majesty's Ship The Endeavour*, London: Stansfield Parkinson.
Parry, W. J. and Kelly, R. L. 1987 Expedient core technology and sedentism. In J. K. Johnson and C. A. Morrow (eds) *The Organization of Core Technology*, pp.285-304. Boulder: Westview Press.
Pawley, A. K. 1998 The trans New Guinea Phyllum hypothesis: A reassessment. In J. Miedema, C. Ode and R. A. C. Dam (eds) *Perspectives on the Bird's Head of Irian Jaya, Indonesia: Proceedings of the Conference 13 – 17 October 1997*, pp.655-690. Amsterdam: Rodopi.
Pawley, A. K. 2002 The Austronesian dispersal: Languages, technology, people. In P. Bellwood and C Renfrew (eds) *Examining the Farming/Language Dispersal Hypothesis*, pp.251-273. Cambridge: McDonald Institute for Archaeological Research, University of Cambridge.
Peake, H. 1928 *The Origins of Agriculture*, London: Ernest Benn Ltd.
Pearsall, D. M. 1995 Domestication and agriculture in the New World Tropics. In T. D. Price and A. Gebauer (eds) *Last Hunters – First Farmers: New Perspectives on the Prehistoric Transition to Agriculture*, pp.157-192. Santa Fe: School of American Research Press.
Pearson, S. E. 1949 In the Kalkadoon country the habitat and habits of a Queensland Aboriginal tribe. *Journal of the Historical Society of Queensland* 4(2):190-205.
Peel, L. (ed.) 1996 *The Henty Journals: A Record of Farming, Whaling and Shipping at Portland Bay 1834-39.* Melbourne: Miegunyah Press.
Peggs, A. J. 1903 Notes on the Aborigines of Roebuck Bay, Western Australia. *Folklore* 14:324- 367
Pei, A. 1998 Notes on new advancements and revelations in the archaeology of early rice domestication in the Dong Ting Lake region. *Antiquity* 72(278):878-885.
Pei, A. 2003 Rice paddy agriculture and pottery from the middle reaches of the Yangtze River. In Y. Yasuda (ed.) *The Origins of Pottery and Agriculture*, pp.167-181. New Delhi: Roli Books.
Pelsaert, F 1629 The journals of Francisco Pelsaert. In H. Drake-Brockman (comp.) 1982 *Voyage to Disaster*, pp.122-254. London: Angus and Robertson.
Penney, J. and Rhodes, D. 1990 *Lake Condah Archaeological Project: Post-Contact Archaeological Component. 2 Parts*. Melbourne: Dept. of Conservation and Environment.
Penney, R. 1991 The spirit of the Murray. Manuscripts recently discovered on the banks of the Murray. *Journal of the Anthropological Society of South Australia* 28(1):27-87.
Peron, F. and de Freycinet, L. 2003 *Voyage of Discovery to the Southern Lands*, C. Cornell (trans.) Adelaide: Friends of the State Library of South Australia.
Perry, L. 2002 Starch granule size and the domestication of manioc (*Manihot esculenta*) and sweet potato (*Ipomoea batatas*). *Economic Botany* 56(4):335-349.
Perry, L., Sandweiss, D. H., Piperno, D. R., Rademaker, K. Malpass, M. A., Umire, A. and de la Vera, P. 2006 Early maize agriculture and interzonal interaction in southern Peru. *Nature* 440(7080):76-79.
Perry, S. A. 1839 Report by Deputy-Surveyor-General on the Clarence River. In J. D. Lang 1847 *Cooksland in North-East Australia*, pp.39-44. London: Longman, Brown, Green and Longmans.
Peterson, J. 2002 *Sexual Revolutions: Gender and Labor at the Dawn of Agriculture*, Walnut Creek: Altamira Press.
Peterson, N. 1976 Ethno-archaeology in the Australian Iron Age. In G. de G. Sieveking, I. H. Longworth and K. E. Wilson (eds) *Problems in Economic and Social Archaeology*, pp.265- 275. London: Duckworth.
Peterson, N. 1979 Aboriginal uses of Australian Solanaceae. In J. G. Hawkes, R. N. Lester and A. D. Skelding (eds) *The Biology and Taxonomy of the Solanaceae*, pp.171-189. London: Academic Press.
Peterson, N. 1993 Demand sharing: Reciprocity and the pressure for generosity among foragers. *American Anthropologist* 95(4):860-874.
Petrie, C. C. 1904 *Tom Petrie's Reminiscences of Early Queensland*, Brisbane: Watson, Ferguson and Co.
Phillips, G. 1856 Letter from Mr. George Phillips to the Hon. Surveyor-General, Western Australia, Perth, July 6, 1855. *Journal of the Royal Geographical Society* 26:269-272.
Piddington, M. and Piddington, R. 1932 Report of fieldwork in north-western Australia. *Oceania* 2(3):342-358.
Piddocke, S. 1968 The potlatch system of the southern Kwakiutl: A new perspective. In E. E. St Clair and H. K. Schneider (eds) *Economic Anthropology: Readings in Theory and Analysis*, pp.283-299. New York: Holt, Rinehart and Winston.
Piperno, D. R and Flannery, K. V. 2001 The earliest archaeological maize (*Zea mays* L.) from highland Mexico: New accelerator mass spectrometry dates and their implica-

tions. *Proceedings of the National Academy of Sciences of the United States of America* 98(4):2101-2103.
Piperno, D. R. and Holst, I. 1998 The presence of starch grains on prehistoric stone tools from the humid Neotropics: Indications of early tuber use and agriculture in Panama. *Journal of Archaeological Science* 25(8):765-776.
Piperno, D. R. and Pearsall, D. M. 1998 *The Origins of Agriculture in the Lowland Neotropics*, San Diego: Academic Press.
Pires-Ferreira, J. W. 1976a Shell and iron-ore mirror exchange in Formative Mesoamerica, with comments on other commodities. In K. V. Flannery (ed.) *The Early Mesoamerican Village*, pp.311-328. New York: Academic Press.
Pires-Ferreira, J. W. 1976b Obsidian exchange in Formative Mesoamerica. In K. V. Flannery (ed.) *The Early Mesoamerican Village*, pp.292-306. New York: Academic Press.
Pires-Ferreira, J. W. and Flannery, K. V. 1976 Ethnographic models for Formative exchange. In K. V. Flannery (ed.) *The Early Mesoamerican Village*, pp.286-292. New York: Academic Press.
Plattner, S. 1989 Introduction. In S. Plattner (ed.) *Economic Anthropology*, pp.1-20. Stanford: Stanford University Press.
Playford, P. 1996 *Carpet of Silver*, Nedlands: University of Western Australia Press.
Polanyi, K. 1968 The economy as instituted process. In E. E. Le Clair and H. K. Schneider (eds) *Economic Anthropology: Readings in Theory and Analysis*, pp.122-143. New York: Holt, Rinehart and Winston.
Pretty, G. L. 1977 The cultural chronology of the Roonka Flat. In R. V. S. Wright (ed.) *Stone Tools as Cultural Markers: Change, Evolution and Complexity*, pp.288-331. Canberra: Australian Institute of Aboriginal Studies.
Price, T. D. 1995 Social inequalities at the origins of agriculture. In T. D. Price and G. M. Feinman (eds) *Foundations of Social Inequality*, pp.129-151. New York: Plenum Press.
Price, T. D and Brown, J. A. 1985 Aspects of hunter-gatherer complexity. In T. D Price and J. A. Brown (eds) *Prehistoric Hunter-Gatherers: The Emergence of Cultural Complexity*, pp.3- 20. Orlando: Academic Press.
Price, T. D. and Gebauer, A. 1995 New perspectives on the transition to agriculture. In T. D. Price and A. Gebauer (eds) *Last Hunters – First Farmers: New Perspectives on the Prehistoric Transition to Agriculture*, pp.3-19. Santa Fe: School of American Research Press.
Pringle, H. 1997 Oldest mound complex found at Louisiana site. *Science* 277:1761-1762.
Pringle, H. 1998 The slow birth of agriculture. *Science* 282(5393):1446-1450.
Proeve, E. H. and Proeve, H. F. W. 1952 *A Work of Love and Sacrifice: The Story of the Mission Among the Dieri Tribe at Coopers Creek*, Point Pass: Auricht's Printing Office.
Pryor, F. L. 1983 Causal theories about the origins of agriculture. In P. Uselding (ed.) *Research in Economic History 8*, pp.93-124. Greenwich (Connecticut): JAI Press.
Pryor, F. L. 1986 The adoption of agriculture: Some theoretical and empirical evidence. *American Anthropologist* 88(4):879-897.
Puckridge, J. 1992 Lake Eyre Basin: Land of paradox. *Habitat* 20(3):18-21.
Pulleine, R. H. 1921 Old native camps at Commodore Point, Encounter Bay. *Transactions and Proceedings of the Royal Society of South Australia* 45:278-280.
Pumpelly, R. 1908 *Explorations in Turkestan. Expedition of 1904. Prehistoric Civilizations of Anau: Origins, Growth and Influence of Environment*, Washington: Carnegie Institution.
Radiocarbon CONTEXT Database 2002-2006 U. Bohner and D. Schyle http://context- database.uni-koeln.de/
Rafferty, J.E. 1985 The archaeological record on sedentariness: Recognition, development, and implications. *Advances in Archaeological Method and Theory* 8:113-156.
Rankin, T. 1898 Brewarrina native fish traps. *Science of Man and the Journal of the Royal Anthropological Society of Australasia* 1(9):202.
Rathe, G. 1990 *The Wreck of the Barque 'Stefano' Off the North West Cape of Australia in 1875*, Victoria Park: Hesperian Press.
Read, P. J. 1983 A history of the Wiradjuri people, 1883 – 1969. Unpublished PhD thesis, Australian National University, Canberra.
Redman, C. L. 1978 *The Rise of Civilization*, San Francisco: W. H. Freeman.
Reed, C. A. 1977 The origins of agriculture: Prologue. In C. A. Reed (ed.) *The Origins of Agriculture*, pp.9-21. The Hague: Mouton.
Renfrew, C. 1973 Beyond a subsistence economy: The evolution of social organization in prehistoric Europe. In C. B. Moore (ed.) *Reconstructing Complex Societies: An Archaeological Colloquium*, pp.69-93. *Supplement to the Bulletin of the American School of Oriental Research No.20.*
Renker, A. and Gunther, E. 1990 Makah. In W. Suttles, (ed.) *Handbook of North American Indians, Volume 7: Northwest Coast*, pp.422-430. Washington: Smithsonian Institution.
Renouf, M. A. P. 1984 Northern coastal hunter-fishers: An archaeological model. *World Archaeology* 16(1):18-27.
Reuther, J. G. 1981 *The Diari* [microform], P. A. Scherer (trans.), Canberra: Australian Institute of Aboriginal Studies.
Reynolds, H. 1978 Before the instant of contact: Some evidence from nineteenth-century Queensland. *Aboriginal History* 1(2):63-69.
Reynolds, P. J. 1992 Crop yields of the prehistoric cereal types emmer and spelt: The worst option. In P. C. Anderson[-Gerfaud] (ed.) *Prehistoire de L'Agriculture: Nouvelles Approches Experimentales et Ethnographiques*,

pp.383-393. Paris: Editions du CNRS. *Monographie du CRA No. 6.*

Reynolds, R. G. 1986 An adaptive computer model for the evolution of plant collecting and early agriculture in the eastern Valley of Oaxaca. In K. V. Flannery (ed.) *Guila Naquitz: Archaic Foraging and Early Agriculture in Oaxaca, Mexico*, pp.439-500. Orlando: Academic Press.

Reynolds, R. G. D. 1976 Linear settlement systems on the Upper Grijalva River. The application of a Markovian model. In K. V. Flannery (ed.) *The Early Mesoamerican Village*, pp.180-194. New York: Academic Press.

Ridley, W. 1864 Journal of a missionary tour among the aborigines of the western interior of Queensland in the year 1855. In J. D. Lang *Queensland Australia*, pp435-445. London: Edward Stanford.

Ridley, W. 1875 *Kamilaroi and Other Australian Languages*, Sydney: Government Printer.

Ridley, W. 1878 Wailwun language and traditions. *Journal of the Anthropological Institute of Great Britain and Ireland* 7:246-254.

Riley, T. J. 1991 Comments [on Blumler and Byrne 1991]. *Current Anthropology* 32(1):44.

Rindos, D. 1984 *The Origins of Agriculture: An Evolutionary Perspective*, Orlando: Academic Press.

Rindos, D. 1996 Symbiosis, instability, and the origins and spread of agriculture: A new model. In M. J. O'Brien (ed.) *Evolutionary Archaeology: Theory and Application*, pp.208-235. Salt Lake City: University of Utah Press.

Robbins, L. H., Campbell, A. C., Murphy, M. I., Brook, G. A., Srivastava, P. and Badenhorst, S. 2005 The advent of herding in Southern Africa: Early AMS dates on domestic livestock from the Kalahari Desert. *Current Anthropology* 46(4):671-677.

Robinson, G. A. 1844a Extracts from a report made by the Chief Protector of Aborigines (G. A. Robinson Esq,) of his proceedings on a visit to the country lying to the north of 'Portland Bay', in the months from August to March 1842. In *Great Britain Parliamentary Papers* [relative to the] *Aborigines 1844, Volume 3: Despatches Illustrating the Condition of the Aboriginal Population of the Australian Colonies*. Encl. 1 in No.61, pp.239-240.

Robinson, G. A. 1844b Report: Chief Protector's Office, Melbourne, 9 April, 1842. Enclosure in No. 49 (No. 42/539). In *Great Britain Parliamentary Papers* [relative to the] *Aborigines 1844, Volume 3: Despatches Illustrating the Condition of the Aboriginal Population of the Australian Colonies*, pp.208-211.

Robinson, G. A. 1845 Reply to a circular letter from G. A Robinson, Chief Protector of Aborigines, Melbourne, Port Phillip. In *New South Wales: Votes and Proceedings of the Legislative Council During the Session of the Year 1845 - New South Wales, Aborigines. Report from the Select Committee in the condition of the Aborigines with Appendix, Minutes of Evidence and replies to a circular letter*, pp.45-51.

Robinson, G. A. 1971 *Friendly Mission: The Tasmanian Journals and Papers of George Augustus Robinson 1829-1834*. N. J. B. Plomley (ed.), Bellerive: Tasmanian Historical Research Association.

Rocek, T. R. 1998 Pithouses and pueblos on two continents: Interpretations of sedentism and mobility in the southwestern United States and south west Asia. In T. R. Rocek and O. Bar- Yosef (eds) *Seasonality and Sedentism: Archaeological Perspectives From Old and New World Sites*, pp.199-216. Cambridge (Mass.): Peabody Museum of Archaeology and Ethnology.

Roe, J. S. 1847 Report on Champion Bay and its vicinity by J. S. Roe, Surveyor-General, July 12, 1847. In *Exploration Diaries* 4:20-33. Battye Library: PR 5441.

Roheim, G. 1945 *The Eternal Ones of the Dream*, New York: International Universities Press.

Rolland, N. 1985 Social inequality and the Pleistocene record. In M. Thompson, M. T. Garcia and F. J. Kense (eds) *Status, Structure and Stratification: Current Archaeological Reconstructions. Proceedings of the Sixteenth Annual Conference of the Archaeological Association of the University of Calgary Archaeological Association*, pp.255-268. Calgary: University of Calgary Archaeological Association.

Roosevelt, A. C. 1984 Population, health and evolution of subsistence: Conclusions from the conference. In M. N. Cohen and G. J. Armelagos (eds) *Paleopathology at the Origins of Agriculture*, pp. 559-583. Orlando: Academic Press.

Roscoe, P. 2005 Foraging, ethnographic analogy, and Papuan pasts: Contemporary models for the Sepik-Ramu past. In A. Pawley, R. Attenborough, J. Golson and R. Hide (eds) *Papuan Pasts: Cultural, Linguistic and Biological Histories of Papuan-Speaking Peoples*, pp.555- 584. Canberra: Pacific Linguistics, Research School of Pacific and Asian Studies, Australian National University.

Rosenberg, M. 1990 The Mother of Invention: Evolutionary theory, territoriality, and the origins of agriculture. *American Anthropologist* 92(2):399-415.

Ross, A. 1985 Archaeological evidence for population change in the middle to late Holocene in southeastern Australia. *Archaeology in Oceania* 20(3):81-89.

Ross, A. 2001 The Aboriginal Moreton Bay fishery: Using archaeological evidence to reply to Walters. *Australian Aboriginal Studies* 2001/2:63-65.

Ross, J. 1871 Letter from Mr John Ross concerning country north-west of Cooper's Creek, Australia. *Proceedings of the Royal Geographical Society* 15:96-100.

Roth, H. L. 1887 On the Origin of Agriculture. *Journal of the Anthropological Institute of Great Britain and Ireland* 16:102-136.

Roth, H. L. 1899 *The Aborigines of Tasmania*, Halifax: King and Sons.

Roth, W. E. 1897 *Ethnological Studies Among the North-West-Central Queensland Aborigines*, Brisbane: Government Printer.

Roth W. E. 1901 *Food: Its Search, Capture and Preparation*, Brisbane, Acting Government Printer. *North Queensland Ethnography Bulletin No. 3.*

Roth, W. E. 1903 Notes on a savage life in the early days of Western Australia settlement. *Proceedings of the Royal Society of Queensland,* 17(2):45-69.

Roth, W. E. 1904 *Domestic Implements, Arts and Manufactures*, Brisbane: Government Printer. *North Queensland Ethnography Bulletin No. 7.*

Roth, W. E. 1908 Australian canoes and rafts. *Man* 8:161-162.

Roth, W. E. 1910a *Transport and Trade*, Sydney: Australian Museum. *North Queensland Ethnography Bulletin No. 14.*

Roth, W. E 1910b *Huts and Shelters*, Sydney: Australian Museum. *North Queensland Ethnography Bulletin No. 16.*

Roush, W. 1997 Squash seeds yield new view of early American farming. *Science* 276(5314):894- 895.

Rowland, M. J. 1980 The Keppel Islands – preliminary investigations. *Australian Archaeology* 11:1-17.

Rowland, M. J. 1981 Radiocarbon dates for a shell fishhook and disc from Mazie Bay, North Keppel Island. *Australian Archaeology* 12:63-69.

Rowland, M. J. 1983 A note on corrections to radiocarbon dates from the Keppel Islands. *Australian Archaeology* 17:134-5,

Rowland, M. J. 1984 A long way in a bark canoe: Aboriginal occupation of the Percy Islands. *Australian Archaeology* 18:17-31.

Rowland, M. J. 1986 The Whitsunday Islands: Initial historical and archaeological observations and implications for future work. *Queensland Archaeological Research* 3:72-87.

Rowland, M. J. 1987 The distribution of watercraft on the east coast of Queensland: Implications for culture contact. *Australian Aboriginal Studies* 1987/2:38-45.

Rowland, M. J. 2002 'Crows', swimming logs and auditory exostoses: Isolation on the Keppel Islands and broader implications. *Tempus* 7:61-73.

Rowlands, R. J. and Rowlands, J. M. 1969 An Aboriginal dam in northwestern New South Wales. *Mankind* 7(2):132-136.

Rowley, C. D. 1972 *The Destruction of Aboriginal Society*, Ringwood: Penguin Books.

Rowley-Conwy, P. 1986 Between cave painters and crop planters: Aspects to the temperate European mesolithic. In M. Zvelebil (ed.) *Hunters in Transition: Mesolithic Societies of Temperate Eurasia and Their Transition to Farming*, pp.17-32. Cambridge: Cambridge University Press.

Rowley-Conwy, P. 1999 Prehistory in southern Scandinavia. In J. Coles, R. Bewley and P Mellars (eds) *World Prehistory: Studies in Memory of Grahame Clark*, pp.125-159. Oxford: Oxford University Press.

Rowley-Conwy, P. 2001 Time, change and the archaeological of hunter-gatherers: How original is the Original Affluent Society? In C. Panter-Brick, R. H. Layton and P. Rowley-Conwy (eds) *Hunter-Gatherers: An Interdisciplinary Perspective*. pp.39-72. Cambridge: Cambridge University Press.

Ruby, R. 1995 *Jericho: Dreams, Ruins, Phantoms*, New York: Henry Holt and Co,

Ruff, C. 1987 Sexual dimorphism in human lower limb bone structure: Relationship to subsistence strategy and sexual division of labor. *Journal of Human Evolution* 16(5):391-416.

Russell, H. S. 1845 Exploring excursions in Australia. *Journal of the Royal Geographical Society of London* 15:305-327.

Russell, N. 1986 *Like Engendering Like*, Cambridge: Cambridge University Press.

Russo, M. 1998 Measuring sedentism from fauna: Archaic cultures along the southwest Florida coast. In T. R. Rocek and O. Bar-Yosef (eds) *Seasonality and Sedentism: Archaeological Perspectives From Old and New World Sites*, pp.143-164. Cambridge (Mass.): Peabody Museum of Archaeology and Ethnology.

Ryan, L. 1981 *The Aboriginal Tasmanians*, St Lucia: University of Queensland Press.

Ryan, W. M. 1969 *White Man, Black Man*, Milton: Jacaranda Press.

Sahlins, M. 1972 *Stone Age Economics*, Chicago: Aldine, Atherton.

Salvado, R. 1977 *The Salvado Memoirs*, E. J. Stormon (ed. and trans.) Nedlands: University of Western Australia Press.

Sanger, E. B. 1883 Notes on the Aborigines of Coopers Creek. *American Naturalist* 17:1220-1225.

Sassaman, K. E. 1992 Lithic technology and the hunter-gatherer sexual division of labour. *North American Archaeologist* 13(3):249-262.

Sato, Y. 2003 Origin of rice cultivation in the Yangtze River basin. In Y. Yasuda (ed.) *The Origins of Pottery and Agriculture*, pp.143-150. New Delhi: Roli Books.

Satterthwait, L. D. 1986 Aboriginal net hunting. *Mankind* 16(1):31-48.

Satterthwait, L. D. 1987 Socioeconomic implications of Australian Aboriginal net hunting. *Man* 22: 613-636.

Saville-Kent, W. 1897 *The Naturalist in Australia*, London: Chapman and Hall.

Savolainen, P., Zhang, Y-P., Luo, J., Lundeberg, J. and Leitner, T. 2002 Genetic evidence for an East Asian origin of domestic dogs. *Science* 298(5598):1610-1613.

Scarry, C. M. 1993 Introduction. In C. M. Scarry (ed.) *Foraging and Farming in the Eastern Woodlands*, pp.1-11. Gainesville: University Press of Florida.

Schalk, R. F. 1977 The structure of an anadromous fish resource. In L. R. Binford (ed.) *For Theory Building in Archaeology: Essays on Faunal Remains, Aquatic Resources, Spatial Analysis and Systemic Modelling*, pp.207-249. New York: Academic Press.

Schell, P. 1998 *The Archaeology of the Stony Rises*. Melbourne: Aboriginal Affairs Victoria.

Schrire, C. 1972 Ethno-archaeological models and subsistence behaviour in Arnhem Land. In D. L. Clarke (ed.) *Models in Archaeology*, pp.653-670. London: Methuen and Co.

Schrire, C. 1984 Wild surmises on savage thoughts. In C.

Schrire (ed.) *Past and Present in Hunter Gatherer Studies*, pp.1-25. Orlando: Academic Press.

Schulze, L. 1891 The Aborigines of the upper and middle Finke River: Their habits and customs with introductory notes on the physical and natural-history features of the country. *Transactions and Proceedings of the Royal Society of South Australia* 14:210-246.

Schurmann, C. W. 1844 *Vocabulary of the Parnkalla Language.* Adelaide: George Dehane.

Schurmann, C. W. 1879 The Aboriginal tribes of Port Lincoln. In J. D. Wood (comp.) *Native Tribes of South Australia*, pp.207-252. Adelaide: E. S. Wigg.

Scott, M. P. 1972 Some Aboriginal food plants of the Ashburton District, WA. *The Western Australian Naturalist* 12(4):94-96.

Sharp, L. 1952 Steel axes for stone-age Australians. In E. H. Spicer (ed.) *Human Problems in Technological Change*, pp.69-90. New York: Russell Sage Foundation.

Sharp, L. 1968 Steel axes for stone-age Australians. In E. E. Le Clair and H. K. Schneider (eds) *Economic Anthropology: Readings in Theory and Analysis*, pp.341-353. New York: Holt Rinehart and Winston.

Shaw, A. G. L. 1962 *The Story of Australia*, London: Faber & Faber.

Shelach, G. 2000 The earliest Neolithic cultures of northeast China: Recent discoveries and new perspectives on the beginning of agriculture. *Journal of World Prehistory* 14(4):363-413.

Sherratt, A. 1997 Climatic cycles and behavioural revolutions: The emergence of moderns humans and the beginnings of farming. *Antiquity* 71(272):271-287.

Shipek, F. C. 1989 An example of intensive plant husbandry: The Kumeyaay of southern California. In D. R. Harris and G. C. Hillman (eds) *Foraging and Farming: The Evolution of Plant Exploitation*, pp.159-170. London: Unwin Hyman.

Shipman, P. 1983 Early hominid lifestyles: Hunting and gathering or foraging and scavenging. In J. Clutton-Brock and C. Grigson (eds) *Animals and Archaeology: Volume 1, Hunters and Their Prey*, pp.31-50. Oxford: British Archaeological Reports. *BAR International Series No. 163.*

Shnirelman, V. A. 1992 Complex hunter-gatherers: Exception or common phenomenon. *Dialectical Anthropology* 17(2):183-196.

Sievwright, C. 1842 Journal of the proceedings of Assistant Protector Sievwright, Western District, from 1st March to 31st March 1842. Archival estrays: Official Papers of New South Wales, List 11 CSIL/11. [Dixson Library, Sydney.]

Silverstein, M. 1990 Chinookans of the Lower Columbia. In W. Suttles, (ed.) *Handbook of North American Indians, Volume 7: Northwest Coast*, pp.533-546. Washington: Smithsonian Institution.

Simons, M. 2003 *The Meeting of the Waters: The Hindmarsh Island Affair*, Sydney: Hodder Headline Australia.

Simpson, K. N. G. and Blackwood, R. 1973 An Aboriginal cache of freshwater mussels at Lake Victoria, New South Wales. *Memoirs of the National Museum of Victoria* 34:217.

Smart, D. J. Q. 2003 *Later Mesolithic Fishing Strategies and Practices in Denmark*, Oxford: BAR Publishing. *BAR International Series No. 1119.*

Smith, B. D. 1995a *The Emergence of Agriculture*, New York: Scientific American Library.

Smith, B. D. 1995b Seed plant domestication in Eastern North America. In T. D. Price and A. Gebauer (eds) *Last Hunters – First Farmers: New Perspectives on the Prehistoric Transition to Agriculture*, pp.193-213. Santa Fe: School of American Research Press.

Smith, B. D. 1997 The initial domestication of *Cucurbita pepo* in the Americas 10,000 years ago. *Science* 276(5314):932-934.

Smith, B. D. 2001 Documenting plant domestication: The consilience of biological and archaeological approaches. *Proceedings of the National Academy of Sciences of the United States of America* 98(4):1324-1326.

Smith, M. 1999 Southwest Australian coastal economies: A new review. In J. Hall and I. McNiven (eds) *Australian Coastal Archaeology*, pp.15-24. Canberra: ANH Publications, Department of Archaeology and Natural History, Research School of Pacific and Asian Studies.

Smith, M. A. 1985 A morphological comparison of Central Australian seedgrinding implements and Australian Pleistocene-Age grindstones. *The Beagle* 2(1):23-38.

Smith, M. A. 1986 The antiquity of seedgrinding in central Australia. *Archaeology in Oceania* 21(1):29-39.

Smith, M. A. 1988 Central Australian seed grinding implements and Pleistocene grindstones. In B. Meehan and R. Jones (eds) *Archaeology With Ethnography: An Australian Perspective*, pp.94-108. Canberra: Dept. of Prehistory, Research School of Pacific Studies, Australian National University.

Smith, M. A. 1989a The case for a resident human population in the Central Australian Ranges during full glacial aridity. *Archaeology in Oceania* 24(3):93-105.

Smith, M. A. 1989b Seed gathering in inland Australia: Current evidence from seed-grinders on the antiquity of the ethnohistorical pattern of exploitation. In D. R. Harris and G. C. Hillman (eds) *Foraging and Farming: The Evolution of Plant Exploitation*, pp.305-317. London: Unwin Hyman.

Smith, M. A. 1998 Biogeography, human ecology and prehistory in the sandridge deserts. In T. Murray (ed.) *Archaeology of Aboriginal Australia: A Reader*, pp.119-140. St. Leonards: Allen and Unwin.

Smith, M. A. and Veth, P. 2004 Radiocarbon dates for baler shell in the Great Sandy Desert. *Australian Archaeology* 58:37-38.

Smithyman, K. and Kalman, B. 2004 *Nations of the Northwest Coast*, New York: Crabtree Publishing.

Smyth, R. B. 1878 *The Aborigines of Victoria*, 2 volumes, Melbourne: Govt. Printer.

Solow, R. 1956 A contribution to the theory of economic growth. *Quarterly Journal of Economics* 70(1):65-94.

Solow, R. 1957 Technical change and the aggregate production function. *Review of Economics and Statistics* 39(3):312-320.

South Australian Parks and Wildlife Service 1988 *Innamincka Regional Reserve. Management Plan South Australia: Draft.*

Sowunmi, M. A. 1985 The beginnings of agriculture in West Africa. *Current Anthropology* 26(1):127-129.

Spencer, B. 1918-19 What is nardoo. *Victorian Naturalist* 34(11):170-171; 35(1):8-15.

Spencer, B. and Gillen, F. J. 1899 *The Native Tribes of Central Australia*, London: Macmillan.

Spencer, B. and Gillen, F. J. 1904 *The Northern Tribes of Central Australia*, London: Macmillan.

Spencer, B. and Gillen, F. J. 1912 *Across Australia*, 2 volumes, London: Macmillan.

Spriggs, M. 1996 Early agriculture and what went before in Island Melanesia: Continuity or intrusion. In D. R. Harris (ed.) *The Origins and Spread of Agriculture and Pastoralism in Eurasia*, pp.524-537. London: UCL Press.

Srejovic, D. 1972 *Europe's First Monumental Sculpture: New Discoveries at Lepenski Vir*. F. Edwards (trans.) NY: Stein and Day.

Srisuwan, S., Sihachakar, D. and Siljak-Yakovlev, S. 2006 The origin and evolution of sweet potato (*Ipomoea batatas* Lam.) and its wild relatives through the cytogenetic approaches. *Plant Science* 171(3):424-433.

Stahle, J. H. 1880 The Gounrditch-Mara Tribe. In L. Fison and A. W. Howitt, *Kamilaroi and Kurnai*, pp.274-278. Melbourne: George Robertson.

Stanley, A. 1981 Daughters of Isis, daughters of Demeter: When women sowed and reaped. *Women's Studies International Quarterly* 4(3):289-304.

Stanner, W. E. H. 1933-34a Ceremonial economics of the Mulluck Mulluck and Madngella tribes of the Daly River, north Australia. A preliminary paper. *Oceania* 4:156-175,458-471.

Stanner, W. E. H. 1933-34b The Daly River tribes: A report of field work in northern Australia. *Oceania* 4:10-36.

Stark, B. L. 1981 The rise of sedentary life. In J. A. Sabloff (ed.) *Supplement to the Handbook of Middle American Indians, Volume 1: Archaeology*, pp.345-372. Austin: University of Texas Press.

Stekelis, M. and Yizraely, T. 1963 Excavations at Nahal Oren: Preliminary report. *Israel Exploration Journal* 13(1):1-12.

Stephens, J. 1839 *The History of the Rise and Progress of the New British Province of South Australia*, London: Smith, Elder and Co.

Stiner, M. C., Munro, N. D., Surovell, T. A., Tchernov, E. and Bar-Yosef, O. 1999 Paleolithic population growth pulses evidenced by small animal exploitation. *Science* 283(5399):190- 194.

Stirling, E. C. 1896 Anthropology. In W. B. Spencer (ed.) *Report of the Work of the Horn Scientific Expedition: Volume 4, Anthropology*, 4 volumes, London: Dalau and Co.

Stirling, E. C. and Waite, E. R. 1918-21 Description of toas, or Australian Aboriginal direction signs. *Records of the South Australian Museum* 1:105-155.

Stockton, J. 1975 An Aboriginal fishtrap from southern Queensland? *Mankind* 10(1):44-47.

Stokes, J. L. 1841 Visit to the country adjacent to Champion Harbour, reported to be Port Grey, and examination of the coastline to the northward. *Perth Gazette* 25 December 1841 pp.2-3.

Stokes, J. L. 1846 *Discoveries in Australia*, 2 volumes, London: T. and W. Boone.

Stow, C. [K. L. Parker] 1930 *Woggheeguy: Australian Aboriginal Legends*, Adelaide: F. W. Preece and Sons.

Stow, J. P. 1894 *The Voyage of the Forlorn Hope*, Melbourne: George Robertson and Co.

Strangways, G 1838 Mr G. Strangways account. *South Australian and Colonial Register* 25 August 1838, pp.27-30.

Strassmann, B. I. and Gillespie, B. 2002 Life-history theory, fertility and reproductive success in humans. *Proceedings of the Royal Society B: Biological Sciences* 269(1491):553-562.

Strehlow, T. G. H. 1965 Culture, social structure and environment in Aboriginal central Australia. In R. M. and C. H. Berndt (eds) *Aboriginal Man In Australia*, pp.121-145. Sydney: Angus and Robertson.

Stretton, W. G. 1893 Customs, rites and superstitions of the Aboriginal tribes of the Gulf of Carpentaria with a vocabulary. *Transactions and Proceedings of the Royal Society of South Australia* 17:227-253.

Stuart, E. J. 1923 *A Land of Opportunities*, London: Bodley Head.

Stuart, J. M. 1864 *The Journals of John McDouall Stuart During the Years 1858-1862*, London: Saunders, Otley and Co.

Sturt, C. 1833 *Two Expeditions Into the Interior of Southern Australia*, 2 volumes, London: Smith, Elder.

Sturt, C. 1849 *Narrative of an Expedition into Central Australia ... In the Years 1844-46*, 2 volumes, London: T. and W. Boone.

Sullivan, M. E. 1982 Exploitation of offshore islands along the New South Wales coastline. *Australian Archaeology* 15:8-19.

Sullivan, M. E. 1987 The recent prehistoric exploitation of edible mussel in Aboriginal shell middens in southern New South Wales. *Archaeology in Oceania* 22(3):97-106.

Suttles, W. 1968 Coping with abundance: Subsistence on the Northwest Coast. In R. B. Lee and I. DeVore (eds) *Man the Hunter*, pp.56-68. Chicago: Aldine.

Suttles, W. (ed.) 1990a *Handbook of North American Indians, Volume 7: Northwest Coast*, Washington: Smithsonian Institution.

Suttles, W. 1990b Introduction. In W. Suttles, (ed.) *Handbook of North American Indians, Volume 7: Northwest Coast*, pp.1-15. Washington: Smithsonian Institution.

Suttles, W. 1990c Environment. In W. Suttles, (ed.) *Handbook of North American Indians, Volume 7: Northwest Coast*, pp.16-29. Washington: Smithsonian Institution.

Suttles, W. 1990d Central Coast Salish. In W. Suttles, (ed.) *Handbook of North American Indians, Volume 7: Northwest Coast*, pp.453-475. Washington: Smithsonian Institution.

Suttles, W. and Jonaitis, J. A. 1990 History of research in ethnology. In W. Suttles, (ed.) *Handbook of North American Indians, Volume 7: Northwest Coast*, pp.73-87. Washington: Smithsonian Institution.

Syrett, M. D. 1993 Questioning the relationship between hunter-gatherer social complexity and sedentism. A re-evaluation of the sedentary foragers of Lepenski Vir. Unpublished Masters Paper, UCSB Department of Anthropology, Santa Barbara

Swaardecroon, H., Chasselijn, C. and Craine, J. S. 1705 Report to Jan van Horn and Council of India. In R. H. Major 1859 *Early Voyages to Terra Australis*, pp.165-173. London: Hakluyt Society.

Swan, T. W. 1956 Economic growth and capital accumulation. *The Economic Record* 32:334-361.

Tacon, P. 1993 Art of the escarpments: Image-making in western Arnhem Land. In G. Burenhult (ed.) *People of the Stone Age: Hunter-Gatherers and Early Farmers*, pp.218-225. St. Lucia: University of Queensland Press.

Taplin, G. 1878 *The Narrinyeri: An Account of the Tribes of South Australian Aborigines*, Adelaide: E. S. Wigg

Taplin, G. 1879a The Narrinyeri. In J. D. Woods (comp.) *The Native Tribes of South Australia*, pp.1-156. Adelaide: E. S. Wigg.

Taplin, G. 1879b Narrinyeri tribe (Murray River) In G. Taplin (ed.) *The Folklore, Manners, Customs and Languages of the South Australian Aborigines*, pp.33-50. Adelaide: Acting Government Printer.

Tainter, J. A. 1978 Mortuary practices and the study of prehistoric social systems. *Advances in Archaeological Method and Theory* 1:105-141.

Tauber, H. 1981 $^{13}C$ evidence for dietary habits of prehistoric man in Denmark. *Nature* 292:332- 333.

Tchernov, E. 1991 Biological evidence for human sedentism in Southwest Asia during the Natufian. In O. Bar-Yosef and F. R. Valla (eds) *The Natufian Culture in the Levant*, pp.315-340. Ann Arbor: International Monographs in Prehistory.

Teichelmann, C. G. and Schuermann, C. W. 1923 Outline of grammar, vocabulary and phraseology of the Aboriginal language of South Australia. In T. A. Parkhouse (ed.) *Reprints and Papers Relating to the Autochthones of Australia*, pp.57-68. Woodville: Parkhouse.

Testart, A. 1982 The significance of food storage among hunter-gatherers: Residence patterns, population densities and social inequalities. *Current Anthropology* 23(5):523-537.

Thomas, N. W. 1905 Australian canoes and rafts. *Journal of the Anthropological Institute of Great Britain and Ireland* 35:56-79.

Thomas, N. W. 1906 *Native Races of the British Empire: Natives of Australia*, London: Archibald Constable.

Thomas, N. W. 1907 Addenda to 'Australian canoes and rafts'. *Journal of the Anthropological Institute of Great Britain and Ireland* 36:409-412.

Thomas, W. n.d.a. Papers of William Thomas, MSS 214, Box 24, Item 22. [manuscript] Mitchell Library, Sydney, New South Wales MS 214.

Thomas, W. n.d.b. Manuscript notes and attributed sketches in R. Brough Smyth Papers. [manuscript] State Library of Victoria MS 8781.

Thomas, W. 1844 Journal of Proceedings 2.3.1844 for period 1.12.1843 - 1.3.1844. PROV VA512 Aboriginal Protectorate Records: VPRS 4467, Reel 2, Unit 3 No. 79, Aboriginal Protectorate Records Weekly, Quarterly and Annual Reports. [Victorian Public Records Office, Laverton, Victoria.]

Thomas, W. 1858 Papers of William Thomas, MSS 214, Box 24, Item 11: "Aborigines Superior Race" 1858. [manuscript] Mitchell Library, Sydney, New South Wales MS 214.

Thomas, W. 1898 Letter. In T. F. Bride (ed.) *Letters from Victorian Pioneers*, pp.84-100. Melbourne: Government Printer.

Thomson, D. F. 1934 Notes on a hero cult from the Gulf of Carpentaria, North Queensland. *Journal of the Royal Anthropological Institute of Great Britain and Ireland* 64:217-235.

Thomson, D. F. 1936 Notes on some bone and stone implements from North Queensland. *Journal of the Royal Anthropological Institute of Great Britain and Ireland* 66:71-74.

Thomson, D. F. 1938 A new type of fish trap from Arnhem Land, Northern Territory of Australia. *Man* 38:193-198.

Thomson, D. F. 1939a The tree dwellers of the Arafura Swamps: A new type of bark canoe from central Arnhem Land. *Man* 39:121-126.

Thomson, D. F. 1939b The seasonal factor in human culture. *Proceedings of the Prehistoric Society* 5(10):209-221.

Thomson, D. F. 1949 *Economic Structure and Ceremonial Exchange in Arnhem Land*, Melbourne: Macmillan.

Thomson, D. F. 1952 Notes on some primitive watercraft in north Australia. *Man* 52:1-5.

Thomson, D. F. 1954 The moulding of clay in Arnhem Land. *Man* 54:117-119.

Thorley, P. B. 1998 Pleistocene settlement in the Australian arid zone: Occupation of an inland riverine lands landscape in the central Australian ranges. *Antiquity* 72(275):34-45.

Thornly, N. 1878 Some words of the language of the Western tribes of Victoria. In R. B. Smyth *The Aborigines of Victoria, Volume 2*, pp.60-62. 2 volumes, Melbourne: Government Printer.

Tibbett, K. 2002 Archaeological analysis of stone axe exchange networks in the Lake Eyre Basin during the mid- to late Holocene. *Australian Archaeology* 55:22-29.

Tiechelmann [Teichelmann], C. G. 1841 *Aborigines of South Australia*, Adelaide: Committee of the South Australian Wesleyan Methodist Auxiliary Mission Society.

Tindale, N. B. 1951 Aboriginal net making. *Mankind* 4(6):257-258.

Tindale, N. B. 1957 Culture succession in south eastern Australia from Late Pleistocene to the present. *Records of the South Australia Museum* 13:1-49.

Tindale, N. B. 1974 *Aboriginal Tribes of Australia: Their Terrain, Environmental Controls, Distribution, Limits and Proper Names*, Berkley: University of California Press.

Tindale, N. B. 1977 Adaptive significance of the Panara or grass seed culture of Australia. In R. V. S. Wright (ed.) *Stone Tools as Cultural Markers: Change, Evolution and Complexity*, pp.345-349. Canberra: Australian Institute of Aboriginal Studies.

Tindale, N. B. 1981 Desert Aborigines and the southern coastal peoples: Some comparisons. In A. Keast (ed.) *Ecological Biogeography of Australia*, pp.1855- 1884. The Hague: Dr W. Junk bv Publishers.

Tolcher, H. M. 1990 *Innamincka: The Town With Two Lives*, Innamincka: Innamincka Progress Association.

Tolcher, H. M 2003 *Seed of the Coolibah: A History of the Yandruwandha and Yawarrawarrka People*, Adelaide: H. M. Tolcher.

Tolmer, A. 1882 *Reminiscences of an Adventurous and Chequered Career at Home and at the Antipodes*, 2 volumes, London: Sampson Low, Marston, Searle and Rivington.

Tonkinson, R. 1978 *The Mardudjara Aborigines: Living the Dream in Australia's Desert*, Fort Worth: Holt, Rinehart and Winston.

Tonkinson, R. 1991 *The Mardu Aborigines: Living the Dream in Australia's Desert*, Fort Worth: Holt, Rinehart and Winston.

Tonkinson, R. 1993 Foreword. In R. M. Berndt, C. H. Berndt and J. E. Stanton, *A World That Was: The Yaraldi of the Murray River and the Lakes, South Australia*, pp.xvi–xxxi. Carlton: Melbourne University Press.

Torrence, R. 1989 Retooling: Towards a behavioural theory of stone tools. In R. Torrence (ed.) *Time, Energy and Stone Tools*, pp.57-66. Cambridge: Cambridge University Press.

Towle, C. C. 1944 Quarries used by the Aborigines of the Paroo River. *Victorian Naturalist* 60(5):73-75.

Towner, E. T. 1954-56 Lake Eyre and its tributaries. *Queensland Geographical Journal* 57(43):65- 94.

Toyama, S. 2003 The origins and spread of rice cultivation as seen from rice remains. In Y. Yasuda (ed.) *The Origins of Pottery and Agriculture*, pp.263-272. New Delhi: Roli Books.

Trigger, B. 1985 *Natives and Newcomers: Canada's 'Heroic Age' Reconsidered*, Kingston: McGill University Press.

Trigger, B. 1987 Inland, coast and islands: Traditional Aboriginal society and material culture of the southern Gulf of Carpentaria. *Records of the South Australian Museum* 21(2):69-84.

Tsutsumi, T. 2003 Origins of pottery and human strategies for adaptation during the termination of the Last-Glacial Period in the Japanese Archipelago. In Y. Yasuda (ed.) *The Origins of Pottery and Agriculture*, pp.241-262. New Delhi: Roli Books.

Tully, A. M. 1896 Communications from correspondents. *The Australasian Anthropological Journal* 1(2):22-23.

Turbet, P. 1989 *The Aborigines of the Sydney District Before 1788*, Kenthurst: Kangaroo Press.

Turnbull, C. 1984 *The Forest People*, Oxford: Isis Large Print.

Uerpmann, H-P. 1996 Animal domestication – accident or intention? In D. R. Harris (ed.) *The Origins and Spread of Agriculture and Pastoralism in Eurasia*, pp.227-237. Washington DC: Smithsonian Institute Press.

Ulm, J. 2002 Reassessing marine fishery intensification in south east Queensland. *Queensland Archaeological Research* 13:79-96.

Ulm, S. and Hall, J. 1996 Radiocarbon and cultural chronologies in southeast Queensland prehistory. In S. Ulm, I. Lilley and A. Ross (eds) *Australian Archaeology 95: Proceedings of the Australian Archaeological Association Annual Conference*, pp.45-62. St Lucia: Anthropology Museum, University of Queensland.

Underhill, A. P. 1997 Current issues in Chinese Neolithic archaeology. *Journal of World Prehistory* 11(2):103-160.

Underhill, A. P. and Habu, J. 2006 Early communities in East Asia: Economic and sociopolitical organisation at the local and regional level. In M. T. Stark (ed.) *Archaeology of Asia*, pp.121-148. Malden: Blackwell.

Unger-Hamilton, R. 1989 The epi-paleolithic southern Levant and the origins of cultivation. *Current Anthropology* 30(1):88-103.

Unger-Hamilton, R. 1992 Experiments in harvesting wild cereals and other plants. In P. C. Anderson[-Gerfaud] (ed.) *Prehistoire de L'Agriculture: Nouvelles Approches Experimentales et Ethnographiques*, pp.211-224. Paris: Editions du CNRS. *Monographie du CRA No. 6.*

Uniacke, J. 1825 Narrative of Mr Oxley's expedition to survey Port Curtis, Moreton Bay and Port Bowen: With a view to form convict penal establishments there. In B. Field (ed.) *Geographical Memoirs of New South Wales*. pp.1-86. London: John Murray.

United States Department of Agriculture 1936 *Food Plants of the North American Indians*, Washington: USDA Miscellaneous Publication No. 237.

Valla, F. R. 1998 Natufian seasonality: A guess. In T. R. Rocek and O. Bar-Yosef (eds) *Seasonality and Sedentism: Archaeological Perspectives From Old and New World Sites*, pp.93-108. Cambridge (Mass.): Peabody Museum of Archaeology and Ethnology.

Vancouver, G. 1798 *A Voyage of Discovery to the North Pacific Ocean and Round the World ... 1790-95*, 3 volumes, London: G. G. and J. Robinson.

Vanderwal, R. and Horton, D. 1984 *Coastal Southwest Tasmania*, Canberra: Department of Prehistory, Research School of Pacific Studies, Australian National University. *Terra Australis 9.*

Varien, M. D. 1999 *Sedentism and Mobility in a Social*

*Landscape: Mesa Verde and Beyond*, Tucson: University of Arizona Press.
Vavilov, N. I. 1997 *Five Continents*, D. Love (trans.), S. Reznik and P. Stapleton (eds). St Petersberg: International Plant Genetics Resources Institute.
Veth, P. M 1987 Martu prehistory: Variation in arid zone adaptations. *Australian Archaeology* 25:102-111.
Veth, P. M. 1989 Islands in the interior: A model for the colonization of Australia's arid zone. *Archaeology in Oceania* 24(3):81-92.
Veth, P. M. 1993 *Islands of the Interior: The Dynamics of Prehistoric Adaptations Within the Arid Zone of Australia*, Ann Arbor: International Monographs in Prehistory.
Veth, P. M. 1995 Aridity and settlement in northwest Australia. *Antiquity* 69(265):733-746.
Veth, P. M. 2005 Cycles of aridity and human mobility: Risk minimization amongst late Pleistocene foragers of the Western Desert, Australia. In P. Veth, M. A. Smith and P. Hiscock (eds) *Desert Peoples: Archaeological Perspectives*, pp.100-115. Oxford, Blackwell Publishing.
Veth, P. M., Hamm, G. and Lampert, R. J. 1990 The archaeological significance of the lower Cooper Creek. *Records of the South Australian Museum* 24(1):43-66.
Victoria: Legislative Council, Votes and Proceedings 1858-9 *Report of the Select Committee of the Legislative Council on the Aborigines, together with the Proceedings of the Committee, Minutes of Evidence and Appendices.*
Vinsrygg, S. 1987 Sex-roles and the division of labour in hunting-gathering societies. In R. Bertelsen, A. Lillehammer and J. R. Naess (eds) *Were They All Men?: An Examination of Sex Roles in Prehistoric Society*, Stvanger: Arkeologisk Museum.
Vivienne, M. 1901 *Travels in Western Australia*, London: Heinemann.
Voorhies, B. 1996 The transformation from foraging to farming in Lowland Mesoamerica. In S. L. Fedick (ed.) *The Managed Mosaic: Ancient Maya Agriculture and Resource Use*, pp.17-29. Salt Lake City: University of Utah Press.
Wadley, L. 2001 What is cultural modernity?: A general view and a South African perspective from Rose Cottage Cave. *Cambridge Archaeological Journal* 11:201-221.
Walcott, P. 1884 Vocabulary of the Aboriginal language at Nickol Bay. In A. C. and F. T. Gregory (comps.) *Journals of Australian Expeditions*, pp.97-98. Brisbane: Government Printers.
Walker, J. 1977 *Joe's Book: Facts and Theories on the Bunya Mountains*, Bell: J. Walker.
Walsh, F. J. 1990 An ecological study of traditional Aboriginal use of 'country': Martu in the Great and Little Sandy Deserts, Western Australia. In D. A. Saunders, A. J. M. Hopkins and R. A. How (eds) *Australian Ecosystems: 200 Years of Utilization, Degradation and Reconstruction*, pp.23-37. Chipping Norton: Ecological Society of Australia. *Proceedings of the Ecological Society of Australia, Volume 16.*
Walters, I. 1985a The Toorbul Point Aboriginal fish trap. *Queensland Archaeological Research* 2:38-49.
Walters, I. 1985b Some observations on the material culture of Aboriginal fishing in the Moreton Bay area: Implications for archaeology. *Queensland Archaeological Research* 2:50-57.
Walters, I. 1988 Fish hooks: Evidence for dual social systems in southeastern Australia. *Australian Archaeology* 27:98-114.
Wang, X., Sun, C., Cai, H. and Zhang, J. 1999 Origin of the Chinese cultivated rice (*Oryza sativa* L). *Chinese Science Bulletin* 44(4).
Warner, W. L. 1958 *A Black Civilization: A Social Study of an Australian Tribe*, New York: Harper Bros.
Watanabe, H. 1968 Subsistence and ecology of northern food gatherers with special reference to the Ainu. In R. B. Lee and I. DeVore (eds) *Man the Hunter*, pp.69-77. Chicago: Aldine.
Watkins, G. 1890-91 Notes on the Aborigines of Stradbroke and Moreton Islands. *Proceedings of the Royal Society of Queensland* 8:40-50.
Watson, P. J. and Kennedy, M. C. 1991 The development of horticulture in the Eastern Woodlands of North America: Women's Role. In J. M. Gero and M. W. Conkey (eds) *Engendering Archaeology: Women and Prehistory*, pp.255-275. Oxford: Blackwell.
Watson, P. L. 1983 *This Precious Foliage*, Sydney: University of Sydney. *Oceania Monograph No. 26.*
Watson, P. L. 1998: *Frontier Lands and Pioneer Legends*, Allen and Unwin. St. Leonards.
Weiss, E., Wetterstrom, W. and Bar-Yosef, O. 2004 The broad spectrum revisited: Evidence from plant remains. *Proceedings of the National Academy of Sciences of the United States of America* 101(26):9551-9555.
Wells, F. H. 1894 The habits, customs and ceremonies of the Aboriginals on the Diamantina, Herbert and Eleanor Rivers in eastern central Australia. *Report of the Australasian Association for the Advancement of Science* 5(1893):515-522.
Wells, L. A. 1930 The Northern Territory and Queensland border survey expedition 1883/6. Typescript, State Library of South Australia.
Welsby, T. 1913 *The Discoverers of the Brisbane River*, Brisbane: Didams and Co.
Wesson, J. P. 1981 Excavations of stone structures in the Condah area. Unpublished Masters Preliminary thesis, La Trobe University, Melbourne.
West, A. L. 1999 *Aboriginal String Bags, Nets and Cordage*, Melbourne: Museum Victoria. *Occasional Papers, Anthropology and History No. 2.*
Westgarth, W. 1846 *A Report on the Conditions, Capabilities and Prospects of the Australian Aborigines*. Melbourne: W. Clarke.
Westgarth, W.1848 *Australia Felix*. London: Oliver and Boyd.
Westgarth, W.1851 Report on the condition and prospects of the Aborigines of Australia. *Journal of the Indian Archipelago and Eastern Asia* 5(2):704-28.

Westgarth, W.1888/1970 *Personal Recollections of Early Melbourne and Victoria.* Geelong: Rippleside Press.

Wetterstrom, W. 1998 The origins of agriculture in Africa: With particular reference to sorghum and pearl millet. *Review of Archaeology* 19(2):30-46.

Whalen, M. E. 1976 Zoning within an Early Formative community in the Valley of Oaxaca. In K. V. Flannery (ed.) *The Early Mesoamerican Village*, pp.75-79. New York: Academic Press.

Whalen, M. E. 1988 House and household in Formative Oaxaca. In R. R. Wilk and W. Ashmore (eds) *Household and Community in the Mesoamerican Past.* pp.249-272. Albuquerque: University of New Mexico Press.

White, J. C. 1989 Ethnological observations on wild and cultivated rice and yams in northeastern Thailand. In D. R. Harris and G. C. Hillman (eds) *Foraging and Farming: The Evolution of Plant Exploitation*, pp.152-158. London: Unwin Hyman.

White, J. P. 1971 New Guinea and Australian prehistory: The Neolithic problem. In D. J. Mulvaney and J. Golson (eds) *Aboriginal Man and Environment in Australia*, pp.182-195. Canberra: ANU Press.

White, J. P. 1993 Australia: The different continent. In G. Burenholt (ed.) *People of the Stone Age: Hunter-Gatherers and Early Farmers*, pp.207-209. St. Lucia: University of Queensland Press.

White, J. P. and O'Connell, J. F. 1982 *A Prehistory of Australia, New Guinea and Sahul*, Sydney: Academic Press.

White, S. A. 1914 *Into the Dead Heart: An Ornithological Trip Through Central Australia*, Adelaide: W. K. Thomas.

Widlok, T. 2005 Theoretical shifts in the anthropology of desert hunter-gatherers. In P. Veth, M. A Smith and P. Hiscock (eds) *Desert Peoples: Archaeological Perspectives*, pp.17-33. Oxford: Blackwell Publishing.

Widmer, R. J. 1988 *The Evolution of the Calusa: A Non-Agricultural Chiefdom on the Southwest Florida Coast*, Tuscaloosa: University of Alabama Press.

Wilde, W. H. 1988 *Courage a Grace: A Biography of Dame Mary Gilmore*, Melbourne: Melbourne University Press.

Wilke, P. J., Bettinger, R., King, T. F. and O'Connell, J. F. 1972 Harvest selection and domestication in seed plants. *Antiquity* 46(183):203-209.

Willcox, G. H. 1992 Archaeobotanical significance of growing Near Eastern progenitors of domestic plants at Jales (France). In P. C. Anderson[-Gerfaud] (ed.) *Prehistoire de L'Agriculture: Nouvelles Approches Experimentales et Ethnographiques*, pp.159-177. Paris: Editions du CNRS. *Monographie du CRA No. 6.*

Willcox, G. H. 1999 Agrarian change and the beginnings of cultivation in the Near East: Evidence from wild progenitors, experimental cultivation and archaeobotanical data. In C. Gosden and J. Hather (eds) *The Prehistory of Food: Appetites for Change*, pp.478-500. London: Routledge.

Willey, G. R., Ekholm, G. F. and Millon, R. F. 1964 The patterns of farming life and civilization. In R. C. West (ed.) *Handbook of Middle American Indians, Volume 1*, pp. 446-498. Austin: University of Texas Press.

Williams, E. 1984 Documentation and archaeological investigation of an Aboriginal 'village' site in south western Victoria. *Aboriginal History* 8(2):173-188.

Williams, E. 1985 Estimation of prehistoric populations of archaeological sites in south western Victoria: Some problems. *Archaeology in Oceania* 20(3):73-80.

Williams, E. 1987 Complex hunter-gatherers: A view from Australia. *Antiquity* 61(232):310-21.

Williams, E. 1988 *Complex Hunter-Gatherers: A Late Holocene Example from Temperate Australia.* Oxford: British Archaeological Reports. *BAR International Series No. 423.*

Williams, W. 1923 Vocabulary of the languages of Adelaide. In T. A. Parkhouse (ed.) *Reprints and Papers Relating to the Autochones of Australia*, pp.57-68. Woodville: Parkhouse.

Williams, W. 1950 Archaeological sites – Somerset. *The North Queensland Naturalist* 17(94:7.

Williamson, C. 1998 Late Holocene Australia and the writing of Aboriginal history. In T. Murray (ed.) *Archaeology of Aboriginal Australia: A Reader*, pp.141-148. St. Leonards: Allen and Unwin.

Wills, W. H. 1995 Archaic foraging and the beginnings of food production in the American Southwest. In T. D. Price and A. Gebauer (eds) *Last Hunters – First Farmers: New Perspectives on the Prehistoric Transition to Agriculture*, pp.215-242. Santa Fe: School of American Research Press.

Wills, W. J. 1861 Wills' diary from Coopers Creek to Carpentaria and return. In *The Victorian Exploring Expedition: Diary of Burke and Wills, Journal of Howitt and Narrative of King*, pp.14-52. Melbourne: The Age.

Willshire, W. H. 1896 *The Land of Dawning*, Adelaide: W. K. Thomas and Co.

Wimberley, C. C. 1899 Linguistics. *Science of Man* 2(4):69-70.

Wing, E. 1977 Animal domestication in the Andes. In C. A. Reed (ed.) *Origins of Agriculture*, pp. 837-857. The Hague: Mouton.

Winter, M. C. 1976 The archaeological cluster in the Valley of Oaxaca. In K. V. Flannery (ed.) *The Early Mesoamerican Village*, pp.25-31. New York: Academic Press.

Winter, T. 1898 Letter. In T. F. Bride (ed.) *Letters from Victorian Pioneers*, pp.275-279. Melbourne: Government Printer.

Withnell, J. G. 1901 *The Customs and Traditions of the Aboriginal Natives of North Western Australia*, Roebourne.

Witsen, N. 1705 *Noord en Ooste Tartarye*, Amsterdam: Francois Halma.

Wobst, H. M. 1978 The archaeo-ethnology of hunter-gatherers or the tyranny of the ethnographic record in archaeology. *American Antiquity* 43(2):303-309.

Woodburn, J. 1968 An introduction to Hadza ecology. In R.

B. Lee and I. DeVore (eds) *Man the Hunter*, pp.49-55. Chicago: Aldine.

Woodburn, J. 1982 Egalitarian societies. *Man* 17:431-451.

Worsley, P. M. 1961 The utilisation of natural food resources by an Australian Aboriginal tribe. *Acta Ethnographica* 10(1-2):153-190.

Worsnop, T. 1897 *The Prehistoric Arts, Manufactures, Works, Weapons etc. of the Aborigines of Australia*, Adelaide: Government Printer.

Wright, G. A. 1978 Social differentiation in the early Natufian. In C. L. Redman, M. J. Berman, E. V. Curtin, W. T. Langhorne, N. M. Vesaggi and J. C. Wanser (eds) *Social Archaeology: Beyond Subsistence and Dating*, pp.201-223. New York: Academic Press.

Wright, W. 1861-62 Mr Wright's Diary. In *Report of the Commissioners Appointed to Enquire Into and Report Upon the Circumstances Connected With the Suffering and Death of Robert O'Hara Burke and William John Wills, the Victorian Explorers*, pp80-88 (Appendix H), Melbourne: Government Printers.

Wrigley, J. W. 1988 *Australian Native Plants*, Sydney: Collins.

Wyatt, W. 1879 Some account of the manners and superstitions of the Adelaide and Encounter Bay tribes with a vocabulary of their languages, names persons and places. In J. D. Wood (comp.) *Native Tribes of South Australia*, pp.159-181. Adelaide: E. S. Wigg.

Wyndham, W. T. 1889 The Aborigines of Australia. *Journal and Proceedings of the Royal Society of New South Wales* 23:36-42.

Wyrwoll, K. H., McKenzie, N. L., Pederson, B. J., and Tapley, I. J. 1986 The Great Sandy Desert of Northwestern Australia: The last 7000 years. *Search* 17(7-9):208-210.

Yan, W. 1999 Neolithic settlements in China: Latest finds and research. *Journal of East Asian Archaeology* 1:131-147.

Yan, W. 2003 The origins of rice agriculture, pottery and cities. In Y. Yasuda (ed.) *The Origins of Pottery and Agriculture*, pp.151-156. New Delhi: Roli Books.

Yarnell, R. 1985 Review of 'The Origins of Agriculture'. *American Antiquity* 50(3):698-699.

Yasuda, Y. 2003a Origins of pottery and agriculture in east Asia. In Y. Yasuda (ed.) *The Origins of Pottery and Agriculture*, pp.119-142. New Delhi: Roli Books.

Yasuda, Y. 2003b The second east side story: Origin of agriculture in west Asia. In Y. Yasuda (ed.) *The Origins of Pottery and Agriculture*, pp.17-38. New Delhi: Roli Books.

Yen, D. E. 1985 Wild plants and domestication in Pacific Islands. In V. N. Misra and P. Bellwood (eds) *Recent Advances in Indo-Pacific Prehistory*, pp.315-326. New Delhi: Oxford and IBH Publishing Co.

Yen, D. E. 1989 The domestication of environment. In D. R. Harris and G. C. Hillman (eds) *Foraging and Farming: The Evolution of Plant Exploitation*, pp.55-75. London: Unwin Hyman.

Yen, D. E. 1995 The development of Sahul agriculture with Australia as bystander. *Antiquity* 69(265):831-847.

Yengoyan, A. A. 1976 Structure, event and ecology in Aboriginal Australia. In N. Peterson (ed.) *Tribes and Boundaries in Australia*, pp.121-132. Canberra: Australian Institute of Aboriginal Studies.

Yesner, D. R 1980 Maritime hunter-gatherers: Ecology and prehistory. *Current Anthropology* 21(6):727-750.

Yesner, D. R. 1987 Life in the 'Garden of Eden': Causes and consequences of the adaption of marine diets by human societies. In M. Harris and E. Ross (eds) *Food and Evolution*, pp.285-310. Philadelphia: Temple University Press.

Yoffee, N. 1996 Perspectives on trends toward social complexity in prehistoric Australia and New Guinea. In T. Murray (ed.) *Archaeology of Aboriginal Australia*, pp.168-181. St. Leonards: Allen and Unwin.

Yoffee, N. 1998 *Comunidades, Ritualities and Chiefdoms: Social Evolution in the American Southwest and Southeast and Social Evolutionary Theory*. Melbourne: Melbourne University Press. *North American Studies Bernard Bailyn Lecture No. 4.*

Yuan, J. 2003 Rice, pottery 10000 Yrs BP at Yuchanyan, Dao County, Hunan Province. In Y. Yasuda (ed.) *The Origins of Pottery and Agriculture*, pp.157-166. New Delhi: Roli Books.

Yuan, J. and Flad, R. K. 2004 Pig domestication in ancient China. *Antiquity* 76:724-732.

Zhang, C. 2003 Early pottery and rice phytolith remains from Xianrendong and Diaotonghuan sites, Wannian, Jiangxi Province. In Y. Yasuda (ed.) *The Origins of Pottery and Agriculture*, pp.185-191. New Delhi: Roli Books.

Zhang, J. and Wang, X. 1998 Notes on the recent discovery of ancient cultivated rice at Jiahu, Henan Province: A new theory concerning the origin of *Oryza japonica* in China. *Antiquity* 72(278):897-901.

Zhang, W. 2003 The bi-peak-tubercle of rice, the character of ancient rice and the origin of cultivated rice. In Y. Yasuda (ed.) *The Origins of Pottery and Agriculture*, pp.205-216. New Delhi: Roli Books.

Zhao, Z. 1998 The Middle Yangtze region of China is one place where rice was domesticated: Phytolith evidence from Diaotonghuan Cave, Northern Jiangxi. *Antiquity* 72(278):885-897.

Zhimin, A. 1989 Prehistoric agriculture in China. In D. R. Harris and G. C. Hillman (eds) *Foraging and Farming: The Evolution of Plant Exploitation*, pp.643-649. London: Unwin Hyman.

Zohary, D. 1969 The progenitors of wheat and barley in relation to domestication and agricultural dispersal in the Old World. In P. J. Ucko and G. W. Dimbleby (eds) *Domestication and Exploitation of Plants and Animals*, pp.47-66. London: Duckworth and Co.

Zohary, D. 1989 Domestication of the Southwestern Asian Neolithic crop assemblage of cereals, pulses and flax: The evidence from the living plants. In D. R. Harris and G. C. Hillman (eds) *Foraging and Farming: The*

*Evolution of Plant Exploitation*, pp.358-373. London: Unwin Hyman.
Zola, N. and Gott, B. 1992 *Koorie Plants, Koorie People: Traditional Aboriginal Food, Fibre and Healing Plants of Victoria*, Melbourne: Koorie Heritage Trust.
Zvelebil, M. 1986 Postglacial foraging in the forests of Europe. *Scientific American* 254(5):86-93.
Zvelebil, M. and Lillie, M. 2000 Transitions to agriculture in eastern Europe. In T. D. Price (ed.) *Europe's First Farmers*, pp.57-92. Cambridge: Cambridge University Press.
Zvelebil, M. and Rowley-Conwy, P. 1986 Foragers and farmers in Atlantic Europe. In M. Zvelebil (ed.) *Hunters in Transition: Mesolithic Societies of Temperate Eurasia and Their Transition to Farming*, pp.67-93. Cambridge: Cambridge University Press.

www.ingramcontent.com/pod-product-compliance
Ingram Content Group UK Ltd.
Pitfield, Milton Keynes, MK11 3LW, UK
UKHW061848041026
15591UKWH00015B/964